ARTHROPOD VECTOR: CONTROLLER OF DISEASE TRANSMISSION

VOLUME 1: VECTOR MICROBIOME AND INNATE IMMUNITY OF ARTHROPODS

ARTHROPOD VECTOR: CONTROLLER OF DISEASE TRANSMISSION

VECTOR MICROBIOME AND INNATE IMMUNITY OF ARTHROPODS

VOLUME 1

Edited by

STEPHEN K. WIKEL
Emeritus Professor and Chair of Medical Sciences, School of Medicine, Quinnipiac University, CT, United States

SERAP AKSOY
Professor of Epidemiology, Yale School of Public Health, Yale University School of Medicine, New Haven, CT, United States

GEORGE DIMOPOULOS
Professor, Department of Molecular Microbiology and Immunology, Johns Hopkins Bloomberg School of Public Health, Baltimore, MD, United States

Academic Press is an imprint of Elsevier
125 London Wall, London EC2Y 5AS, United Kingdom
525 B Street, Suite 1800, San Diego, CA 92101-4495, United States
50 Hampshire Street, 5th Floor, Cambridge, MA 02139, United States
The Boulevard, Langford Lane, Kidlington, Oxford OX5 1GB, United Kingdom

Notices
Knowledge and best practice in this field are constantly changing. As new research and experience broaden our understanding, changes in research methods, professional practices, or medical treatment may become necessary.

Practitioners and researchers must always rely on their own experience and knowledge in evaluating and using any information, methods, compounds, or experiments described herein. In using such information or methods they should be mindful of their own safety and the safety of others, including parties for whom they have a professional responsibility.

To the fullest extent of the law, neither the Publisher nor the authors, contributors, or editors, assume any liability for any injury and/or damage to persons or property as a matter of products liability, negligence or otherwise, or from any use or operation of any methods, products, instructions, or ideas contained in the material herein.

Library of Congress Cataloging-in-Publication Data
A catalog record for this book is available from the Library of Congress

British Library Cataloguing-in-Publication Data
A catalogue record for this book is available from the British Library

ISBN: 978-0-12-805350-8

Publisher: Sara Tenney
Acquisition Editor: Kristi Gomez
Editorial Project Manager: Pat Gonzalez
Production Project Manager: Julia Haynes
Designer: Matthew Limbert

Typeset by TNQ Books and Journals

Contents—Volume 1

5. Molecular Mechanisms Mediating Immune Priming in *Anopheles gambiae* Mosquitoes

JOSE L. RAMIREZ, ANA BEATRIZ F. BARLETTA AND CAROLINA V. BARILLAS-MURY

6. The Mosquito Immune System and Its Interactions With the Microbiota: Implications for Disease Transmission

FAYE H. RODGERS, MATHILDE GENDRIN AND GEORGE K. CHRISTOPHIDES

7. Using an Endosymbiont to Control Mosquito-Transmitted Disease

ERIC P. CARAGATA AND LUCIANO A. MOREIRA

8. Effect of Host Blood–Derived Antibodies Targeting Critical Mosquito Neuronal Receptors and Other Proteins: Disruption of Vector Physiology and Potential for Disease Control

JACOB I. MEYERS AND BRIAN D. FOY

9. Role of the Microbiota During Development of the Arthropod Vector Immune System

AURÉLIEN VIGNERON AND BRIAN L. WEISS

10. Host–Microbe Interactions: A Case for *Wolbachia* Dialogue

SASSAN ASGARI

Contents—Volume 2

5. Basic and Translational Research on Sand Fly Saliva: Pharmacology, Biomarkers, and Vaccines

WALDIONÊ DE CASTRO, FABIANO OLIVEIRA, ILIANO V. COUTINHO-ABREU, SHADEN KAMHAWI AND JESUS G. VALENZUELA

6. Unique Features of Vector-Transmitted Leishmaniasis and Their Relevance to Disease Transmission and Control

TIAGO D. SERAFIM, RANADHIR DEY, HIRA L. NAKHASI, JESUS G. VALENZUELA AND SHADEN KAMHAWI

7. Early Immunological Responses Upon Tsetse Fly–Mediated Trypanosome Inoculation

GUY CALJON, BENOÎT STIJLEMANS, CARL DE TREZ AND JAN VAN DEN ABBEELE

8. Mosquito Modulation of Arbovirus–Host Interactions

STEPHEN HIGGS, YAN-JANG S. HUANG AND DANA L. VANLANDINGHAM

9. Tick Saliva: A Modulator of Host Defenses

STEPHEN WIKEL

10. Tick Saliva and Microbial Effector Molecules: Two Sides of the Same Coin

DANA K. SHAW, ERIN E. MCCLURE, MICHAIL KOTSYFAKIS AND JOAO H.F. PEDRA

11. Tsetse Fly Saliva Proteins as Biomarkers of Vector Exposure

JAN VAN DEN ABBEELE AND GUY CALJON

12. Epidemiological Applications of Assessing Mosquito Exposure in a Malaria-Endemic Area

ANDRE SAGNA, ANNE POINSIGNON AND FRANCK REMOUE

13. *Ixodes* Tick Saliva: A Potent Controller at the Skin Interface of Early *Borrelia burgdorferi* Sensu Lato Transmission

SARAH BONNET AND NATHALIE BOULANGER

14. Translation of Saliva Proteins Into Tools to Prevent Vector-Borne Disease Transmission

SUKANYA NARASIMHAN, TYLER R. SCHLEICHER AND EROL FIKRIG

List of Contributors

Sassan Asgari The University of Queensland, Brisbane, QLD, Australia

Carolina V. Barillas-Mury National Institutes of Health, Rockville, MD, United States

Ana Beatriz F. Barletta National Institutes of Health, Rockville, MD, United States

Carol D. Blair Colorado State University, Fort Collins, CO, United States

Eric P. Caragata Centro de Pesquisas René Rachou – Fiocruz, Belo Horizonte, Brazil

George K. Christophides Imperial College London, London, United Kingdom

Adriana Costero-Saint Denis National Institutes of Health, Bethesda, MD, United States

Brian D. Foy Colorado State University, Fort Collins, CO, United States

Mathilde Gendrin Imperial College London, London, United Kingdom

Marcelo Jacobs-Lorena Johns Hopkins Bloomberg School of Public Health, Baltimore, MD, United States

Wolfgang W. Leitner National Institutes of Health, Bethesda, MD, United States

Shirley Luckhart University of California, Davis, CA, United States

Jacob I. Meyers Texas A&M University, College Station, TX, United States

Kristin Michel Kansas State University, Manhattan, KS, United States

Luciano A. Moreira Centro de Pesquisas René Rachou – Fiocruz, Belo Horizonte, Brazil

Ken E. Olson Colorado State University, Fort Collins, CO, United States

Xiaoling Pan Michigan State University, East Lansing, MI, United States

Jose E. Pietri University of California, Santa Cruz, CA, United States

Jose L. Ramirez National Institutes of Health, Rockville, MD, United States; U.S. Department of Agriculture, Peoria, IL, United States

Victoria L.M. Rhodes Kansas State University, Manhattan, KS, United States

Michael A. Riehle University of Arizona, Tucson, AZ, United States

Faye H. Rodgers Imperial College London, London, United Kingdom

Michael R. Strand University of Georgia, Athens, GA, United States

Suzanne Thiem Michigan State University, East Lansing, MI, United States

Aurélien Vigneron Yale School of Public Health, New Haven, CT, United States

Tonu M. Wali National Institutes of Health, Bethesda, MD, United States

Sibao Wang Chinese Academy of Sciences, Shanghai, China

Brian L. Weiss Yale School of Public Health, New Haven, CT, United States

Zhiyong Xi Michigan State University, East Lansing, MI, United States

Preface

These two volumes bring together in one place an up-to-date, multidisciplinary examination, by leading authorities, of factors that make the vector arthropod the controller of pathogen transmission. Arthropod vector ability to transmit infectious agents is increasingly recognized as being impacted by the themes addressed in these two volumes: vector microbiome, arthropod innate immunity, and vector saliva stimulation and modulation of host defenses. The three areas addressed in these two volumes are increasingly active areas of investigation that are resulting in significant new insights for understanding vector competence of a variety of arthropod vector species. These research areas increasingly represent opportunities for translation of basic findings into novel approaches for controlling arthropod vectors and vector-borne diseases.

The first volume examines arthropod factors that determine vector competence in the context of gut and reproductive organ–associated microbiomes and complex interactions of vector arthropod immune defenses with the microbiome and vector-borne infectious agents. The introductory chapter identifies knowledge gaps in the biology of vector transmission of disease causing microbes that are addressed in the chapters of these volumes. First volume vector microbiome–related topics encompass: vector microbiome–mediated immune inductions; microbiome influence on shaping arthropod immunity; impact of vector microbiota on disease transmission; and engineered microbiome in control applications. Arthropod vector immune defenses include signaling pathways; priming of innate defenses during vector infection; and developing vector immunity-based novel control strategies for arthropod transmitted pathogens.

The second chapter of the first volume explores the relationships of vector arthropod immune signal transduction pathways with mitochondrial regulation. Microbiome vector immune responses are reviewed in chapters that focus on *Wolbachia*-mediated immune inductions; *Wolbachia* in host and microbe dialogue; role of microbiota in development of the arthropod immune system; and interactions of vector and microbe in mosquito immunity and development. Chapter themes focusing on novel vector-borne disease control approaches include use of symbionts to control dengue; modulation of mosquito immune defenses as a control strategy; targeting dengue virus replication in the mosquito vector; paratransgenesis-mediated pathogen control utilizing engineered symbionts; mosquitocidal activity of antibodies that target chloride channels; and modulation of mosquito immune defenses to control dengue. The first volume closes with a chapter addressing the implication of insulin-like peptides in modulating *Plasmodium* infection in the mosquito.

The second volume provides expert reviews of the complex interactions occurring between hosts, disease vectors, and vector-borne pathogens. Topics addressed in this volume include: vector saliva composition; saliva stimulation, modulation, and suppression of host defenses; influence of vector saliva on transmission and establishment of infectious agents; use of host antibody responses to vector saliva as biomarkers of exposure and risk of vector-borne pathogen infection; and development of saliva molecule–based disease transmission blocking vaccines. This volume concludes with a chapter examining multiple considerations that must be addressed in translating basic arthropod disease vector research described in these two volumes into commercial products to control vectors and the diseases they transmit.

The scope of second volume chapter themes provides a comprehensive examination of the multiple, interrelated, and complex interactions occurring at the host interface with the vector arthropod. The first chapter explores the cutaneous innate and adaptive immune defenses that confront the blood-feeding vector arthropod. The next three chapters focus on how arthropod disease vectors counteract the challenges posed by the host defenses of hemostasis, itch and pain responses, and wound healing, respectively. The fifth chapter examines the genomics and proteomics of vector saliva to provide the underpinnings for subsequent chapters that address the interactions of sandflies, tsetse flies, mosquitoes, and ticks with host defenses and how those interactions create environments favorable for pathogen transmission and establishment of infections agents. These chapters describe both conserved and unique host defense countermeasures across the range of arthropod disease vectors. A subsequent chapter is an excellent example of increasingly detailed molecular characterizations of the interplay between vectors and hosts by examining how tick saliva interacting with Nod-like receptors modulates host innate immune response signaling. Additional chapters describe the use of saliva biomarkers as indicators of vector exposure and epidemiological role in assessing potential risk of infection with malaria and African trypanosomiasis. Multiple authors provide insights as to how basic knowledge of the interactions between vectors and hosts can be used to develop immunologically based strategies to control vector-borne protozoa, bacteria, and arboviruses.

The idea for this two-volume work emerged from the 2015 Keystone Symposium entitled, "The Arthropod Vector: The Controller of Transmission." Our hope is that information contained in these two volumes will stimulate further research and encourage both interdisciplinary collaborations and new avenues of research on arthropod vectors of disease.

Stephen K. Wikel
Serap Aksoy
George Dimopoulos

CHAPTER

1

The Site of the Bite: Addressing Knowledge Gaps in Vector Transmission of Diseases

Wolfgang W. Leitner, Adriana Costero-Saint Denis, Tonu M. Wali

National Institutes of Health, Bethesda, MD, United States

Of the more than 200 pathogens that can cause disease in humans, more than 17% are transmitted by vectors (WHO, 2016). Every year, more than one billion people are infected by vector-borne diseases (WHO, 2014) and more than a million deaths annually are attributed to pathogens transmitted by vectors (WHO, 2016). Despite impressive progress made in certain elimination programs, for example malaria (Newby et al., 2016), sustained success depends on continued investment in control strategies and a strong surveillance system. However, regional political instability, as well as climate change, threatens to undo any progress. A well-documented example of spread of malaria by war is the movement of troops from island to island in the Pacific during WWII, which distributed new strains of *Plasmodium* to these islands while mosquitoes thrived in a devastated ecosystem (Masterson, 2014). Gains against vector-borne diseases are difficult to sustain due to the following: the sheer number of pathogens and their vectors; the need to get buy-in from all nations in an affected area; and the transient efficacy of the control measures (e.g., pesticides, drugs) when resistant vectors or pathogens emerge. Bed nets only protect from infectious bites when intact and properly used, and even then lose their efficacy when mosquitoes change their feeding routine to adjust to the availability of their host. Therefore, durable solutions whose efficacy persists beyond a short window of time are urgently needed. These strategies include vaccines rather than drugs, and vector control-strategies beyond pesticides and bed nets.

VECTORS: THE NEGLECTED PART OF THE EQUATION

The evolution of blood feeding is a fascinating adaptation to a unique biological niche. It has independently evolved multiple times among more than 14,000 species of hematophagous arthropods (reviewed in Ribeiro, 1995) and is used by insects such as mosquitoes, flies, and lice, but also arachnids such as ticks and mites. Obtaining a blood meal requires that the first lines of defense of the

Arthropod Vector: Controller of Disease Transmission, Volume 1
http://dx.doi.org/10.1016/B978-0-12-805350-8.00001-5

host are penetrated, namely the mechanical and chemical barrier of the epidermis. Blood-feeding arthropods accomplish this either through a brute-force approach (pool feeders "dig" through the skin until blood vessels are damaged to release blood) or by using a proboscis that has evolved to act like a hypodermic needle. Either approach represents a unique opportunity for pathogens to enter a host through the back door, which a blood-feeding vector has opened in the skin. Numerous pathogens have adapted to this lifestyle with some going through elaborate—and distinct—developmental stages in both the vector and human hosts (such as *Plasmodium* or *Leishmania*).

While blood-feeding arthropods transmit the majority of vector-borne pathogens, some are transmitted by vectors indirectly, such as *Schistosomes* parasitic flatworms that are released by snails into freshwater. While such animals technically qualify as vectors, they do not interact directly with the host and, thus, the transmission of the pathogen is guided by different rules than those involved in direct introduction into a host. For this reason, such vectors are excluded from the discussions later.

This chapter describes the path that led to—and culminated in—the May 2015 keystone meeting "The Arthropod Vector: the Controller of Transmission."

IDENTIFYING THE RESEARCH GAPS

The National Institute of Allergy and Infectious Diseases (NIAID) supports a broad research portfolio on a large number of vector-borne diseases.[1] One aspect that remains underrepresented, however, is the vector vertebrate host interaction. It has long been recognized that "blood-feeding arthropods are not simply flying or crawling hypodermic needles and syringes" (Wikel and Alarcon-Chaidez, 2001). Many researchers have noted that delivery of vector-borne pathogens to a host through a blood-feeding vector produces significantly different outcomes than injecting isolated pathogens with a needle and syringe, even when purified pathogens are injected into the same anatomical location targeted by the vector,that is, the skin [as shown for example for malaria (Kebaier et al., 2009; Leitner et al., 2010) https://www.ncbi.nlm.nih.gov/pubmed/20507620, *Leishmania* (Samuelson et al., 1991), Cache Valley virus (Edwards et al., 1998), or West Nile virus (Styer et al., 2006)]. What does the vector inadvertently contribute to the infectivity of the microbial payload during a blood meal? A number of molecules in the saliva of blood feeders, some of which have potent pharmacological activity, have already been identified (reviewed in Abdeladhim et al., 2014 for sand flies, or in Kotal et al., 2015 for ticks, or in Tsujimoto et al., 2012 for black flies) and are being explored as drugs (Chudzinski-Tavassi et al., 2016; Francischetti et al., 2005). A subset of these molecules also have immunomodulatory activity (Araujo-Santos et al., 2014; Boppana et al., 2009; Carregaro et al., 2015). But overall, surprisingly little is known about the molecular and cellular events at the site of blood feeding, where three distinct organisms interact: the arthropod vector, the vertebrate host, and the pathogen(s) being transmitted.

[1] The National Institute of Allergy and Infectious Diseases conducts and supports basic and applied research to better understand, treat, and ultimately prevent infectious, immunologic, and allergic diseases (NIAID, 2013b). A number of globally important infectious disease agents are transmitted to humans by the bite of blood-feeding arthropods. NIAID currently supports basic and translational research on mosquitoes, ticks, sand flies, triatomine bugs, tsetse flies, and other arthropods that transmit diseases to humans. For more information on the research NIAID is currently supporting, please visit https://www.niaid.nih.gov/about/mission-planning-overview;https://www.niaid.nih.gov/research/vector-bio.

To begin defining the research gaps of this triad in more detail and to explore, as well as highlight novel approaches for combating vector-borne diseases, the NIAID Extramural Program has hosted a series of workshops and conferences with the objective of bringing together researchers in the areas of vector biology, vaccinology, (skin) immunology, and microbiology (NIAID, 2016). Information about these workshops, including agendas and publications, can be found at https://www.niaid.nih.gov/research/vector-host-interactions.

In 2011, the "Immunologic Consequences of Vector Derived Factors" (ICVDF, 2011) were explored (summarized in Leitner et al., 2011) to take the discussion beyond well-described pharmacological agents in vector saliva that interfere with the three arms of the vertebrate hemostatic system (platelet aggregation, coagulation, and vasoconstriction) and begin to address the impact these proteins have on the skin's immune system. While salivary proteins have well-defined roles during a blood meal and directly benefit the feeding vector, it is less clear how they affect a pathogen's ability to establish an infection. On the other hand, the vector also delivers molecules that actively interfere with, or modulate, the host's immune system. While it is less clear how such saliva factors could benefit the vector, especially when the blood meal lasts for seconds to minutes, the benefits for a vector-delivered pathogen are much easier to understand, albeit insufficiently studied in vivo at this point. The establishment of an immunosuppressive environment at the bite site would, however, provide a compelling explanation for how small numbers of viruses, bacteria, or parasites can successfully infect a host when they are delivered by a vector, while similar numbers of pathogens *injected* into the same tissue with a syringe are often rejected successfully by the host's immune system.

During the discussion period of this initial workshop, it became evident that a significant gap in our understanding of the immunogenicity of vector saliva antigens exists: while vector saliva [e.g., salivary proteins from *Aedes aegypti* (Bizzarro et al., 2013; Wasserman et al., 2004) and *Culex* sp. (Wanasen et al., 2004) or black flies (Tsujimoto et al., 2010)] has been shown to interfere with lymphocyte activity (proliferation, cytokine secretion); at least certain saliva proteins are surprisingly immunogenic, despite the incredibly small amount of antigen delivered during a blood meal. The antibody response to saliva proteins, however, provides a useful and highly sensitive tool to determine exposure to vector bites, which can be used to objectively measure the efficacy of vector control strategies or to estimate the risk of infection by vector-borne pathogens (Ali et al., 2012; Dama et al., 2013). Prototypes for analytical test kits were introduced at the conference, and while the development of such kits for a variety of vector species would be desirable, the immunogenicity of vector saliva varies widely in terms of antibody titers induced, duration of the response, and what proteins are predominantly recognized.

The immunogenicity of vector saliva proteins and their presumed involvement in the successful establishment of infections with vector-borne pathogens has also resulted in a new area within the vaccine development field: vaccination with vector saliva proteins with the objective of neutralizing immunomodulatory saliva proteins and, thus, allowing the host's immune system to eliminate the infectious inoculum without interference by vector factors. The most advanced vaccines based on this approach are those that target sand fly saliva and successfully prevent *Leishmania* infection in a variety of species (Oliveira et al., 2015). Because of the advanced stage of such novel *Leishmania* vaccines, they were further discussed at another workshop (see later) in the context of product development.

Since the workshop generated significant interest and highlighted gaps, as well as new research opportunities in the area of immunological effects of vector molecules, the topic was further explored at a symposium at the ASTMH (2011) conference. The discussions were aimed at attracting more investigators with diverse scientific backgrounds to the research area and to promote collaborations.

ROLE OF IMMUNE CELL SUBSETS IN THE ESTABLISHMENT OF VECTOR-BORNE INFECTIONS

Recognizing the need to bridge the gap between vector and infectious disease researchers and immunologists, the NIAID Program developed a new format for its next workshop in 2012: instead of inviting established investigators to present their ongoing or completed research studies, junior investigators were tasked with proposing ideas for novel, unconventional approaches to address gaps in the understanding of how immune responses are altered and affected by vector saliva factors. The concepts were developed with the input from senior investigators from a complementary research field to promote and encourage "outside-the-box" ideas. The panel of speakers made a number of recommendations (published in Leitner et al., 2012) for moving the field forward more efficiently.

1. Improve model systems, such as humanized mice or chimeric mice with human skin to reduce artifacts based on significant differences between species. The call for better model systems extends to vectors and vector-borne pathogens: While laboratory-adapted species may be easier to work with and better characterized, they often differ from their wild-caught counterparts significantly enough to limit the results obtained with them. An only recently recognized difference is a vector's microbiome. Similar to laboratory mice, a laboratory raised arthropod's microbiome is not only different, but also lacks the diversity of that found in corresponding arthropods in the field (Dong et al., 2009).
2. Standardize experimental design and protocols which differ significantly between laboratories. Although numerous reports have demonstrated immunomodulatory effects of various immune cell subsets of the vertebrate host, it is impossible to compare these results when some labs use saliva extract while others work with isolated or recombinant saliva proteins and at different concentrations.
3. More studies by multidisciplinary teams would better address the challenges involved in studying a process that involves three very distinct species and immune events in a tissue—the skin—with an immune system that is by far not as well understood as the central immune system.
4. More focus on immune cell subpopulations in the skin: While the effects of various arthropod saliva components on cells—often cell lines—in vitro has been well documented, how this translates to primary cells, such as the different dermal and epidermal dendritic cell populations, remains to be shown.
5. Blood feeding is a process of two-way communication, but when studying the process of blood feeding, most researchers focus on the consequence of this process on the vertebrate host. Biological information is, however, also transferred to the vector through the blood. Cytokines, hormones, antibodies, and other molecules in the blood meal have a significant effect, for example, on the vector's immune system, but also on the pathogens they carry. This has, most prominently, been studied in the context of transmission-blocking vaccines, which exert their effect in the blood-fed vector, but little

has been done beyond the examination of effector antibodies.

EFFECT OF VECTOR INNATE IMMUNITY AND HUMAN-DERIVED IMMUNE MOLECULES ON THE TRANSMISSION OF VECTOR-BORNE PATHOGENS

While the two initial NIAID workshops focused primarily on how vector-derived factors affect the vertebrate host and its ability to deal with vector-borne pathogens, we also wanted to focus on how vector factors can affect transmission. For example, how the vertebrate host can affect the vector through "information" contained within the blood meal (Pakpour et al., 2014); how a vector's immune system may open new opportunities for combating vector-borne pathogens; and how the vector's microbiome may be contributing to transmission by affecting the vector's immune system and the pathogens within the vector. To further explore these topics, a third workshop was convened, entitled "Effect of Vector Innate Immunity and Human-Derived Immune Molecules on the Transmission of Vector-Borne Pathogens" (NIAID, 2013a), which was based on concept talks by junior investigators, a format successfully used in the previous workshop. A variety of vector biology aspects were discussed, which could be exploited for the development of novel vector control strategies.

1. Vector Immunity:
   Arthropod vectors employ a variety of immune mechanisms when infected, such as phagocytosis, agglutination, melanization, or the production of microbicidal free radicals. An approach that has already been deployed is the release of genetically modified mosquitoes. Various types of genetically altered mosquitoes are in the pipeline. Some strains are based on the introduction of "suicide genes" to significantly limit the life span of the released male animals and their offspring in the absence of nutrients or drugs not available to them outside a laboratory. While highly attractive from a safety standpoint, a major drawback of this approach is the need to continuously produce and release GM animals to maintain the suppression of the population by "crowding out" wild-type males. Alternatively, if the mosquito's immune system is manipulated, it may be able to eliminate pathogens acquired from an infectious blood meal before they are passed to the next host during a subsequent blood meal. Various innate immune defense genes have been identified that can be upregulated to "harden" mosquitoes against infections. However, the constitutive overexpression of immune pathways is frequently associated with a reduced life span and/or fitness of the GM mosquitoes, which would prevent them from successfully competing with their wild-type counterparts thus requiring the frequent and expensive re-release of animals. This issue has been addressed by placing the immune defense genes [such as the Imd pathway-controlled NF-kB REl2 transcription factor (Dong et al., 2011)] under promoters activated by a blood meal, which restricts expression to the critical time period when pathogens may be acquired.

   Workshop participants discussed a number of factors that impact the immune response of vectors:

   a. *The nutritional status of a vector*: The amount of nutrition available to a vector determines the strength of its immune system, the integrity of its tissue barriers (such as the gut barrier) and its ability to efficiently repair tissues, all factors that determine its susceptibility to pathogens and, thus, the likelihood that it will transmit a pathogen.
   b. *Immune tolerance*: The nonresponsiveness of the immune system to certain

antigens has been extensively studied in mammalians, especially in the context of transplantation (preventing response to non-self antigen), autoimmunity (suppressing response to self-antigen), or allergy (suppressing response to environmental antigen). Tolerance has also been recognized as an important mechanism during infection. For example, while strong immune responses against malaria parasites are associated with mortality, "natural immunity" against the pathogen is characterized by persistent parasitemia and an acquired nonresponsiveness of the host's immune system. Amazingly little is known about tolerance mechanisms in arthropod vectors, but evidence exists in *Drosophila* that immune tolerance—mediated by the nutritional status of the animal—is a critical factor in the survival after infection with certain pathogens (Ayres and Schneider, 2009). How could immune tolerance of vectors be used in a strategy to interfere with pathogen transmission? As described later, changing the microbiome of a blood-feeding arthropod can significantly affect its ability to act as a vector, but may require induction of tolerance to the newly introduced commensals.

c. *Circadian rhythm*: While the circadian rhythm of vector species has been studied in the context of their feeding behavior (Rund et al., 2016), much less is known about its influence on other biological functions. It is known to affect immune responses, and the infectivity of *Drosophila* with pathogens varies with the time of the day and the corresponding changes in the expression of crucial immune regulatory genes. It has been proposed that the feeding behavior of mosquitoes (i.e., preferential feeding during dusk and dawn) also coincides with the up/downregulation of immune response genes and that pathogens may have adapted to this rhythm to minimize immune destruction by the vector.

2. The Vector Microbiome:
While microbiome research in mammals has seen a boom, parallel research in arthropods is still in its infancy. New research has shown that a blood-feeding vector's microbiome is highly dynamic and changes dramatically during different developmental stages (Duguma et al., 2015; Wang et al., 2011), as well as in response to environmental stimuli (Pennington et al., 2016) or host-derived elements in the blood meal (Gendrin et al., 2015). Nevertheless, the impact of commensals on the vector's susceptibility to pathogens has been clearly established (Jupatanakul et al., 2014), as well as its effect on the vector's innate immune system (reviewed in Dennison et al., 2014; Weiss and Aksoy, 2011). "Beneficial" commensals either compete out pathogens directly (Bahia et al., 2014) or prevent transmission of pathogens through stimulating the vector's immune system (Eappen et al., 2013; Pan et al., 2012). *Wolbachia*, which is a common symbiont in many arthropods, has been modified to interfere with pathogen development in the *Aedes* mosquito (Aliota et al., 2016; Moreira et al., 2009).

Another approach being developed using vector microbiome is the genetic modification of symbiotic bacteria to express antipathogenic molecules, for example, antimalarial effector molecules (Wang et al., 2012). While workshop participants acknowledged the importance of understanding the microbiome of vectors, they cited several obstacles such as the significant heterogeneity of arthropod's commensals and the inability to grow many commensal species in vitro for further study. Additionally, significant differences in the microbiome of lab-raised and wild-caught mosquitoes complicates translational studies.

3. Altering Vector Biology With Factors in the Blood Meal:
   The blood meal a vector obtains during feeding contains a significant amount of information that affects the subsequent biology and behavior of the vector. Altered behavior can be the result of a pathogen in the blood meal which, in case of mosquitoes infected with malaria parasites, can change their feeding persistence (Leitner et al., 2010), duration, and probing behavior (reviewed in Cator et al., 2012). How much of the altered behavior is truly triggered by the pathogen remains to be determined after it was shown that the vector's immune response to a microbial challenge was the cause for altered behavior (Cator et al., 2012, 2015). Host-derived factors in the blood meal that affect the vector include insulin/insulin-like growth factors, which are sufficiently conserved to be recognized by arthropod receptors (reviewed in Luckhart et al., 2015).

DROSOPHILA—A USEFUL MODEL FOR VECTORS?

While the manipulation of the mosquitoes' immune system is already being explored as a strategy to interfere with pathogen transmission as discussed earlier, this discussion also highlights the importance of studying the immune system of arthropods in general. Insects such as *Drosophila* are not only valuable model organisms to decipher conserved innate immune pathways and receptors [the discovery of Toll-like receptors being the most famous example (Rock et al., 1998)], but a better understanding of arthropod immunity also provides new potential targets for innovative strategies to combat vector-borne pathogens. *Drosophila* researchers have developed a huge arsenal of reagents, technologies, and protocols, which could be applied to vector research without the need to "reinvent the wheel." The relevance of *Drosophila* research to vector biology was underscored by the successful development of *Wolbachia*-modified *Anopheles* (Bian et al., 2013), which are poor vectors for several pathogens not only because of their reduced life span, but also the constitutive stimulation of innate immune responses, a phenomenon first demonstrated in *Drosophila* in which *Wolbachia* is a naturally occurring symbiont (Teixeira et al., 2008).

ARTHROPOD VECTORS AND DISEASE TRANSMISSION: TRANSLATIONAL ASPECTS

Discussions at these workshops revealed the enormous potential for translation of the basic research being discussed. Although investigators are aware of the potential to turn their research into novel approaches for vector and transmission control, few understand the process and challenges to accomplish this goal. The fourth workshop, entitled "Arthropod Vectors and Disease Transmission: Translational Aspects," focused on the translational aspect of vector research (Leitner et al., 2015). Four potential translational areas were explored:

1. *Vaccines based on vector factors*: As explored already in previous workshops, vector saliva is a rich source of potential new vaccine targets. Targeting immunomodulatory proteins, which facilitate—or enable—the small infectious inoculum to establish a systemic infection, with neutralizing antibodies may allow the host's immune system to eliminate pathogens directly at the bite site. Such vaccines may be developed as a stand-alone strategy or formulated with a vaccine targeting pathogen-derived antigen to deliver a two-tiered hit. Despite the attractiveness of this approach, surprisingly few vaccine candidates that are based on this concept are in the development pipeline. The most advanced vector-based vaccines

under development are those using sand fly saliva antigens to prevent the transmission of *Leishmania* (reviewed in Reed et al., 2016). Interestingly, their mechanism of action is not based on neutralizing antibodies, but the induction of a local Th1 (delayed-type-hypersensitivity) response (Oliveira et al., 2015). Such an immune environment has long been known to interfere with parasite development although in this instance, it represents a bystander response to antigens invariably associated with, but not derived from parasites. Commercial development of such vaccines may first be pursued for veterinary applications to target the animal reservoir of the parasite. This drastically reduces regulatory hurdles and would help establish the usefulness and safety of the approach. However, from a funding standpoint, finding a sponsor for a vaccine that indirectly protects humans in endemic areas by immunizing animals will likely be very challenging.

2. *Bugs-to-drugs*: The fact that vector saliva contains bioactive molecules designed to exert their effects at very small concentrations makes these molecules attractive drug candidates. They act as vasodilators, anti-coagulants, and immunomodulators, just to name a few functions of the few proteins that have been characterized so far. Their independent evolution in different vector species has resulted in proteins that are functionally related, but share no other similarities, which would allow long-term application in humans without risk of losing efficacy due to neutralizing antibody responses to the proteins, by simply switching between proteins from different species, a trick ticks are already using when exposing a host to their saliva proteins for an extended period of time. Workshop participants discussed a variety of potential applications for vector-derived saliva proteins, such as the suppression of transplant rejection by immunoregulatory proteins in an effort to reduce side effects of currently employed drugs.
3. *Biomarkers*: Antibodies induced by immunogenic saliva proteins can serve as highly sensitive biomarkers of exposure to vectors. Kits are already being developed and will be useful in determining the efficacy of vector control strategies, or the mapping of pathogen transmission hot spots and represent an enormous leap forward from having to conduct this type of analysis manually by trapping vectors. Developing such analytical assays is complicated by numerous factors such as the selection of the optimal saliva antigen and the need to understand the immune response to those antigens, particularly the duration of the antibody response. The fact that the composition of saliva proteins is altered by the presence of vector-borne pathogens may allow the design of next-generation analytical kits capable of detecting not only the frequency of blood meals, but also the ratio of bites by uninfected and pathogen-carrying vectors.
4. *The role of the microbiome during vector-borne pathogen transmission*: After establishing that the microbiome of a vector (and potentially even the microbiome of complex pathogens) significantly affects the pathogen burden in a vector and, thus, its ability to transmit it (reviewed in Finney et al., 2015), the logical question is whether this knowledge could be applied to the development of a product aimed at blocking the transmission of vector-borne pathogens. Together with transgenic mosquitoes, which are resistant to many pathogens, paratransgenic approaches have been explored for triatomine bugs and, currently, sand flies (Bian et al., 2013; Taracena et al., 2015). This approach has also been extended to mosquitoes (Bongio and Lampe, 2015; Wang et al., 2012). The most prominent among

them is *Wolbachia*-transfected mosquitoes that, depending on the strain of *Wolbachia*, either prevent virus from developing in the mosquito (Zhang et al., 2013) or interfere with mosquito reproduction (Calvitti et al., 2012), or reduces longevity (Kambris et al., 2009; Yeap et al., 2011). Considering the current lack of vaccines or drugs against important arboviruses such as dengue, chikungunya and, most recently, Zika, the development of paratransgenic approaches may represent the most effective way to combat the transmission of arboviruses in affected areas.

TRANSLATIONAL CONSIDERATIONS FOR NOVEL VECTOR MANAGEMENT APPROACHES

The workshops and conferences organized by NIAID between 2011 and 2015 focused on basic research aspects related to the interface between vector, pathogen, and mammalian host, on novel ideas to combat the transmission of vector-borne pathogens, as well as translational potential to accomplish this goal. While most of these new products are still at various stages of the development pipeline, a small number have started to be deployed outside the United States, revealing the lack of a regulatory and legal framework necessary for deployment within the United States. Therefore, the NIAID in collaboration with the Foundation for the NIH (FNIH) hosted a workshop in 2015 with the intention to consider the outcome of the discussions for developing guidance for investigators seeking NIAID funding to support translational research in vector biology.

Experts in the areas of risk assessment, environmental, clinical and ethical/social/cultural considerations presented the various challenges an investigator faces when moving new approaches from the laboratory to the field. Among those challenges (summarized in Costero-Saint Denis et al., 2017), one of the most important ("make-or-break") issues is community outreach, an aspect that tends not to be on the radar of many laboratory researchers. However, all of the speakers who presented successful field studies of novel strategies to control vector-borne pathogen transmission cited their local outreach and education efforts as major factors responsible for the smooth and successful completion of their field studies.

"New" products to combat vector-borne diseases include more than those based on truly novel approaches, such as genetically modified or paratransgenic vectors. They also extend to the repurposing of previously established products. A prototypic example of the latter is the mass administration of ivermectin, a drug used to treat neglected tropical diseases such as onchocerciasis, lymphatic filariasis, and strongyloidiasis, and now being tested to kill malaria vectors. While a single dose of the drug already makes a blood toxic to *Anopheles*, the current challenge is to generate a formulation that creates a long-term depot (Chaccour et al., 2015).

KEYSTONE SYMPOSIA ON MOLECULAR AND CELLULAR BIOLOGY–THE ARTHROPOD VECTOR: THE CONTROLLER OF TRANSMISSION

The NIAID recognized that the discussions at the workshops held between 2011 and 2014 were only accessible to a small number of investigators. To reach the broader community and to attract more investigators from outside the vector biology and vector-borne disease community, a keystone symposium was convened in Taos, New Mexico, in 2015. The topics presented and explored at this conference are summarized in the subsequent chapters.

CONCLUSIONS

Significant gaps in the understanding of how arthropod vectors contribute to the transmission of vector-borne pathogens other than by simply acting as "flying syringes" prompted the NIAID to host a series of workshops leading to a keystone conference to explore ways to address these gaps and explore new, unconventional strategies to combat vector-borne disease transmission. The chapters in this book are representative of the areas of vector biology that were explored in the workshops and the keystone meeting, and which represent the cutting edge of vector research now. A theme that emerged from the discussions at these conferences was the need for—and benefit of—multidisciplinary research teams. Research on the transmission of a pathogen during an arthropod vector's blood-feeding represents a unique ecosystem (i.e., the bite site), and thus denotes a unique research challenge. The transmission process is complex as it involves the interaction of several organisms: the blood-feeding arthropod vector, the mammalian host, the pathogen (a virus, bacterium, or parasite), and—as only currently appreciated—the vector's microbiome. The intricacy of this system necessitates the interaction of multiple and distinct types of expertise, namely entomologists, immunologists (ideally with experience in skin immunology), and virologists/bacteriologists/parasitologists. While reductionist approaches, which focus on select aspects of one of the organisms involved in the transmission of vector-borne pathogens, have yielded many insights in the past, they have also reached their limit and shed limited light on the overall transmission process. By not acknowledging the complexity of this multispecies interaction as it exists in nature, progress in finding novel approaches to interfere with pathogen transmission is slowed or stalled. A textbook example is how the relatively recent recognition of the vector microbiome's impact on the vector competence of an arthropod was quickly translated into novel transmission-blocking strategies. Additional approaches based on the biology of these interactions will yield novel rational ways to prevent and control vector-borne diseases and the heavy burden they pose to most of the global population.

We hope that the following chapters will help underline the need to not lose sight of the site of the bite.

References

Abdeladhim, M., Kamhawi, S., Valenzuela, J.G., 2014. What's behind a sand fly bite? The profound effect of sand fly saliva on host hemostasis, inflammation and immunity. Infect Genet. Evol. 28, 691–703.

Ali, Z.M.I., Bakli, M., Fontaine, A., Bakkali, N., Hai, V.V., Audebert, S., Boublik, Y., Pages, F., Remoue, F., Rogier, C., Fraisier, C., Almeras, L., 2012. Assessment of *Anopheles* salivary antigens as individual exposure biomarkers to species-specific malaria vector bites. Malar. J. 11.

Aliota, M.T., Peinado, S.A., Velez, I.D., Osorio, J.E., 2016. The wMel strain of *Wolbachia* reduces transmission of zika virus by *Aedes aegypti*. Sci. Rep. 6, 28792.

Araujo-Santos, T., Prates, D.B., Franca-Costa, J., Luz, N.F., Andrade, B.B., Miranda, J.C., Brodskyn, C.I., Barral, A., Bozza, P.T., Borges, V.M., 2014. Prostaglandin E_2/leukotriene B_4 balance induced by *Lutzomyia longipalpis* saliva favors *Leishmania infantum* infection. Parasit Vectors 7, 601.

ASTMH, 2011. Immunologic Consequences of Vector-Derived Factors. http://www.abstractsonline.com/Plan/ViewSession.aspx?sKey=86e860fb-c405-4599-8c9d-4f36e1ceb0c6&mKey={6BADBAB5-8298-4B1A-B5C2-0EEFA5120BB0}.

Ayres, J.S., Schneider, D.S., 2009. The role of anorexia in resistance and tolerance to infections in *Drosophila*. PLoS Biol. 7, e1000150.

Bahia, A.C., Dong, Y., Blumberg, B.J., Mlambo, G., Tripathi, A., BenMarzouk-Hidalgo, O.J., Chandra, R., Dimopoulos, G., 2014. Exploring *Anopheles* gut bacteria for *Plasmodium* blocking activity. Environ. Microbiol. 16, 2980–2994.

Bian, G., Joshi, D., Dong, Y., Lu, P., Zhou, G., Pan, X., Xu, Y., Dimopoulos, G., Xi, Z., 2013. *Wolbachia* invades *Anopheles stephensi* populations and induces refractoriness to *Plasmodium* infection. Science 340, 748–751.

Bizzarro, B., Barros, M.S., Maciel, C., Gueroni, D.I., Lino, C.N., Campopiano, J., Kotsyfakis, M., Amarante-Mendes, G.P., Calvo, E., Capurro, M.L., Sa-Nunes, A., 2013. Effects of *Aedes aegypti* salivary components on dendritic cell and lymphocyte biology. Parasit Vectors 6, 329.

Bongio, N.J., Lampe, D.J., 2015. Inhibition of *Plasmodium berghei* development in mosquitoes by effector proteins secreted from *Asaia* sp. Bacteria using a novel native secretion signal. PLoS One 10, e0143541.

Boppana, V.D., Thangamani, S., Adler, A.J., Wikel, S.K., 2009. SAAG-4 is a novel mosquito salivary protein that programmes host CD4(+) T cells to express IL-4. Parasite Immunol. 31, 287–295.

Calvitti, M., Moretti, R., Skidmore, A.R., Dobson, S.L., 2012. *Wolbachia* strain wPip yields a pattern of cytoplasmic incompatibility enhancing a *Wolbachia*-based suppression strategy against the disease vector *Aedes albopictus*. Parasit Vectors 5, 254.

Carregaro, V., Ribeiro, J.M., Valenzuela, J.G., Souza-Junior, D.L., Costa, D.L., Oliveira, C.J., Sacramento, L.A., Nascimento, M.S., Milanezi, C.M., Cunha, F.Q., Silva, J.S., 2015. Nucleosides present on phlebotomine saliva induce immunossuppression and promote the infection establishment. PLoS Negl. Trop. Dis. 9, e0003600.

Cator, L.J., Lynch, P.A., Read, A.F., Thomas, M.B., 2012. Do malaria parasites manipulate mosquitoes? Trends Parasitol. 28, 466–470.

Cator, L.J., Pietri, J.E., Murdock, C.C., Ohm, J.R., Lewis, E.E., Read, A.F., Luckhart, S., Thomas, M.B., 2015. Immune response and insulin signalling alter mosquito feeding behaviour to enhance malaria transmission potential. Sci. Rep. 5, 11947.

Chaccour, C.J., Rabinovich, N.R., Slater, H., Canavati, S.E., Bousema, T., Lacerda, M., Ter Kuile, F., Drakeley, C., Bassat, Q., Foy, B.D., Kobylinski, K., 2015. Establishment of the Ivermectin Research for Malaria Elimination Network: updating the research agenda. Malar. J. 14, 243.

Chudzinski-Tavassi, A.M., Morais, K.L., Pacheco, M.T., Pasqualoto, K.F., de Souza, J.G., 2016. Tick salivary gland as potential natural source for the discovery of promising antitumor drug candidates. Biomed. Pharmacother. 77, 14–19.

Costero-Saint Denis, A., Leitner, W.W., Wali, T., James, S., 2016. Translational considerations of novel vector management approaches. PLoS Negl. Trop. Dis. 10(8), e0004800.

Dama, E., Cornelie, S., Bienvenu Somda, M., Camara, M., Kambire, R., Courtin, F., Jamonneau, V., Demettre, E., Seveno, M., Bengaly, Z., Solano, P., Poinsignon, A., Remoue, F., Belem, A.M., Bucheton, B., 2013. Identification of *Glossina palpalis gambiensis* specific salivary antigens: towards the development of a serologic biomarker of human exposure to tsetse flies in West Africa. Microbe. Infect. 15, 416–427.

Dennison, N.J., Jupatanakul, N., Dimopoulos, G., 2014. The mosquito microbiota influences vector competence for human pathogens. Curr. Opin. Insect Sci. 3, 6–13.

Dong, Y., Manfredini, F., Dimopoulos, G., 2009. Implication of the mosquito midgut microbiota in the defense against malaria parasites. PLoS Pathog. 5, e1000423.

Dong, Y., Das, S., Cirimotich, C., Souza-Neto, J.A., McLean, K.J., Dimopoulos, G., 2011. Engineered anopheles immunity to *Plasmodium* infection. PLoS Pathog. 7, e1002458.

Duguma, D., Hall, M.W., Rugman-Jones, P., Stouthamer, R., Terenius, O., Neufeld, J.D., Walton, W.E., 2015. Developmental succession of the microbiome of *Culex* mosquitoes. BMC Microbiol. 15, 140.

Eappen, A.G., Smith, R.C., Jacobs-Lorena, M., 2013. Enterobacter-activated mosquito immune responses to *Plasmodium* involve activation of SRPN6 in *Anopheles stephensi*. PLoS One 8, e62937.

Edwards, J.F., Higgs, S., Beaty, B.J., 1998. Mosquito feeding-induced enhancement of Cache Valley Virus (Bunyaviridae) infection in mice. J. Med. Entomol. 35, 261–265.

Finney, C.A., Kamhawi, S., Wasmuth, J.D., 2015. Does the arthropod microbiota impact the establishment of vector-borne diseases in mammalian hosts? PLoS Pathog. 11, e1004646.

Francischetti, I.M., Mather, T.N., Ribeiro, J.M., 2005. Tick saliva is a potent inhibitor of endothelial cell proliferation and angiogenesis. Thromb. Haemost 94, 167–174.

Gendrin, M., Rodgers, F.H., Yerbanga, R.S., Ouedraogo, J.B., Basanez, M.G., Cohuet, A., Christophides, G.K., 2015. Antibiotics in ingested human blood affect the mosquito microbiota and capacity to transmit malaria. Nat. Commun. 6, 5921.

Jupatanakul, N., Sim, S., Dimopoulos, G., 2014. The insect microbiome modulates vector competence for arboviruses. Viruses 6, 4294–4313.

Kambris, Z., Cook, P.E., Phuc, H.K., Sinkins, S.P., 2009. Immune activation by life-shortening *Wolbachia* and reduced filarial competence in mosquitoes. Science 326, 134–136.

Kebaier, C., Voza, T., Vanderberg, J., 2009. Kinetics of mosquito-injected *Plasmodium* sporozoites in mice: fewer sporozoites are injected into sporozoite-immunized mice. PLoS Pathog. 5, e1000399.

Kotal, J., Langhansova, H., Lieskovska, J., Andersen, J.F., Francischetti, I.M., Chavakis, T., Kopecky, J., Pedra, J.H., Kotsyfakis, M., Chmelar, J., 2015. Modulation of host immunity by tick saliva. J. Proteomics 128, 58–68.

Leitner, W.W., Bergmann-Leitner, E.S., Angov, E., 2010. Comparison of *Plasmodium berghei* challenge models for the evaluation of pre-erythrocytic malaria vaccines and their effect on perceived vaccine efficacy. Malar. J. 9, 145.

Leitner, W.W., Costero-Saint Denis, A., Wali, T., 2011. Immunological consequences of arthropod vector-derived salivary factors. Eur. J. Immunol. 41, 3396–3400.

Leitner, W.W., Costero-Saint Denis, A., Wali, T., 2012. Role of immune cell subsets in the establishment of vector-borne infections. Eur. J. Immunol. 42, 3110–3115.

Leitner, W.W., Wali, T., Kincaid, R., Costero-Saint Denis, A., 2015. Arthropod vectors and disease transmission: translational aspects. PLoS Negl. Trop. Dis. 9, e0004107.

Luckhart, S., Pakpour, N., Giulivi, C., 2015. Host-pathogen interactions in malaria: cross-kingdom signaling and mitochondrial regulation. Curr. Opin. Immunol. 36, 73–79.

Masterson, K.M., 2014. The Malaria Project: The U.S. Government's Secret Mission to Find a Miracle Cure. Penguin Group, New York, NY.

Moreira, L.A., Iturbe-Ormaetxe, I., Jeffery, J.A., Lu, G.J., Pyke, A.T., Hedges, L.M., Rocha, B.C., Hall-Mendelin, S., Day, A., Riegler, M., Hugo, L.E., Johnson, K.N., Kay, B.H., McGraw, E.A., van den Hurk, A.F., Ryan, P.A., O'Neill, S.L., 2009. A *Wolbachia* symbiont in *Aedes aegypti* limits infection with dengue, chikungunya, and *Plasmodium*. Cell 139, 1268–1278.

Newby, G., Bennett, A., Larson, E., Cotter, C., Shretta, R., Phillips, A.A., Feachem, R.G., 2016. The path to eradication: a progress report on the malaria-eliminating countries. Lancet 387, 1775–1784.

NIAID, 2013a. Effect of Vector Innate Immunity and Human-Derived Immune Molecules on the Transmission of Vector-Borne Pathogens. https://www.niaid.nih.gov/news/events/meetings/vectorBorne/Pages/default.aspx.

NIAID, 2013b. NIAID Mission and Planning. https://www.niaid.nih.gov/about/whoweare/Pages/default.aspx.

NIAID, 2016. Effect of Arthropod Blood Feeding on Infectious Diseases. https://www.niaid.nih.gov/topics/vector/Pages/arthropodBloodFeeding.aspx.

Oliveira, F., Rowton, E., Aslan, H., Gomes, R., Castrovinci, P.A., Alvarenga, P.H., Abdeladhim, M., Teixeira, C., Meneses, C., Kleeman, L.T., Guimaraes-Costa, A.B., Rowland, T.E., Gilmore, D., Doumbia, S., Reed, S.G., Lawyer, P.G., Andersen, J.F., Kamhawi, S., Valenzuela, J.G., 2015. A sand fly salivary protein vaccine shows efficacy against vector-transmitted cutaneous leishmaniasis in nonhuman primates. Sci. Transl. Med. 7.

Pakpour, N., Riehle, M.A., Luckhart, S., 2014. Effects of ingested vertebrate-derived factors on insect immune responses. Curr. Opin. Insect Sci. 3, 1–5.

Pan, X.L., Zhou, G.L., Wu, J.H., Bian, G.W., Lu, P., Raikhel, A.S., Xi, Z.Y., 2012. *Wolbachia* induces reactive oxygen species (ROS)-dependent activation of the Toll pathway to control dengue virus in the mosquito *Aedes aegypti*. Proc. Natl. Acad. Sci. U.S.A. 109, E23–E31.

Pennington, M.J., Prager, S.M., Walton, W.E., Trumble, J.T., 2016. *Culex quinquefasciatus* larval microbiomes vary with instar and exposure to common wastewater contaminants. Sci. Rep. 6, 21969.

Reed, S.G., Coler, R.N., Mondal, D., Kamhawi, S., Valenzuela, J.G., 2016. Leishmania vaccine development: exploiting the host-vector-parasite interface. Expert Rev. Vaccines 15, 81–90.

Ribeiro, J.M., 1995. Blood-feeding arthropods: live syringes or invertebrate pharmacologists? Infect Agents Dis. 4, 143–152.

Rock, F.L., Hardiman, G., Timans, J.C., Kastelein, R.A., Bazan, J.F., 1998. A family of human receptors structurally related to *Drosophila* Toll. Proc. Natl. Acad. Sci. U S A 95, 588–593.

Rund, S.S., O'Donnell, A.J., Gentile, J.E., Reece, S.E., 2016. Daily rhythms in mosquitoes and their consequences for malaria transmission. Insects 7.

Samuelson, J., Lerner, E., Tesh, R., Titus, R., 1991. A mouse model of *Leishmania braziliensis* braziliensis infection produced by coinjection with sand fly saliva. J. Exp. Med. 173, 49–54.

Styer, L.M., Bernard, K.A., Kramer, L.D., 2006. Enhanced early West Nile virus infection in young chickens infected by mosquito bite: effect of viral dose. Am. J. Trop. Med. Hyg. 75, 337–345.

Taracena, M.L., Oliveira, P.L., Almendares, O., Umana, C., Lowenberger, C., Dotson, E.M., Paiva-Silva, G.O., Pennington, P.M., 2015. Genetically modifying the insect gut microbiota to control Chagas disease vectors through systemic RNAi. PLoS Negl. Trop. Dis. 9, e0003358.

Teixeira, L., Ferreira, A., Ashburner, M., 2008. The bacterial symbiont *Wolbachia* induces resistance to RNA viral infections in *Drosophila melanogaster*. PLoS Biol. 6, e2.

Tsujimoto, H., Gray, E.W., Champagne, D.E., 2010. Black fly salivary gland extract inhibits proliferation and induces apoptosis in murine splenocytes. Parasite Immunol. 32, 275–284.

Tsujimoto, H., Kotsyfakis, M., Francischetti, I.M., Eum, J.H., Strand, M.R., Champagne, D.E., 2012. Simukunin from the salivary glands of the black fly *Simulium vittatum* inhibits enzymes that regulate clotting and inflammatory responses. PLoS One 7, e29964.

Wanasen, N., Nussenzveig, R.H., Champagne, D.E., Soong, L., Higgs, S., 2004. Differential modulation of murine host immune response by salivary gland extracts from the mosquitoes *Aedes aegypti* and *Culex quinquefasciatus*. Med. Vet. Entomol. 18, 191–199.

Wang, Y., Gilbreath 3rd, T.M., Kukutla, P., Yan, G., Xu, J., 2011. Dynamic gut microbiome across life history of the malaria mosquito *Anopheles gambiae* in Kenya. PLoS One 6, e24767.

Wang, S., Ghosh, A.K., Bongio, N., Stebbings, K.A., Lampe, D.J., Jacobs-Lorena, M., 2012. Fighting malaria with engineered symbiotic bacteria from vector mosquitoes. Proc. Natl. Acad. Sci. U.S.A. 109, 12734–12739.

Wasserman, H.A., Singh, S., Champagne, D.E., 2004. Saliva of the Yellow Fever mosquito, *Aedes aegypti*, modulates murine lymphocyte function. Parasite Immunol. 26, 295–306.

Weiss, B., Aksoy, S., 2011. Microbiome influences on insect host vector competence. Trends Parasitol. 27, 514–522.

WHO, 2014. A Global Brief on Vector-Borne Diseases. http://apps.who.int/iris/bitstream/10665/111008/1/WHO_DCO_WHD_2014.1_eng.pdf.

WHO, 2016. Vector-Borne Diseases Fact Sheet. http://www.who.int/mediacentre/factsheets/fs387/en/.

Wikel, S.K., Alarcon-Chaidez, F.J., 2001. Progress toward molecular characterization of ectoparasite modulation of host immunity. Vet. Parasitol. 101, 275–287.

Yeap, H.L., Mee, P., Walker, T., Weeks, A.R., O'Neill, S.L., Johnson, P., Ritchie, S.A., Richardson, K.M., Doig, C., Endersby, N.M., Hoffmann, A.A., 2011. Dynamics of the "popcorn" *Wolbachia* infection in outbred *Aedes aegypti* informs prospects for mosquito vector control. Genetics 187, 583–595.

Zhang, G., Hussain, M., O'Neill, S.L., Asgari, S., 2013. *Wolbachia* uses a host microRNA to regulate transcripts of a methyltransferase, contributing to dengue virus inhibition in *Aedes aegypti*. Proc. Natl. Acad. Sci. U.S.A. 110, 10276–10281.

CHAPTER

2

Conservation and Convergence of Immune Signaling Pathways With Mitochondrial Regulation in Vector Arthropod Physiology

Shirley Luckhart[1], *Michael A. Riehle*[2]

[1]University of California, Davis, CA, United States; [2]University of Arizona, Tucson, AZ, United States

HISTORICAL IMPORTANCE OF INSECTS IN OUR UNDERSTANDING OF DISEASE

To fully understand the conservation of immune signaling in the context of this chapter, it is worthwhile to reflect on our historical understanding of insect–pathogen interactions. Domesticated insects—silkworms and honeybees—were sources of some of the earliest reported observations of insect infection. These observations date back to Chinese reports of fungal pathogens (*Cordyceps* spp.) in silkworms in the 7th century BC (Wang, 1965) and to "rusts" of honeybees, a reddish pigmentation imparted by bacterial infections described by Aristotle in *Historia Animalium* (Steinhaus, 1956). Today, we know these honeybees infections as "foulbrood." Given the economic importance of these organisms, the control of disease and prevention of their spread was of obvious importance. However, it was not until Louis Pasteur was commissioned by the French government in 1865 to save the cultivation of *Bombyx mori* (mulberry silkworm) and *Bombyx yamamai* (oak silkworm) for silk production that the causes of these diseases were identified as microbial in origin (Takeda, 2014). Pasteur's studies and implementation of hygiene practices to separate overtly diseased silkworms from healthy ones saved the French sericulture industry from ruin (Ewan and Steinhaus, 1976; Steinhaus, 1956). These studies are thought to have inspired the field of insect pathology, but perhaps more importantly they were foundational for the germ theory of disease, attributable to both Pasteur and Agostino Bassi, whose 25 years of work on silkworm diseases preceded that of Pasteur's but was not as widely recognized (Ewan and Steinhaus, 1976; Steinhaus, 1956).

In addition to microbial infections, honeybees can be afflicted by a hive scavenger, the beemoth

Arthropod Vector: Controller of Disease Transmission, Volume 1
http://dx.doi.org/10.1016/B978-0-12-805350-8.00002-7

or waxworm, which we now know as *Galleria mellonella* in the family *Pyralidae*. This organism has been documented for nearly two-and-a-half millennia, with perhaps the first written description by Aristotle (384–322 BC), followed by the Roman agricultural author Lucius Columella in the 1st century AD and by Jan Swammerdam and Carl Linnaeus in Europe in the 1600s and 1700s, respectively (Paddock, 1918). Today, this curious pest is easily controlled with routine hygiene practices, but its hive-dwelling behavior at elevated temperatures (optimally 35°C) has contributed to interest in *G. mellonella* as a temperature-matched model host for human pathogens (Cook and McArthur, 2013).

Contributions to our understanding of infection and immunity in arthropods have risen exponentially over the past five decades, with nearly 6100 papers identifiable as "insect immunity" in PubMed since 1970. In contrast, prior to 1970 only 50 or so publications on insect immunity can be found. This is likely a reflection of the rise of model insects, not the least of which are honeybees and a large variety of silkworms, following centuries of domestication and handling, but also fruit flies which have become domesticated as "human commensals" (Keller, 2007). This moniker derives from the fact that all extant populations of this species feed on domesticated fruits or other food sources from human activity. This characteristic has allowed *Drosophila melanogaster* to spread throughout the world from its Afrotropical origins, including to the United States in 1875 in a "jar of pickled plums" (New York (State). State Entomologist and Lintner, 1882). The rapid rise of local populations, along with its ease of rearing and its short and predictable generation time, quickly endeared *D. melanogaster* to biologists, including William E. Castle and Thomas Hunt Morgan. Castle and Morgan started to work with this fruit fly in 1900–01, just a few years after its introduction into the United States following the suggestion of Charles Woodworth, a Harvard entomologist who was the first to breed *D. melanogaster* in captivity. While genetic discoveries were made rapidly with this "human commensal," Swedish microbiologists Hans Boman and Bertil Rasmusson published in *Nature* in 1972 the first observations on humoral immunity in *D. melanogaster* (Boman et al., 1972). This immune response, initiated by the injection of nonpathogenic bacteria, protected the flies from a second lethal dose of *Pseudomonas*. Although the response was robust and potent, the prospects of biochemical purification of the active factor from thousands of tiny flies set Boman on a path to use Saturniid moths introduced to him by Harvard entomologist Carroll Williams as "one-test-tube animals" (Boman et al., 1974). In 1981, Boman and his colleagues (Steiner et al., 1981) not only identified novel small antimicrobial peptides (AMPs)—including the "P9" peptides or cecropin A and B, famously lost from initial gels due to their small sizes—but also established the conservation of this biology across highly diverse species from Lepidoptera to Diptera (Faye and Lindberg, 2016).

In brief, insects in agriculture and other aspects of the human landscape have informed our fundamental understanding of infection and disease, first in an economic context and now as easily manipulated model organisms. In addition to domesticated models, however, arthropods that have exploited vertebrates for their nutrient-rich blood have concurrently and unwittingly become important couriers of pathogens between themselves and the vertebrates they feed on. As humans, we have sought for centuries to better understand our own biology of infection and mechanisms of disease. Here, we argue that considerable insight into these phenomena can be gained by exploring the biology of arthropods at the blood-feeding interface.

THE BLOOD-FEEDING INTERFACE

Arthropod vectors of disease-causing pathogens consume blood primarily as a nutritional resource for reproduction and self-maintenance.

We endeavor to show here, however, that protein and chemical factors from vertebrate blood also signal response pathways in the vector to regulate an array of biological responses, most of which are relevant to vectorial capacity. Because these blood factors have distinct biological functions in the vertebrate host, vector arthropods give us the opportunity to study the evolutionary origins of signaling mediated by these factors and to directly query the associated functional implications for human and animal health. To understand this biology—and the importance of it to vector–pathogen interactions—we review relevant arthropod evolution and the challenges facing those arthropods that evolved to consume vertebrate blood.

Coincident and rapid radiation in plants and arthropods approximately 350–450 million years ago yielded a diversity of terrestrial, aquatic, and marine arthropods that fed on plants and other invertebrates in these environments (Misof et al., 2014). These lifestyles facilitated the acquisition of bacteria, viruses, and eukaryotic microorganisms that could be regurgitated during feeding, acquired internally or externally by developing offspring or consumed via predation of the arthropod or arthropod feces by other invertebrates. With the multiple independent origins of blood feeding across divergent arthropod groups—at least 20 times in arthropods, with the Prostigmata and Astigmata broadly estimated at 400–100 MYA and the Insecta estimated between late Jurassic (200 MYA) to the Tertiary (65 MYA)—came radical new biology on many fronts (Mans, 2011). In this new host interface, the survival of previously acquired microorganisms depended on adaptation to cell biology and signaling pathways that were similar enough between arthropods and vertebrates to facilitate their transition to vector-borne pathogens over evolutionary time. In this context, we argue that a closer examination of vector-borne pathogens in their primeval arthropod hosts can reveal the most fundamental aspects of host–pathogen interactions. While blood feeding provided new opportunities for microorganisms to interact with arthropod and vertebrate hosts, however, it also presented unique challenges.

For small arthropods, obtaining and digesting blood for nutrition and reproduction is a challenging way of life. For many hematophagous insects, including mosquitoes, midges, and blackflies, blood feeding is female specific and the derived nutrients are used to successfully complete a reproductive cycle. In these insects, nectar or sugar is acquired as a nutritional supplement in females and the sole food source for males. However, other arthropods, including tsetse flies, kissing bugs, bedbugs, lice, fleas, and ticks, utilize vertebrate blood as their sole nutritional source and thus both males and females blood-feed. Regardless of sex specificity, acquiring a blood meal requires the arthropod to locate and identify a mobile host using a broad range of visual and chemosensory cues (Cork et al., 1996; Gibson and Torr, 1999; Lehane, 2005; Takken and Verhulst, 2013). These can include heat, carbon dioxide, and visual cues to identify the vertebrate host over considerable distances, and chemosensory and gustatory cues for host identification at close range. Once a host is identified and feeding begins, the arthropod must evade detection and mitigate a wide range of host defenses including vasoconstriction, coagulation, and immune responses. These challenges are overcome by a variety of salivary compounds, including anticoagulants, vasodilators, anesthetics, and immunomodulators. For example, orthologous apyrases in the saliva of mosquitoes, sand flies, bed bugs, tsetse flies, kissing bugs, and ticks function as anticoagulants (Hughes, 2013). In contrast to the apyrases, vasodilators and anesthetics tend to be more diverse and species specific. This pharmacology has been extensively studied to identify the sialome or the sum of salivary gland products, with hard ticks having among the most complex collections due to their prolonged periods of host attachment.

Once a host has been identified and fed upon, the physiological challenges of hematophagy must be overcome. First, most blood-feeding arthropods take discrete blood meals and engorge themselves when they are able to acquire blood. Thus, blood meals are often large in proportion to body size, doubling or tripling the blood-feeder's weight and limiting mobility (Gwadz, 1969; Pietrokovsky et al., 1996). Accordingly, mechanisms have evolved to quickly reduce the mass of the blood meal through diuresis and digestion of the meal itself (Benoit and Denlinger, 2010). The diuretic hormones serotonin and corticotropin can induce the rapid secretion of excess water through the Malpighian tubules and digestion is accomplished through the rapid synthesis and secretion of proteases, lipases, and glycosylases into the midgut lumen. The latter enzymes allow for the rapid breakdown of blood meal nutrients for absorption. Despite these adaptations, blood itself can have varying quantities of essential nutrients, possibly affecting fitness and reproductive capacity in arthropods with narrow host ranges (Takken and Verhulst, 2013).

In addition to water elimination and the need for a complex array of digestive enzymes for a meal that can be nutritionally challenging, blood-feeding arthropods must manage a variety of toxic and bioactive compounds in the blood meal. As hemoglobin is digested, the toxic prooxidant heme is released, increasing oxidative stress and membrane destabilization in the midgut. To prevent this damage, some blood-feeding insects such as mosquitoes and sandflies sequester free heme in the peritrophic matrix surrounding the blood meal (Devenport et al., 2006; Pascoa et al., 2002). The kissing bug *Rhodnius prolixus* utilizes a heme degradation pathway to detoxify this molecule (Galay et al., 2015; Paiva-Silva et al., 2006). Vertebrate blood also includes a wide array of hormones and cytokines, some of which can directly interact with the midgut of the blood-feeding insect (Pakpour et al., 2013a). These can include growth factors such as insulin and insulin growth factor 1 (IGF-1), as well as cytokines such as transforming growth factor (TGF)-β1. Hormone binding proteins, including IGF-1 Binding Protein (IGFBP), in the saliva and/or blood meal may have broader roles in arthropod physiology (Radulović et al., 2015). Ingested antibodies can also interact with lumenal surface proteins of the midgut epithelium, a fact that has been exploited to control blood-feeding insects or the pathogens they imbibe. For example, when cattle were immunized against hard or soft tick midgut antigens, those ticks that fed on immunized cattle had lyzed midguts, resulting in greatly reduced tick fitness or tick death. Pathogen targeting has also been achieved via delivery of blood meal-derived antibodies: mosquito midgut antibodies can block development of *Plasmodium berghei*, *Plasmodium falciparum*, and *Plasmodium vivax* parasites in the insect host (Atkinson et al., 2015; Lal et al., 1994, 2001; Sandeu et al., 2016), while both arthropod- and pathogen-targeted antibodies have been proposed for development of transmission-blocking vaccines for tick-borne pathogens (Gomes-Solecki, 2014; Merino et al., 2013).

Finally, the blood meal can contain pathogens ranging from simple viruses to multicellular nematodes. While many viruses may have a minimal impact on their blood-feeding vectors, larger parasites can adversely impact arthropod fitness. For example, ingestion of large numbers of *D. immitus* microfilariae can cause significant midgut damage as the nematodes penetrate through the midgut epithelium into the hemocoel. Tissue damage also occurs as the nematodes grow and ultimately exit the Malpighian tubules. In fact, if more than a handful of these parasites develop within the mosquito, the result is often lethal. Nematode damage can have epidemiological consequences as well. Notably, midgut penetration of filarial parasites can enhance passage of mosquito-borne viruses from the midgut into the hemocoel, increasing the potential for salivary gland infection and transmission (Vaughan et al., 1999, 2009, 2012).

Like mosquitoes, mites can suffer reduced fitness from infection with filarial worms. In particular, infection with *Litomosoides sigmodontis* enhances mite secretion of factors that stimulate vertebrate host immunity, which can limit blood-feeding success (Nieguitsila et al., 2013). Even single-celled parasites, such as *Plasmodium* in mosquitoes, can negatively impact vector fitness through direct effects on reproduction (Ahmed and Hurd, 2006; Hogg and Hurd, 1997; Hogg et al., 1997; Jahan and Hurd, 1998).

Considering the numerous challenges to acquiring a blood meal, it is surprising that arthropods have so successfully taken advantage of this resource. To be certain blood is nutrient rich, we would argue that the costs associated with blood-feeding are balanced not only by the nutritional and reproductive potential of blood, but also by the physiological advantages gained from decoding the signaling "information" of sugars, small molecules, peptides, proteins, and lipids that are present in blood. Importantly, the fundamental pathways that respond to these highly conserved signals appeared with emergence of multicellularity in animal species (Böhm et al., 2002; Huminiecki et al., 2009; Skorokhod et al., 1999) and are present in all extant animals, indicating that these ancient pathways are critical to animal success, including perhaps the regulation of information transferred with ingested blood (Fig. 2.1).

ANCIENT REGULATORY PATHWAYS OF HOMEOSTASIS: IIS, TGF-β, MAPK

Processing of blood takes time—4h even in the fastest feeders (lice; Lehane, 1991) to 7–10 days in ticks—providing ample time for signaling factors in blood to bind to midgut epithelial receptors and activate cell signaling pathways in this tissue. Indeed, these interactions can occur very quickly: we have demonstrated that protein kinase phosphorylation in the mosquito midgut epithelium can occur within minutes of blood ingestion (Drexler et al., 2013, 2014; Kang et al., 2008; Pietri et al., 2016; Surachetpong et al., 2009). Coincident translocation of blood factors out of the lumen (Drexler et al., 2013) and subsequent synthesis of new gene products and metabolites (Luckhart et al., 2013) from signaling activated transcription can amplify this biology via tissue-to-tissue communication. Accordingly, the midgut tissue of vector arthropods is significantly endowed with receptors and kinase cascades to decode the signaling "information" of biomolecules present in

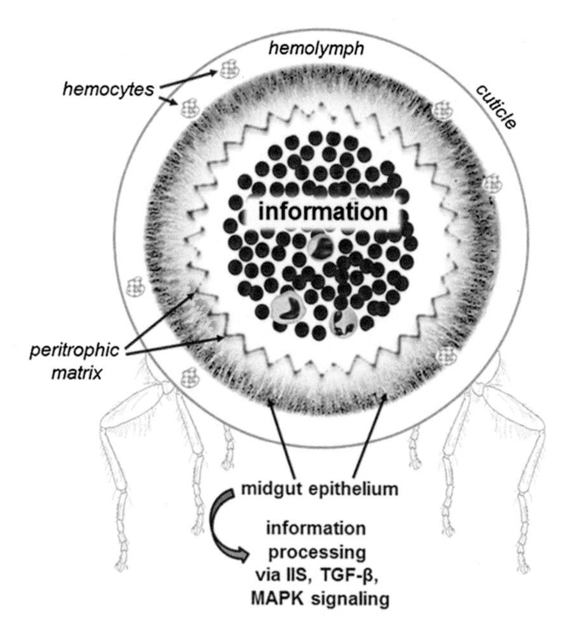

FIGURE 2.1 Processing information from the blood meal. The blood meal–midgut interface represents a key signaling center that coordinates information from the bloodmeal to regulate broad physiological effects including pathogen resistance and fitness traits. Information derived from exogenous blood meal components interact with key receptors at the midgut epithelium. This activates ancient, highly conserved signaling pathways including the insulin/insulin growth factor 1 signaling (IIS) cascade, the transforming growth factor β (TGF- β) signaling pathway, and mitogen-activated protein kinase (MAPK) cascades. These cascades in turn act both locally and systemically to regulate resistance and fitness through mitochondrial activity and turnover.

blood (Pakpour et al., 2013a, 2014). The strong conservation of both the structure and function of these signaling pathways allows us to leverage model invertebrates to posit that the vector arthropod gut is an information processing center for both endogenous and exogenous signals.

In both *D. melanogaster* and *Caenorhabditis elegans*, the midgut and intestine, respectively, are information processing and signaling centers that integrate communication among gut cells (autonomous signaling) and between the gut and other tissues (nonautonomous signaling) for the regulation of diverse physiological processes (Biteau et al., 2010; Hur et al., 2013; Jasper, 2015; Li et al., 2013; Li and Jasper, 2016; Okumura et al., 2014; Regan et al., 2016; Reiff et al., 2015; Rera et al., 2012b; Rogers and Rogina, 2014; Ulgherait et al., 2014; Wang et al., 2014). Transduction via three ancient cell signaling pathways—the TGF-β pathway, the mitogen-activated protein kinase (MAPK) pathway, and insulin/insulin-like growth factor signaling (IIS)—are networked to regulate innate immunity, stress resistance, longevity, and metabolism in these organisms. Central to signaling regulation of these processes are mitochondrial function and bioenergetics (Hur et al., 2013; Li and Jasper, 2016; Pellegrino et al., 2014; Regan et al., 2016; Rogers and Rogina, 2014; Ulgherait et al., 2014; Wang et al., 2014). We first review these signaling pathways and their key involvement in mitochondrial function in more detail, then move on to mitochondrial regulation of life history traits with particular emphasis on vector arthropods.

The Pathways: IIS, TGF-β, MAPK

IIS plays a central role in the regulation of lifespan, resistance to infection, metabolism, and oxidative stress responses in mammals and the model invertebrates *C. elegans* and *D. melanogaster*. We and others demonstrated that activation of IIS in mosquitoes by ingested human insulin and IGF-1, as well as by endogenous mosquito insulin-like peptides (ILPs) regulates egg production (Arik et al., 2009; Brown et al., 2008; Graf et al., 1997; Gulia-Nuss et al., 2011; Riehle and Brown, 1999, 2002, 2003; Roy et al., 2007), longevity (Corby-Harris et al., 2010; Hauck et al., 2013; Kang et al., 2008), resistance (Corby-Harris et al., 2010; Drexler et al., 2014; Horton et al., 2010; Lim et al., 2005; Marquez et al., 2011; Pakpour et al., 2012; Pietri et al., 2016; Surachetpong et al., 2011), metabolism (Drexler et al., 2014; Pietri et al., 2016; Sim and Denlinger, 2008) and oxidative stress (Kang et al., 2007; Luckhart et al., 2013). This work verifies that the full range of IIS phenotypes is, in fact, conserved in mosquitoes. Binding of an ILP (i.e., insulin, IGF, invertebrate ILP) to the insulin receptor (IR) (Luckhart and Riehle, 2007); induces IR autophosphorylation. IR ligands differentially activate two major downstream pathways. Activation of phosphatidylinositol 3-kinase (PI-3K)/Akt phosphorylates forkhead box class O family member protein (FOXO), excluding it from the nucleus and preventing FOXO-dependent transcription from regulating metabolism, redox [nitrogen oxide (NO)/reactive oxygen species (ROS)] levels, longevity, and resistance(Murphy et al., 2003; van der Vos and Coffer, 2011). Activation of the MAPK (MEK/ERK) branch regulates tissue health and resistance (Horton et al., 2011; Wymann et al., 2003).

The TGF-β signaling pathway controls wound healing and immunity, metabolism, and oxidative stress responses in both vertebrates and invertebrates (Morikawa et al., 2016). Further, there is ample evidence of varying levels of conservation of the signaling ligands and pathways across a diversity of vector arthropods (Pakpour et al., 2013a, 2014). The TGF-β superfamily is comprised of prototypical TGF-β ligands, the decapentaplegic (Dpp)-Vg-related (DVR) subfamily [growth and differentiation factors and the bone morphogenetic proteins (BMPs)], and the activins and inhibins (Morikawa et al., 2016). Canonical signaling by these ligands is mediated by heterodimeric complexes of type I and type II receptors and the Smad family of signaling proteins (Morikawa et al., 2016). In general, ligand-bound TGF-β/Activin and Dpp/BMP receptor complexes recruit and phosphorylate receptor Smads (R-Smads), which are released to form complexes with Co-Smads. These complexes move into the nucleus and interact with other DNA-binding proteins to regulate gene transcription. In *D. melanogaster*, BMP signals are transduced by the R-Smad, Mad, and the Co-Smad Medea, while TGF-β/Activin-like signals are transduced by the R-Smad, dSmad2, and Medea (Peterson and O'Connor, 2014). Signaling through both pathways appears to be terminated by the inhibitory Smad Dad. In addition to this canonical signaling, ligand-bound TGF-β receptor activation can transduce the activation of the stress-associated MAPKs, including TAK-1-mediated activation of p38 MAPK and c-Jun N-terminal kinase (JNK) (Heldin and Moustakas, 2016; Mu et al., 2012).

IIS, TGF-β, and MAPK Regulation of Mitochondrial Function

In model invertebrates (Mukherjee et al., 2014; Sadagurski and White, 2013; Tiefenbock et al., 2010; Tóth et al., 2008), vector mosquitoes (Drexler et al., 2014; Luckhart et al., 2013; Pietri et al., 2016), and mammals (Cheng et al., 2010; Sadagurski and White, 2013), many of the physiological effects of both branches of IIS are mediated through the regulation of mitochondrial function. In *D. melanogaster* and *C. elegans*, IIS and associated changes in mitochondrial function contribute to an interorgan signaling network (Droujinine and Perrimon, 2014) that underlies essential physiological processes. In this context, direct insulin signaling (i.e., IIS-to-IIS, IIS-to-ILP secretion) across tissues appears to impact a narrow, but important, range of cell processes (i.e., proteostasis in the fruit fly (Alic et al., 2014; Demontis and Perrimon, 2010), while the larger impact of IIS interorgan communication appears to be mediated via local mitochondrial changes that can alter systemic mitochondrial function. The fruit fly gut and nematode intestine tissues that are perhaps the most sensitive to stress are intimately involved in organ-to-organ IIS communication (Alic et al., 2014; Owusu-Ansah and Perrimon, 2015). For example, a decline in IIS-associated mitochondrial biogenesis in many fruit fly tissues is associated with aging, but systemic enhancement of biogenesis does not increase fly longevity. However, increased gut-specific mitochondrial biogenesis in fruit flies can enhance longevity through finely tuned effects on systemic mitochondrial function (Rera et al., 2012b; Ulgherait et al., 2014; Vellai, 2009). Further, increased gut mitochondrial biogenesis improves the local epithelial barrier, preventing gut microbes from causing systemic infection (Rera et al., 2012b). Mitochondrial changes in *D. melanogaster* fat body and muscle can also signal non-autonomously to independently contribute to IIS phenotypes (Owusu-Ansah and Perrimon, 2015), indicating that mitochondrial changes in these tissues can functionally extend mitochondrial changes in the gut.

The TGF-β pathway can also directly and indirectly influence mitochondrial activity and in turn regulate cell survival and ROS production. TGF-β signaling is a key regulator of apoptosis, in part through disruption of the mitochondrial membrane and through increased ROS production as demonstrated in mink lung epithelial (Mv1Lu) cells (Casalena et al., 2012; Yoon et al., 2005). In these cells, increased ROS levels are induced through specific suppression of complex IV activity in the electron transport chain (ETC). Interestingly, in mouse podocyte cells treated with TGF-β, increased ROS production was also observed, but this effect was not due to suppression of the ETC (Abe et al., 2013). Instead these cells had increased mitochondrial activity, as assessed by increased oxygen consumption rates, which led to the observed increase in ROS production. It was determined that TGF-β-induced mitochondrial activity was regulated by the mTOR pathway, as demonstrated through both phosphorylation and activation of key mTOR signaling molecules, tuberin and mTOR, and through suppression of the mTOR pathway using rapamycin. In addition to increased ROS production, studies in both Mv1Lu cells and hepatocytes demonstrated that TGF-β signaling can also reduce the membrane potential across the inner membrane of the mitochondria (Herrera et al., 2001; Rodrigues et al., 1999; Yoon et al., 2005). It is thought that ROS, and the associated damage and membrane permeability it induces, is responsible for this reduction in membrane potential. However, studies in podocytes revealed the opposite effect with TGF-β increasing mitochondrial membrane potential (Abe et al., 2013). This effect could be reversed by the mTOR inhibitor rapamycin. While TGF-β influences mitochondrial activity and ROS production, mitochondrial ROS has also been shown to affect TGF-β regulation of gene expression in human lung fibroblasts (Jain et al., 2013). In sum, TGF-β signaling to the mitochondria increases mitochondrial ROS production and membrane potential either through disruption of the ETC or increased mitochondrial respiration, and in turn the increased ROS production can positively increase TGF-β transcriptional activity.

JNK- and p38 MAPK-dependent signaling have both been implicated in regulating mitochondrial activity as well, although the mechanisms are less well understood. In particular, both p38 MAPK and JNK are typically found in the cytoplasm and nucleus, but several studies have found these signaling molecules to also be localized in mitochondria (Baines et al., 2002; Ito et al., 2001; Lim et al., 2016). JNK activity in the mitochondria can regulate apoptosis through phosphorylation of Bcl-2-associated death promoter (BAD) protein, but depending on cell type can either activate or inhibit BAD and in turn apoptosis. Similarly, p38 MAPK can also stimulate apoptosis through its effects on mitochondria, but the mode of action remains elusive. Thus, although these two stress response signaling cascades are involved in mitochondrial activity, it is clear that additional research into their precise roles and regulation is essential to understanding the interplay between these signaling cascades and their regulation of mitochondrial activity.

IIS, TGF-β, and MAPK Regulation of Mitochondrial Biogenesis and Turnover

The balance between mitochondrial biogenesis and the clearance of damaged mitochondria via a selective autophagic process called mitophagy is dynamic and necessary for mitochondrial "quality control." Hence, this balance is highly responsive to metabolic changes, energy needs, resistance to infection, and other stresses to ensure tissue homeostasis and appropriate cellular responses (Fig. 2.2; Youle and van der Bliek, 2012). Autophagy can be initiated by oxidative stress and damage to sensitive cellular targets, particularly mitochondria, leading to the initiation of mitochondrial fission or the segregation of damaged mitochondria for elimination by mitophagy (Elgass et al., 2012). In brief, accumulation of phosphatase and tensin homolog (PTEN)-induced putative kinase 1 (PINK1) on the surface of damaged mitochondria results in the recruitment and aggregation of the E3 ubiquitin ligase Parkin, which activates a cascade of Atg proteins to induce autophagosome formation. Signals that promote autophagy include mitochondrial ROS, ERK-dependent signaling, and energy deprivation-induced AMP-activated protein kinase (Huang et al., 2011; Levine et al., 2011). Conversely, the PI-3K/Akt branch of the IIS cascade inhibits autophagy (Denton et al., 2012; Huang et al., 2011; Levine et al., 2011). On the other side of the spectrum from mitophagy is the creation of new mitochondria through mitochondrial biogenesis. While numerous signaling cascades feed into mitochondrial biogenesis, there are a handful of gatekeeper genes that maintain this critical balance, and perhaps the most important is PPAR gamma coactivator-1 alpha (PGC-1). The IIS cascade is a key regulator of mitochondrial biogenesis, acting through

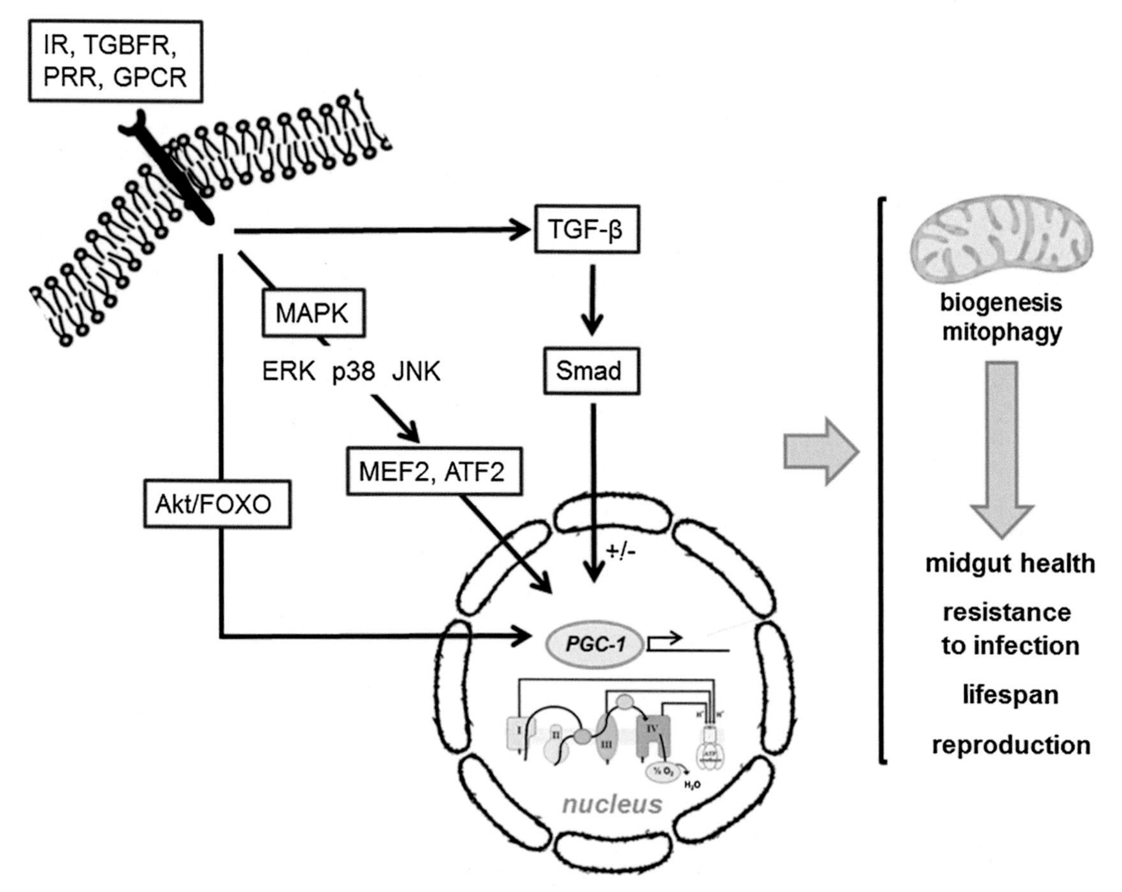

FIGURE 2.2 **Mitochondrial quality control.** The generation of new, healthy mitochondria (biogenesis) and the recycling of old, damaged mitochondria (mitophagy) is highly regulated and responsive to external stimuli. This balance is controlled in large part through PPAR gamma coactivator-1 alpha (PGC-1), a key gatekeeping molecule that regulates mitochondrial biogenesis and mitophagy as dictated by signals from the IIS (Akt/FOXO), MAPK (ERK, p38 MAPK, JNK), and TGF-β signaling cascades. These cascades are activated by membrane receptors (insulin receptor, IR; TGF-β receptor, TGBFR; pattern recognition receptor, PRR; G-protein-coupled receptor, GPCR) that receive input in response to blood meal-components. This mitochondrial balance allows the organism to quickly respond to changing energy needs and also impacts physiologies critical for pathogen transmission, including infection resistance, midgut health, lifespan, and reproduction.

PGC-1. In mammalian cells, when FOXO is excluded from the nucleus, there is no induction of PGC-1 (Fernandez-Marcos and Auwerx, 2011), whereas PTEN overexpression induces expression of PGC-1 (Garcia-Cao et al., 2012). In *D. melanogaster*, overexpression of the FOXO-dependent ortholog of PGC-1 led to increased levels of complexes I, III, IV, and V of the ETC and increased ETC activity, indicating that PGC-1 control of mitochondrial biogenesis is highly conserved (Rera et al., 2011). Increased PGC-1 expression in the *D. melanogaster* midgut also increased midgut barrier integrity (Rera et al., 2011), a phenotype that we have associated with increased resistance to *P. falciparum* in *Anopheles stephensi* in the absence of any change in NF-κB-dependent anti-parasite genes (Hauck et al., 2013; Pakpour et al., 2012, 2013b). PGC-1 is induced by NO-activated guanylate cyclase (Nisoli et al., 2003, 2004, 2005), as well as by AMPK (Fernandez-Marcos and Auwerx, 2011). Hence, IIS-dependent AMPK and PGC-1 control mitochondrial function for tissue homeostasis.

While the role p38 MAPK plays in regulating mitochondrial activity is poorly understood, its role in regulating mitochondrial biogenesis has been better established. Overexpression of p38 MAPK in mouse skeletal muscle led to increased PGC-1 expression (Akimoto et al., 2005). The activation of p38 MAPK leads to the subsequent activation of two molecules, activating transcription factor 2 (ATF2) and myocyte enhancer factor 2 (MEF2), which can regulate the expression of PGC-1 and in turn mitochondrial biogenesis (Han et al., 1997; Handschin et al., 2003; Zhao et al., 1999). In addition, p38 MAPK can directly activate PGC-1 through phosphorylation of key threonine and serine residues, making it resistant to inhibitors (Puigserver et al., 2001; Sano et al., 2007). The MAPK cascades also play important roles in autophagy. JNK signaling regulates autophagy through both transcriptional control of autophagy genes and through the phosphorylation of key autophagy proteins. JNK transcriptional control of autophagy occurs through the activation of downstream transcription factors including FOXO that in turn regulate the expression of Atg genes. JNK signaling can also phosphorylate Bcl-2 which under normal, non-phosphorylated conditions is bound to Beclin1, an important protein involved in nearly all steps of the autophagy process (Zhou et al., 2015). Phosphorylation of Bcl-2 by JNK signaling leads to its dissociation from the Bcl-2/Beclin1 complex, freeing Beclin1 and allowing autophagy to proceed. The role of JNK signaling through both transcriptional regulation and phosphorylation of Bcl-2 is further detailed in a review by Zhou et al. (2015).

As with the IIS cascade, there is evidence that the TGF-β pathway regulates mitochondrial biogenesis via both AMPK and PGC-1. Direct activation of PGC-1 through TGF-β signaling, however, has not been demonstrated. In fact, stimulation of mouse podocytes with TGF-β did not lead to increased mitochondrial numbers or mitochondrial DNA levels (Abe et al., 2013). However as discussed earlier, there is evidence that TGF-β signaling acts in part through the mTOR nutrient sensing pathway, which can activate PGC-1 and initiate mitochondrial biogenesis (Andres et al., 2015), suggesting that TGF-β may indirectly influence this process. TGF-β signaling has also been implicated in autophagy. In *C. elegans*, an RNAi screen to identify genes and pathways involved in regulating autophagy implicated both TGF-β and daf-2 (*C. elegans* IR ortholog) signaling as important regulators of autophagy (Guo et al., 2014). In particular, two *C. elegans* Smads, sma-3 and sma-4, were identified in an initial screen of 139 putative autophagy gene targets. Subsequent analysis found several additional components of the *C. elegans* Sma TGF-β pathway were involved in autophagy.

MITOCHONDRIAL DYNAMICS CONTROLS DIVERSE PHYSIOLOGIES THAT ARE KEY TO VECTOR COMPETENCE

Since the 1990s, it has become clear that mitochondrial contributions to cell fitness extend well beyond biosynthesis and bioenergetics in that these organelles communicate within and across cells to ensure that cellular commitment to biological processes does not occur without assurance that the energy and metabolites are available to support these processes (Chandel, 2014). In this role, the mitochondrial outer membrane can function as a scaffold for signaling protein complexes, thereby directly integrating transduction of environmental stimuli with changes in mitochondrial biogenesis and ETC activity. The scaffolding function regulates the balance between stress/cell death and homeostasis, and this balance can regulate host defense or immunity and lifespan in invertebrates (Chandel, 2014).

In 2012, Cloonan and Choi concluded that agents that improve mitochondrial quality control can be leveraged to develop novel interventions for human diseases. Control of mitochondrial quality is essential for an analogous spectrum of physiologies in model invertebrates as well. In *C. elegans*, reduced IIS alters the rate of aging by an unidentified mechanism with a protective effect on mitochondrial function (Brys et al., 2010). Further, autophagy in *C. elegans* daf-2 (IR) mutants, which are long-lived *and* infection-resistant, is necessary and sufficient for resistance and lifespan extension under reduced IIS (Jia et al., 2009; Melendez et al., 2003).

In *D. melanogaster*, overexpression of AMPK (Rera et al., 2011; Ulgherait et al., 2014) or PGC-1 in the midgut delayed age-related declines in midgut epithelial barrier integrity, indicating that autophagy and mitochondrial biogenesis protect midgut barrier integrity (Rera et al., 2011; Ulgherait et al., 2014). In an analogous fashion, PTEN overexpression in the midgut of transgenic *A. stephensi* enhanced midgut barrier integrity—without changes in NF-κB-dependent anti-parasite gene expression—leading to increased resistance against *P. falciparum* infection (Hauck et al., 2013). Given that Akt inhibits autophagy and PGC-1-mediated mitochondrial biogenesis (Denton et al., 2012; Huang et al., 2011; Levine et al., 2011), these findings are consistent with a scenario in which PTEN overexpression upregulates autophagy and activation of *A. stephensi* PGC-1 to improve midgut barrier integrity, fitness, and resistance to *P. falciparum* infection. Accordingly, we observed increased midgut expression of biomarkers for autophagy, as well as midgut stem cell (MSC) activity in PTEN overexpressing *A. stephensi* relative to nontransgenic controls (Hauck et al., 2013).

An optimal balance of mitochondrial biogenesis and mitophagy is critical for stem cell maintenance and differentiation. Stem cells have long been thought to play an important role in the midgut of insects, including *Aedes aegypti* and *A. stephensi* (Baton and Ranford-Cartwright, 2007; Brown et al., 1985). In *D. melanogaster*, MSCs divide to produce enteroblasts (EBs) that differentiate into digestive/absorptive cells (DCs) or peptidergic endocrine (ECs) cells. MSCs and EBs express escargot (esg) and MSCs also express Delta (Dl), a ligand for the Notch receptor signal pathway. MSCs and EBs are small diploid cells scattered basally in the epithelium that are often associated with ECs expressing the transcription factor prospero (Brown et al., 1985). Upregulation of *D. melanogaster* PGC-1 in the fly midgut prevented MSC overproliferation and accumulation of DC-like esg and misdifferentiated cells (Rera et al., 2012a). This phenotype, along with improved epithelial barrier, confirmed that PGC-1 overexpression prevented MSC dysplasia and increased longevity. We observed that human IGF1 provided to *A. stephensi* in blood at levels consistent with those circulating during malaria induced midgut cytoprotection via homeostatic regulation of ROS, MSC activity, and autophagy, to enhance midgut repair and improve mitochondrial function (Drexler et al., 2014). These changes suggest that improved mitochondrial function improves midgut health for enhanced longevity and resistance (Drexler et al., 2014). Like mitochondrial biogenesis, autophagy is critical for midgut physiology in *D. melanogaster*; but to our knowledge, a critical role for autophagy is known only from larval midgut development (Denton et al., 2009, 2010, 2012) and has not yet been linked to MSC homeostasis in adult flies (Ulgherait et al., 2014).

Resistance to infection is also dependent on mitochondrial dynamics. In mammals, activation of mitochondrial biogenesis during illness (Sweeney et al., 2010, 2011), perhaps in response to host cell damage (Tran et al., 2011), regulates the synthesis of pro- and antiinflammatory cytokines (Dada and Sznajder, 2011; Piantadosi et al., 2011) and promotes host recovery and survival (Carré et al., 2010). Innate immunity in a variety of mammalian cells is dependent on mitochondrial ROS from ETC Complex I and III which potentiate Toll-like receptor signaling, activate NF-kB-dependent gene expression, and function in phagosomal killing (Arnoult et al., 2011). We have shown that IIS-dependent changes to mitochondrial function induces high levels of anti-parasite ROS and NO in *A. stephensi* (Luckhart et al., 2013) and that a shift to glycolysis rather than mitochondrial oxidative phosphorylation (OXPHOS) for ATP synthesis can function to protect the immune-responsive epithelium (Wang et al., 2015). In an effort to establish cause-and-effect of this biology with resistance in *A. stephensi*, we used small molecule inhibitors (SMIs) of glycolysis and mitochondrial function to directly demonstrate that shifts in midgut intermediary metabolism that are specific responses to *A. stephensi* ILP3 and ILP4, in fact, control *P. falciparum* infection (Pietri et al., 2016). Notably, these SMIs had no effect on expression of NF-kB-dependent immune genes (Pietri et al., 2016), indicating that direct manipulation of mitochondrial function can increase resistance to pathogen infection.

The involvement of autophagy in resistance to infection is an area of intense investigation. Indeed, autophagy controls many aspects of innate and adaptive immunity. The most compelling connection between IIS-dependent autophagy and immunity comes from *C. elegans*, in which the daf-2 (IR) mutant phenotype of infection resistance is reversed by mutations that inhibit autophagy (Jia et al., 2009). In *D. melanogaster*, repression of *Atg5*, *7* and *12* reduced autophagy in the midgut and reduced resistance to *E. coli* challenge, as evidenced by increased bacterial titers and reduced survival (Ren et al., 2009). In mammals, autophagy positively controls antibacterial, antiviral, and anti-parasite responses via antimicrobial peptide synthesis, pathogen killing, cytoprotection against microbial toxins and factors, apoptotic corpse clearance, and regulation of inflammatory transcriptional responses (Levine et al., 2011). Further, overexpression of autophagy genes promotes clearance of *Mycobacterium tuberculosis*-containing phagosomes in mouse macrophages and human myeloid cells in vitro (Gutierrez et al., 2004; Singh et al., 2006) protects against fatal Sindbis virus infection in mice (Liang et al., 1998), and localizes anti-pathogen hypersensitive responses in *Arabidopsis* (Liu et al., 2005).

Longevity is one of the most complex phenotypes regulated by mitochondrial physiology. In particular, moderate repression of mitochondrial ETC activity and OXPHOS has been shown to enhance longevity in yeast, *C. elegans*, *D. melanogaster*, and mice (reviewed in Rera et al., 2012a). In particular, enhanced ETC Complex I activity via overexpression of the single-subunit yeast NADH-ubiquinone oxidoreductase (Ndi1) in the *D. melanogaster* midgut can increase longevity and improve midgut barrier integrity with no negative effects on reproduction (Hur et al., 2013). The apparent paradox that longevity can be enhanced by both repression and activation of complex I is resolvable as an example of hormesis in that perturbations in either direction push an organism toward improved mitochondrial homeostasis during aging (Hur et al., 2014). In addition to optimal OXPHOS, autophagy is critical to lifespan extension in *C. elegans* daf2 (IR) mutants (Melendez et al., 2003) and in *D. melanogaster*. In particular, upregulation of Parkin, a key mediator of mitophagy, in *D. melanogaster* significantly increased longevity and egg production. Similarly, mutations in PINK1, which functions with Parkin to regulate mitophagy, have been associated with fly lifespan reduction (Clark et al., 2006). In our studies, we have shown that PTEN overexpression in *A. stephensi* enhances midgut barrier integrity, longevity, and *P. falciparum* resistance with no negative effects on reproduction (Hauck et al., 2013). These observations suggest that PTEN overexpression and predicted effects on PINK1/Parkin (Unoki and Nakamura, 2001) can enhance longevity and resistance in vector arthropods via mitochondrial quality control.

SUMMARY

It is clear that the blood meal conveys a wealth of information to the arthropod vector beyond simple nutritional value. This information is conveyed from the blood meal/midgut interface to the blood-feeding arthropod via a collection of highly conserved signaling cascades that have been fine-tuned to respond to specific signals from the vertebrate blood meal. The remarkable conservation of these fundamental signaling pathways not only allow the arthropod vector to recognize and process signals from the blood meal but provides us the opportunity to utilize our collective knowledge of these signaling cascades across taxa to expand our overall understanding of these pathways and the physiological functions they regulate. At the same time, there will undoubtedly be facets of these signaling cascades that are unique to different blood-feeding arthropods, which will allow us to explore the flexibility of these pathways to respond to both endogenous and exogenous signals.

As our understanding of these pathways has expanded, it has become clear that their central role in regulating mitochondrial physiology drives many of the downstream physiological effects, including regulating the capacity to transmit pathogens and optimization of vector fitness. However, we are just beginning to untangle the complex signaling interactions that regulate mitochondrial activity, biogenesis, and mitophagy, and how these changes to mitochondrial biology impact pathogen transmission, stress responses, and fitness. Elucidating how these complex pathways interact to influence broad insect physiologies in important blood-feeding disease vectors is essential to our ability to manipulate and harness these signaling cascades to mitigate the vectors and the pathogens they transmit. This is particularly true since manipulation of any one pathway will most certainly influence other pathways to respond to or compensate for changes in signaling. Only through acquiring an in-depth understanding of how signals from the blood meal and midgut are interpreted by these entangled signaling pathways can we begin to leverage them for our benefit.

References

Abe, Y., Sakairi, T., Beeson, C., Kopp, J.B., 2013. TGF-beta1 stimulates mitochondrial oxidative phosphorylation and generation of reactive oxygen species in cultured mouse podocytes, mediated in part by the mTOR pathway. Am. J. Physiol. Renal Physiol. 305, F1477–F1490.

Ahmed, A.M., Hurd, H., 2006. Immune stimulation and malaria infection impose reproductive costs in *Anopheles gambiae* via follicular apoptosis. Microbes Infect. 8, 308–315.

Akimoto, T., Pohnert, S.C., Li, P., Zhang, M., Gumbs, C., Rosenberg, P.B., Williams, R.S., Yan, Z., 2005. Exercise stimulates Pgc-1alpha transcription in skeletal muscle through activation of the p38 MAPK pathway. J. Biol. Chem. 280, 19587–19593.

Alic, N., Tullet, J.M., Niccoli, T., Broughton, S., Hoddinott, M.P., Slack, C., Gems, D., Partridge, L., 2014. Cell-nonautonomous effects of dFOXO/DAF-16 in aging. Cell Rep. 6, 608–616.

Andres, A.M., Stotland, A., Queliconi, B.B., Gottlieb, R.A., 2015. A time to reap, a time to sow: mitophagy and biogenesis in cardiac pathophysiology. J. Mol. Cell. Cardiol. 78, 62–72.

Arik, A.J., Rasgon, J.L., Quicke, K.M., Riehle, M.A., 2009. Manipulating insulin signaling to enhance mosquito reproduction. BMC Physiol. 9, 15.

Arnoult, D., Soares, F., Tattoli, I., Girardin, S.E., 2011. Mitochondria in innate immunity. EMBO Rep. 12, 901–910.

Atkinson, S.C., Armistead, J.S., Mathias, D.K., Sandeu, M.M., Tao, D., Borhani-Dizaji, N., Tarimo, B.B., Morlais, I., Dinglasan, R.R., Borg, N.A., 2015. The *Anopheles*-midgut APN1 structure reveals a new malaria transmission-blocking vaccine epitope. Nature Struct. Mol. Biol. 22, 532–539.

Baines, C.P., Zhang, J., Wang, G.W., Zheng, Y.T., Xiu, J.X., Cardwell, E.M., Bolli, R., Ping, P., 2002. Mitochondrial PKCepsilon and MAPK form signaling modules in the murine heart: enhanced mitochondrial PKCepsilon-MAPK interactions and differential MAPK activation in PKCepsilon-induced cardioprotection. Circ. Res. 90, 390–397.

Baton, L., Ranford-Cartwright, L., 2007. Morphological evidence for proliferative regeneration of the *Anopheles stephensi* midgut epithelium following *Plasmodium falciparum* ookinete invasion. J. Invertebr. Pathol. 96, 244–254.

Benoit, J.B., Denlinger, D.L., 2010. Meeting the challenges of on-host and off-host water balance in blood-feeding arthropods. J. Insect Physiol. 56, 1366–1376.

Biteau, B., Karpac, J., Supoyo, S., DeGennaro, M., Lehmann, R., Jasper, H., 2010. Lifespan extension by preserving proliferative homeostasis in *Drosophila*. PLoS Genet. 6, e1001159.

Böhm, M., Gamulin, V., Schröder, H.C., Müller, W.E., 2002. Evolution of osmosensing signal transduction in Metazoa: stress-activated protein kinases p38 and JNK. Cell Tissue Res. 308, 431–438.

Boman, H.G., Nilsson, I., Rasmuson, B., 1972. Inducible antibacterial defence system in *Drosophila*. Nature 237, 232–235.

Boman, H.G., Nilsson-Faye, I., Paul, K., Rasmuson Jr., T., 1974. Insect immunity. I. Characteristics of an inducible cell-free antibacterial reaction in hemolymph of *Samia cynthia* pupae. Infect. Immun. 10, 136–145.

Brown, M.R., Raikhel, A.S., Lea, A.O., 1985. Ultrastructure of midgut endocrine cells in the adult mosquito, *Aedes aegypti*. Tissue Cell 17, 709–721.

Brown, M.R., Clark, K.D., Gulia, M., Zhao, Z., Garczynski, S.F., Crim, J.W., Suderman, R.J., Strand, M.R., 2008. An insulin-like peptide regulates egg maturation and metabolism in the mosquito *Aedes aegypti*. Proc. Natl. Acad. Sci. U.S.A. 105, 5716–5721.

Brys, K., Castelein, N., Matthijssens, F., Vanfleteren, J.R., Braeckman, B.P., 2010. Disruption of insulin signalling preserves bioenergetic competence of mitochondria in ageing *Caenorhabditis elegans*. BMC Biol. 8, 91.

Carré, J.E., Orban, J.C., Re, L., Felsmann, K., Iffert, W., Bauer, M., Suliman, H.B., Piantadosi, C.A., Mayhew, T.M., Breen, P., 2010. Survival in critical illness is associated with early activation of mitochondrial biogenesis. Am. J. Respir. Crit. Care. Med. 182, 745–751.

Casalena, G., Daehn, I., Bottinger, E., 2012. Transforming growth factor-β, bioenergetics, and mitochondria in renal disease. Sem. Nephrol. 32, 295–303.

Chandel, N.S., 2014. Mitochondria as signaling organelles. BMC Biol. 12, 1.

Cheng, Z., Tseng, Y., White, M.F., 2010. Insulin signaling meets mitochondria in metabolism. Trends Endocrinol. Metab. 21, 589–598.

Clark, I.E., Dodson, M.W., Jiang, C., Cao, J.H., Huh, J.R., Seol, J.H., Yoo, S.J., Hay, B.A., Guo, M., 2006. *Drosophila* pink1 is required for mitochondrial function and interacts genetically with parkin. Nature 441, 1162–1166.

Cloonan, S.M., Choi, A.M., 2012. Mitochondria: commanders of innate immunity and disease? Curr. Opin. Immunol. 24, 32–40.

Cook, S.M., McArthur, J.D., 2013. Developing *Galleria mellonella* as a model host for human pathogens. Virulence 4, 350–353.

Corby-Harris, V., Drexler, A., Watkins de Jong, L., Antonova, Y., Pakpour, N., Ziegler, R., Ramberg, F., Lewis, E.E., Brown, J.M., Luckhart, S., Riehle, M.A., 2010. Activation of Akt signaling reduces the prevalence and intensity of malaria parasite infection and lifespan in *Anopheles stephensi* mosquitoes. PLoS Pathog. 6, e1001003.

Cork, A., Bock, G., Cardew, G., 1996. Olfactory basis of host location by mosquitoes and other haematophagous Diptera. Ciba. Found. Symp. 200, 71–88.

Dada, L.A., Sznajder, J.I., 2011. Mitochondrial Ca_2 and ROS take center stage to orchestrate TNF-α–mediated inflammatory responses. J. Clin. Invest. 121, 1683.

Demontis, F., Perrimon, N., 2010. FOXO/4E-BP signaling in *Drosophila* muscles regulates organism-wide proteostasis during aging. Cell 143, 813–825.

Denton, D., Shravage, B., Simin, R., Mills, K., Berry, D.L., Baehrecke, E.H., Kumar, S., 2009. Autophagy, not apoptosis, is essential for midgut cell death in *Drosophila*. Curr. Biol. 19, 1741–1746.

Denton, D., Shravage, B.V., Simin, R., Baehrecke, E.H., Kumar, S., 2010. Larval midgut destruction in *Drosophila*: not dependent on caspases but suppressed by the loss of autophagy. Autophagy 6, 163–165.

Denton, D., Chang, T., Nicolson, S., Shravage, B., Simin, R., Baehrecke, E., Kumar, S., 2012. Relationship between growth arrest and autophagy in midgut programmed cell death in *Drosophila*. Cell Death Differ. 19, 1299–1307.

Devenport, M., Alvarenga, P.H., Shao, L., Fujioka, H., Bianconi, M.L., Oliveira, P.L., Jacobs-Lorena, M., 2006. Identification of the *Aedes aegypti* peritrophic matrix protein AeIMUCI as a heme-binding protein. Biochemistry 45, 9540–9549.

Drexler, A., Nuss, A., Hauck, E., Glennon, E., Cheung, K., Brown, M., Luckhart, S., 2013. Human IGF1 extends lifespan and enhances resistance to *Plasmodium falciparum* infection in the malaria vector *Anopheles stephensi*. J. Exp. Biol. 216, 208–217.

Drexler, A.L., Pietri, J.E., Pakpour, N., Hauck, E., Wang, B., Glennon, E.K., Georgis, M., Riehle, M.A., Luckhart, S., 2014. Human IGF1 regulates midgut oxidative stress and epithelial homeostasis to balance lifespan and *Plasmodium falciparum* resistance in *Anopheles stephensi*. PLoS Pathog. 10, e1004231.

Droujinine, I.A., Perrimon, N., 2014. Defining the interorgan communication network: systemic coordination of organismal cellular processes under homeostasis and localized stress. Front. Cell. Infect. Microbiol. 3, 82.

Elgass, K., Pakay, J., Ryan, M.T., Palmer, C.S., 2012. Recent advances into the understanding of mitochondrial fission. Biochim. Biophys. Acta 1833, 150–161.

Ewan, J., Steinhaus, E.A., 1976. Disease in a Minor Chord. Ohio State University Press, pp. 104–105.

Faye, I., Lindberg, B.G., 2016. Towards a paradigm shift in innate immunity-seminal work by Hans G. Boman and co-workers. Philos. Trans. R. Soc. London. Series B, Biol Sci. 371. http://dx.doi.org/10.1098/rstb.2015.0303.

Fernandez-Marcos, P.J., Auwerx, J., 2011. Regulation of PGC-1α, a nodal regulator of mitochondrial biogenesis. Am. J. Clin. Nutr. 93, 884S–890S.

Galay, R.L., Umemiya-Shirafuji, R., Mochizuki, M., Fujisaki, K., Tanaka, T., 2015. Iron metabolism in hard ticks (Acari: Ixodidae): the antidote to their toxic diet. Parasitol. Int. 64, 182–189.

Garcia-Cao, I., Song, M.S., Hobbs, R.M., Laurent, G., Giorgi, C., de Boer, V.C.J., Anastasiou, D., Ito, K., Sasaki, A.T., Rameh, L., 2012. Systemic elevation of PTEN induces a tumor-suppressive metabolic state. Cell 149, 49–62.

Gibson, G., Torr, S., 1999. Visual and olfactory responses of haematophagous Diptera to host stimuli. Med. Vet. Entomol. 13, 2–23.

Gomes-Solecki, M., 2014. Blocking pathogen transmission at the source: reservoir targeted OspA-based vaccines against *Borrelia burgdorferi*. Front. Cell. Infect. Microbiol. 4, 136.

Graf, R., Neuenschwander, S., Brown, M.R., Ackermann, U., 1997. Insulin-mediated secretion of ecdysteroids from mosquito ovaries and molecular cloning of the insulin receptor homologue from ovaries of bloodfed *Aedes aegypti*. Insect. Mol. Biol. 6, 151–163.

Gulia-Nuss, M., Robertson, A.E., Brown, M.R., Strand, M.R., 2011. Insulin-like peptides and the target of rapamycin pathway coordinately regulate blood digestion and egg maturation in the mosquito *Aedes aegypti*. PLoS One 6, e20401.

Guo, B., Huang, X., Zhang, P., Qi, L., Liang, Q., Zhang, X., Huang, J., Fang, B., Hou, W., Han, J., Zhang, H., 2014. Genome-wide screen identifies signaling pathways that regulate autophagy during *Caenorhabditis elegans* development. EMBO Rep. 15, 705–713.

Gutierrez, M.G., Master, S.S., Singh, S.B., Taylor, G.A., Colombo, M.I., Deretic, V., 2004. Autophagy is a defense mechanism inhibiting BCG and *Mycobacterium tuberculosis* survival in infected macrophages. Cell 119, 753–766.

Gwadz, R.W., 1969. Regulation of blood meal size in the mosquito. J. Insect. Physiol. 15 2039, IN3, 2043–2042–2044.

Han, J., Jiang, Y., Li, Z., Kravchenko, V., Ulevitch, R., 1997. Activation of the transcription factor MEF2C by the MAP kinase p38 in inflammation. Nature 386, 296–299.

Handschin, C., Rhee, J., Lin, J., Tarr, P.T., Spiegelman, B.M., 2003. An autoregulatory loop controls peroxisome proliferator-activated receptor gamma coactivator 1alpha expression in muscle. Proc. Natl. Acad. Sci. U.S.A. 100, 7111–7116.

Hauck, E.S., Antonova-Koch, Y., Drexler, A., Pietri, J., Pakpour, N., Liu, D., Blacutt, J., Riehle, M.A., Luckhart, S., 2013. Overexpression of phosphatase and tensin homolog improves fitness and decreases *Plasmodium falciparum* development in *Anopheles stephensi*. Microbes Infect. 15, 775–787.

Heldin, C.H., Moustakas, A., 2016. Signaling receptors for TGF-beta family members. Cold Spring Harbor Perspect. Biol. 8. http://dx.doi.org/10.1101/cshperspect.a022053.

Herrera, B., Alvarez, A.M., Sanchez, A., Fernandez, M., Roncero, C., Benito, M., Fabregat, I., 2001. Reactive oxygen species (ROS) mediates the mitochondrial-dependent apoptosis induced by transforming growth factor (beta) in fetal hepatocytes. FASEB J. 15, 741–751.

Hogg, J., Hurd, H., 1997. The effects of natural *Plasmodium falciparum* infection on the fecundity and mortality of *Anopheles gambiae sl* in north east Tanzania. Parasitology 114, 325–331.

Hogg, J., Carwardine, S., Hurd, H., 1997. The effect of *Plasmodium yoelii nigeriensis* infection on ovarian protein accumulation by *Anopheles stephensi*. Parasitol. Res. 83, 374–379.

Horton, A.A., Lee, Y., Coulibaly, C.A., Rashbrook, V.K., Cornel, A.J., Lanzaro, G.C., Luckhart, S., 2010. Identification of three single nucleotide polymorphisms in *Anopheles gambiae* immune signaling genes that are associated with natural *Plasmodium falciparum* infection. Malar. J. 9, 160. http://dx.doi.org/10.1186/1475-2875-9-160.

Horton, A.A., Wang, B., Camp, L., Price, M.S., Arshi, A., Nagy, M., Nadler, S.A., Faeder, J.R., Luckhart, S., 2011. The mitogen-activated protein kinome from *Anopheles gambiae*: identification, phylogeny and functional characterization of the ERK, JNK and p38 MAP kinases. BMC Genomics 12, 574.

Huang, J., Lam, G.Y., Brumell, J.H., 2011. Autophagy signaling through reactive oxygen species. Antioxid. Redox Signaling 14, 2215–2231.

Hughes, A.L., 2013. Evolution of the salivary apyrases of blood-feeding arthropods. Gene 527, 123–130.

Huminiecki, L., Goldovsky, L., Freilich, S., Moustakas, A., Ouzounis, C., Heldin, C., 2009. Emergence, development and diversification of the TGF-β signalling pathway within the animal kingdom. BMC Evol. Biol. 9, 1.

Hur, J.H., Bahadorani, S., Graniel, J., Koehler, C.L., Ulgherait, M., Rera, M., Jones, D.L., Walker, D.W., 2013. Increased longevity mediated by yeast NDI1 expression in *Drosophila* intestinal stem and progenitor cells. Aging 5, 662–681.

Hur, J.H., Stork, D.A., Walker, D.W., 2014. Complex-I-ty in aging. J. Bioenerg. Biomembr. 46, 329–335.

Ito, Y., Mishra, N., Yoshida, K., Kharbanda, S., Saxena, S., Kufe, D., 2001. Mitochondrial targeting of JNK/SAPK in the phorbol ester response of myeloid leukemia cells. Cell Death and Differ. 8, 794–800.

Jahan, N., Hurd, H., 1998. Effect of *Plasmodium yoelii nigeriensis* (Haemosporidia: Plasmodiidae) on *Anopheles stephensi* (Diptera: Culicidae) vitellogenesis. J. Med. Entomol. 35, 956–961.

Jain, M., Rivera, S., Monclus, E.A., Synenki, L., Zirk, A., Eisenbart, J., Feghali-Bostwick, C., Mutlu, G.M., Budinger, G.R., Chandel, N.S., 2013. Mitochondrial reactive oxygen species regulate transforming growth factor-beta signaling. J. Biol. Chem. 288, 770–777.

Jasper, H., 2015. Exploring the physiology and pathology of aging in the intestine of *Drosophila melanogaster*. Invertebr. Reprod. Dev. 59, 51–58.

Jia, K., Thomas, C., Akbar, M., Sun, Q., Adams-Huet, B., Gilpin, C., Levine, B., 2009. Autophagy genes protect against *Salmonella typhimurium* infection and mediate insulin signaling-regulated pathogen resistance. Proc. Natl. Acad. Sci. 106, 14564–14569.

Kang, C., You, Y.J., Avery, L., 2007. Dual roles of autophagy in the survival of *Caenorhabditis elegans* during starvation. Genes Dev. 21, 2161–2171.

Kang, M.A., Mott, T.M., Tapley, E.C., Lewis, E.E., Luckhart, S., 2008. Insulin regulates aging and oxidative stress in *Anopheles stephensi*. J. Exp. Biol. 211, 741–748.

Keller, A., 2007. *Drosophila melanogaster*'s history as a human commensal. Curr. Biol. 17, R77–R81.

Lal, A.A., Schriefer, M.E., Sacci, J.B., Goldman, I.F., Louis-Wileman, V., Collins, W.E., Azad, A.F., 1994. Inhibition of malaria parasite development in mosquitoes by anti-mosquito-midgut antibodies. Infect. Immun. 62, 316–318.

Lal, A.A., Patterson, P.S., Sacci, J.B., Vaughan, J.A., Paul, C., Collins, W.E., Wirtz, R.A., Azad, A.F., 2001. Anti-mosquito midgut antibodies block development of *Plasmodium falciparum* and *Plasmodium vivax* in multiple species of *Anopheles* mosquitoes and reduce vector fecundity and survivorship. Proc. Natl. Acad. Sci. U.S.A. 98, 5228–5233.

Lehane, M., 1991. Managing the Blood Meal, Biology of Blood-Sucking Insects. Springer, pp. 79–110.

Lehane, M.J., 2005. The Biology of Blood-sucking in Insects. Cambridge University Press.

Levine, B., Mizushima, N., Virgin, H.W., 2011. Autophagy in immunity and inflammation. Nature 469, 323–335.

Li, H., Jasper, H., 2016. Gastrointestinal stem cells in health and disease: from flies to humans. Dis. Models Mech. 9, 487–499.

Li, H., Qi, Y., Jasper, H., 2013. Dpp signaling determines regional stem cell identity in the regenerating adult *Drosophila* gastrointestinal tract. Cell Rep. 4, 10–18.

Liang, X.H., Kleeman, L.K., Jiang, H.H., Gordon, G., Goldman, J.E., Berry, G., Herman, B., Levine, B., 1998. Protection against fatal Sindbis virus encephalitis by beclin, a novel Bcl-2-interacting protein. J. Virol. 72, 8586–8596.

Lim, J., Gowda, D.C., Krishnegowda, G., Luckhart, S., 2005. Induction of nitric oxide synthase in *Anopheles stephensi* by *Plasmodium falciparum*: mechanism of signaling and the role of parasite glycosylphosphatidylinositols. Infect. Immun. 73, 2778–2789.

Lim, S., Smith, K.R., Lim, S.S., Tian, R., Lu, J., Tan, M., 2016. Regulation of mitochondrial functions by protein phosphorylation and dephosphorylation. Cell Biosci. 6, 1.

Liu, Y., Schiff, M., Czymmek, K., Tallóczy, Z., Levine, B., Dinesh-Kumar, S., 2005. Autophagy regulates programmed cell death during the plant innate immune response. Cell 121, 567–577.

Luckhart, S., Riehle, M.A., 2007. The insulin signaling cascade from nematodes to mammals: insights into innate immunity of *Anopheles* mosquitoes to malaria parasite infection. Dev. Comp. Immunol. 31, 647–656.

Luckhart, S., Giulivi, C., Drexler, A.L., Antonova-Koch, Y., Sakaguchi, D., Napoli, E., Wong, S., Price, M.S., Eigenheer, R., Phinney, B.S., 2013. Sustained activation of Akt elicits mitochondrial dysfunction to block *Plasmodium falciparum* infection in the mosquito host. PLoS Pathog. 9, e1003180.

Mans, B.J., 2011. Evolution of vertebrate hemostatic and inflammatory control mechanisms in blood-feeding arthropods. J. Innate Immun. 3, 41–51.

Marquez, A.G., Pietri, J.E., Smithers, H.M., Nuss, A., Antonova, Y., Drexler, A.L., Riehle, M.A., Brown, M.R., Luckhart, S., 2011. Insulin-like peptides in the mosquito *Anopheles stephensi*: identification and expression in response to diet and infection with *Plasmodium falciparum*. Gen. Comp. Endocrinol. 173, 303–312.

Melendez, A., Talloczy, Z., Seaman, M., Eskelinen, E.L., Hall, D.H., Levine, B., 2003. Autophagy genes are essential for dauer development and life-span extension in *C. elegans*. Science 301, 1387–1391.

Merino, O., Alberdi, P., de la Lastra, J.M.P., de la Fuente, J., 2013. Tick vaccines and the control of tick-borne pathogens. Front. Cell. Infect. Microbiol. 3, 93.

Misof, B., Liu, S., Meusemann, K., Peters, R.S., Donath, A., Mayer, C., Frandsen, P.B., Ware, J., Flouri, T., Beutel, R.G., Niehuis, O., Petersen, M., Izquierdo-Carrasco, F., Wappler, T., Rust, J., Aberer, A.J., Aspock, U., Aspock, H., Bartel, D., Blanke, A., Berger, S., Bohm, A., Buckley, T.R., Calcott, B., Chen, J., Friedrich, F., Fukui, M., Fujita, M., Greve, C., Grobe, P., Gu, S., Huang, Y., Jermiin, L.S., Kawahara, A.Y., Krogmann, L., Kubiak, M., Lanfear, R., Letsch, H., Li, Y., Li, Z., Li, J., Lu, H., Machida, R., Mashimo, Y., Kapli, P., McKenna, D.D., Meng, G., Nakagaki, Y., Navarrete-Heredia, J.L., Ott, M., Ou, Y., Pass, G., Podsiadlowski, L., Pohl, H., von Reumont, B.M., Schutte, K., Sekiya, K., Shimizu, S., Slipinski, A., Stamatakis, A., Song, W., Su, X., Szucsich, N.U., Tan, M., Tan, X., Tang, M., Tang, J., Timelthaler, G., Tomizuka, S., Trautwein, M., Tong, X., Uchifune, T., Walzl, M.G.,

Wiegmann, B.M., Wilbrandt, J., Wipfler, B., Wong, T.K., Wu, Q., Wu, G., Xie, Y., Yang, S., Yang, Q., Yeates, D.K., Yoshizawa, K., Zhang, Q., Zhang, R., Zhang, W., Zhang, Y., Zhao, J., Zhou, C., Zhou, L., Ziesmann, T., Zou, S., Li, Y., Xu, X., Zhang, Y., Yang, H., Wang, J., Wang, J., Kjer, K.M., Zhou, X., 2014. Phylogenomics resolves the timing and pattern of insect evolution. Science 346, 763–767.

Morikawa, M., Derynck, R., Miyazono, K., 2016. TGF-beta and the TGF-beta family: context-dependent roles in cell and tissue physiology. Cold Spring Harbor Perspect. Biol. 8. http://dx.doi.org/10.1101/cshperspect.a021873.

Mu, Y., Gudey, S.K., Landström, M., 2012. Non-Smad signaling pathways. Cell Tissue Res. 347, 11–20.

Mukherjee, S., Basar, M.A., Davis, C., Duttaroy, A., 2014. Emerging functional similarities and divergences between *Drosophila* Spargel/dPGC-1 and mammalian PGC-1 protein. Front. Genet. 5, 216.

Murphy, C.T., McCarroll, S.A., Bargmann, C.I., Fraser, A., Kamath, R.S., Ahringer, J., Li, H., Kenyon, C., 2003. Genes that act downstream of DAF-16 to influence the lifespan of *Caenorhabditis elegans*. Nature 424, 277–283.

New York (State). State Entomologist, Lintner, J., 1882. First Annual Report on the Injurious and Other Insects of the State of New York: Made to the Legislature, Pursuant to Chapter 377 of the Laws of 1881. Weed, Parsons and Company.

Nieguitsila, A., Frutos, R., Moulia, C., Lhermitte-Vallarino, N., Bain, O., Gavotte, L., Martin, C., 2013. Fitness cost of *Litomosoides sigmodontis* filarial infection in mite vectors; implications of infected haematophagous arthropod excretory products in host-vector interactions. BioMed Res. Int. 2013, 584105.

Nisoli, E., Clementi, E., Paolucci, C., Cozzi, V., Tonello, C., Sciorati, C., Bracale, R., Valerio, A., Francolini, M., Moncada, S., 2003. Mitochondrial biogenesis in mammals: the role of endogenous nitric oxide. Sci. Signaling 299, 896.

Nisoli, E., Falcone, S., Tonello, C., Cozzi, V., Palomba, L., Fiorani, M., Pisconti, A., Brunelli, S., Cardile, A., Francolini, M., 2004. Mitochondrial biogenesis by NO yields functionally active mitochondria in mammals. Proc. Natl. Acad. Sci. U.S.A. 101, 16507–16512.

Nisoli, E., Tonello, C., Cardile, A., Cozzi, V., Bracale, R., Tedesco, L., Falcone, S., Valerio, A., Cantoni, O., Clementi, E., 2005. Calorie restriction promotes mitochondrial biogenesis by inducing the expression of eNOS. Sci. Signaling 310, 314.

Okumura, T., Takeda, K., Taniguchi, K., Adachi-Yamada, T., 2014. βν integrin inhibits chronic and high level activation of JNK to repress senescence phenotypes in *Drosophila* adult midgut. PLoS One 9, e89387.

Owusu-Ansah, E., Perrimon, N., 2015. Stress signaling between organs in Metazoa. Annu. Rev. Cell. Dev. Biol. 31, 497–522.

Paddock, F.B., 1918. The Beemoth or Waxworm. Texas Agricultural Experiment Stations No. 231.

Paiva-Silva, G.O., Cruz-Oliveira, C., Nakayasu, E.S., Maya-Monteiro, C.M., Dunkov, B.C., Masuda, H., Almeida, I.C., Oliveira, P.L., 2006. A heme-degradation pathway in a blood-sucking insect. Proc. Natl. Acad. Sci. U.S.A. 103, 8030–8035.

Pakpour, N., Corby-Harris, V., Green, G.P., Smithers, H.M., Cheung, K.W., Riehle, M.A., Luckhart, S., 2012. Ingested human insulin inhibits the mosquito NF-kappaB-dependent immune response to *Plasmodium falciparum*. Infect. Immun. 80, 2141–2149.

Pakpour, N., Akman-Anderson, L., Vodovotz, Y., Luckhart, S., 2013a. The effects of ingested mammalian blood factors on vector arthropod immunity and physiology. Microbes Infect. 15, 243–254.

Pakpour, N., Camp, L., Smithers, H.M., Wang, B., Tu, Z., Nadler, S.A., Luckhart, S., 2013b. Protein kinase C-Dependent signaling controls the midgut epithelial barrier to malaria parasite infection in *Anopheline* mosquitoes. PLoS One 8, e76535.

Pakpour, N., Riehle, M.A., Luckhart, S., 2014. Effects of ingested vertebrate-derived factors on insect immune responses. Curr. Opi. Insect Sci. 3, 1–5.

Pascoa, V., Oliveira, P.L., Dansa-Petretski, M., Silva, J.R., Alvarenga, P.H., Jacobs-Lorena, M., Lemos, F.J., 2002. *Aedes aegypti* peritrophic matrix and its interaction with heme during blood digestion. Insect. Biochem. Mol. Biol. 32, 517–523.

Pellegrino, M.W., Nargund, A.M., Kirienko, N.V., Gillis, R., Fiorese, C.J., Haynes, C.M., 2014. Mitochondrial UPR-regulated innate immunity provides resistance to pathogen infection. Nature 516, 414–417.

Peterson, A.J., O'Connor, M.B., 2014. Strategies for exploring TGF-β signaling in *Drosophila*. Methods 68, 183–193.

Piantadosi, C.A., Withers, C.M., Bartz, R.R., MacGarvey, N.C., Fu, P., Sweeney, T.E., Welty-Wolf, K.E., Suliman, H.B., Piantadosi, C.A., Withers, C.M., 2011. Heme oxygenase-1 couples activation of mitochondrial biogenesis to anti-inflammatory cytokine expression. J. Biol. Chem. 286, 16374–16385.

Pietri, J.E., Pakpour, N., Napoli, E., Song, G., Pietri, E., Potts, R., Cheung, K.W., Walker, G., Riehle, M.A., Starcevich, H., Giulivi, C., Lewis, E.E., Luckhart, S., October 15, 2016. Two insulin-like peptides differentially regulate malaria parasite infection in the mosquito through effects on intermediary metabolism. Biochem. J. 473 (20), 3487–3503 pii:BCJ20160271. PubMed PMID: 27496548.

Pietrokovsky, S., Bottazzi, V., Schweigmann, N., Haedo, A., Wisnivesky-Colli, C., 1996. Comparison of the blood meal size among *Triatoma infestans, T. guasayana* and *T. sordida* (Hemiptera: Reduviidae) of Argentina under laboratory conditions. Memórias do Instituto Oswaldo Cruz 91, 241–242.

Puigserver, P., Rhee, J., Lin, J., Wu, Z., Yoon, J.C., Zhang, C., Krauss, S., Mootha, V.K., Lowell, B.B., Spiegelman, B.M., 2001. Cytokine stimulation of energy expenditure through p38 MAP kinase activation of PPARγ coactivator-1. Mol. Cell 8, 971–982.

Radulović, Ž., Porter, L., Kim, T., Bakshi, M., Mulenga, A., 2015. *Amblyomma americanum* tick saliva insulin-like growth factor binding protein-related protein 1 binds insulin but not insulin-like growth factors. Insect. Mol. Biol. 24, 539–550.

Regan, J.C., Khericha, M., Dobson, A.J., Bolukbasi, E., Rattanavirotkul, N., Partridge, L., 2016. Sex difference in pathology of the ageing gut mediates the greater response of female lifespan to dietary restriction. eLife 5, e10956.

Reiff, T., Jacobson, J., Cognigni, P., Antonello, Z., Ballesta, E., Tan, K.J., Yew, J.Y., Dominguez, M., Miguel-Aliaga, I., 2015. Endocrine remodelling of the adult intestine sustains reproduction in *Drosophila*. eLife 4, e06930.

Ren, C., Finkel, S.E., Tower, J., 2009. Conditional inhibition of autophagy genes in adult *Drosophila* impairs immunity without compromising longevity. Exp. Gerontol. 44, 228–235.

Rera, M., Bahadorani, S., Cho, J., Koehler, C.L., Ulgherait, M., Hur, J.H., Ansari, W.S., Lo, T., Jones, D.L., Walker, D.W., 2011. Modulation of longevity and tissue homeostasis by the *Drosophila* PGC-1 homolog. Cell Metab. 14, 623–634.

Rera, M., Azizi, M.J., Walker, D.W., 2012a. Organ-specific mediation of lifespan extension: more than a gut feeling? Ageing Res. Rev. 12, 436–444.

Rera, M., Clark, R.I., Walker, D.W., 2012b. Intestinal barrier dysfunction links metabolic and inflammatory markers of aging to death in *Drosophila*. Proc. Natl. Acad. Sci. U.S.A. 109, 21528–21533.

Riehle, M.A., Brown, M.R., 1999. Insulin stimulates ecdysteroid production through a conserved signaling cascade in the mosquito *Aedes aegypti*. Insect Biochem. Mol. Biol. 29, 855–860.

Riehle, M.A., Brown, M.R., 2002. Insulin receptor expression during development and a reproductive cycle in the ovary of the mosquito *Aedes aegypti*. Cell Tissue Res. 308, 409–420.

Riehle, M.A., Brown, M.R., 2003. Molecular analysis of the serine/threonine kinase Akt and its expression in the mosquito *Aedes aegypti*. Insect. Mol. Biol. 12, 225–232.

Rodrigues, C.M., Ma, X., Linehan-Stieers, C., Fan, G., Kren, B.T., Steer, C.J., 1999. Ursodeoxycholic acid prevents cytochrome c release in apoptosis by inhibiting mitochondrial membrane depolarization and channel formation. Cell Death Differ. 6, 842–854.

Rogers, R.P., Rogina, B., 2014. Increased mitochondrial biogenesis preserves intestinal stem cell homeostasis and contributes to longevity in Indy mutant flies. Aging 6, 335–350.

Roy, S.G., Hansen, I.A., Raikhel, A.S., 2007. Effect of insulin and 20-hydroxyecdysone in the fat body of the yellow fever mosquito, *Aedes aegypti*. Insect. Biochem. Mol. Biol. 37, 1317–1326.

Sadagurski, M., White, M.F., 2013. Integrating metabolism and longevity through insulin and IGF1 signaling. Endocrinol. Metab. Clin. Am. 42, 127–148.

Sandeu, M.M., Abate, L., Tchioffo, M.T., Bayibéki, A.N., Awono-Ambéné, P.H., Nsango, S.E., Chesnais, C.B., Dinglasan, R.R., de Meeûs, T., Morlais, I., 2016. Impact of exposure to mosquito transmission-blocking antibodies on *Plasmodium falciparum* population genetic structure. Infect. Genet. Evol. 45, 138–144.

Sano, M., Tokudome, S., Shimizu, N., Yoshikawa, N., Ogawa, C., Shirakawa, K., Endo, J., Katayama, T., Yuasa, S., Ieda, M., Makino, S., Hattori, F., Tanaka, H., Fukuda, K., 2007. Intramolecular control of protein stability, subnuclear compartmentalization, and coactivator function of peroxisome proliferator-activated receptor gamma coactivator 1alpha. J. Biol. Chem. 282, 25970–25980.

Sim, C., Denlinger, D.L., 2008. Insulin signaling and FOXO regulate the overwintering diapause of the mosquito *Culex pipiens*. Proc. Natl. Acad. Sci. U.S.A. 105, 6777–6781.

Singh, S.B., Davis, A.S., Taylor, G.A., Deretic, V., 2006. Human IRGM induces autophagy to eliminate intracellular mycobacteria. Sci. Signaling 313, 1438.

Skorokhod, A., Gamulin, V., Gundacker, D., Kavsan, V., Muller, I.M., Muller, W.E., 1999. Origin of insulin receptor-like tyrosine kinases in marine sponges. Bio. Bul. 197, 198–206.

Steiner, H., Hultmark, D., Engstrom, A., Bennich, H., Barman, H., 1981. Sequence and specificity of two antibacterial proteins involved in insect immunity. Nature 292, 246–248.

Steinhaus, E.A., 1956. Microbial Control: The Emergence of an Idea; a Brief History of Insect Pathology Through the Nineteenth Century. University of California Press.

Surachetpong, W., Singh, N., Cheung, K.W., Luckhart, S., 2009. MAPK ERK signaling regulates the TGF-beta1-dependent mosquito response to *Plasmodium falciparum*. PLoS Pathog. 5, e1000366.

Surachetpong, W., Pakpour, N., Cheung, K.W., Luckhart, S., 2011. Reactive oxygen species-dependent cell signaling regulates the mosquito immune response to *Plasmodium falciparum*. Antioxid. Redox Signaling 14, 943–955.

Sweeney, T.E., Suliman, H.B., Hollingsworth, J.W., Piantadosi, C.A., 2010. Differential regulation of the PGC family of genes in a mouse model of *Staphylococcus aureus* sepsis. PLoS One 5, e11606.

Sweeney, T.E., Suliman, H.B., Hollingsworth, J.W., Welty-Wolf, K.E., Piantadosi, C.A., 2011. A Toll-like receptor 2 pathway regulates the Ppargc1a/b metabolic co-activators in mice with *Staphylococcal aureus* sepsis. PLoS One 6, e25249.

Takeda, J.T., 2014. Global insects: silkworms, sericulture, and statecraft in Napoleonic France and Tokugawa Japan. French Hist. cru044.

Takken, W., Verhulst, N.O., 2013. Host preferences of blood-feeding mosquitoes. Annu. Rev. Entomol. 58, 433–453.

Tiefenbock, S.K., Baltzer, C., Egli, N.A., Frei, C., 2010. The *Drosophila* PGC-1 homologue Spargel coordinates mitochondrial activity to insulin signaling. EMBO J. 29, 171–183.

Tóth, M.L., Sigmond, T., Borsos, É., Barna, J., Erdélyi, P., Takács-Vellai, K., Orosz, L., Kovács, A.L., Csikós, G., Sass, M., 2008. Longevity pathways converge on autophagy genes to regulate life span in *Caenorhabditis elegans*. Autophagy 4, 330–338.

Tran, M., Tam, D., Bardia, A., Bhasin, M., Rowe, G.C., Kher, A., Zsengeller, Z.K., Akhavan-Sharif, M.R., Khankin, E.V., Saintgeniez, M., 2011. PGC-1α promotes recovery after acute kidney injury during systemic inflammation in mice. J. Clin. Invest. 121, 4003.

Ulgherait, M., Rana, A., Rera, M., Graniel, J., Walker, D.W., 2014. AMPK modulates tissue and organismal aging in a non-cell-autonomous manner. Cell Rep. 8, 1767–1780.

Unoki, M., Nakamura, Y., 2001. Growth-suppressive effects of BPOZ and EGR2, two genes involved in the PTEN signaling pathway. Oncogene 20, 4457–4465.

van der Vos, K.E., Coffer, P.J., 2011. The extending network of FOXO transcriptional target genes. Antioxid. Redox Signaling 14, 579–592.

Vaughan, J.A., Trpis, M., Turell, M.J., 1999. Brugia malayi microfilariae (Nematoda: Filaridae) enhance the infectivity of Venezuelan equine encephalitis virus to *Aedes* mosquitoes (Diptera: Culicidae). J. Med. Entomol. 36, 758–763.

Vaughan, J.A., Focks, D.A., Turell, M.J., 2009. Simulation models examining the effect of *Brugian filariasis* on dengue epidemics. Am. J. Trop. Med. Hyg. 80, 44–50.

Vaughan, J.A., Mehus, J.O., Brewer, C.M., Kvasager, D.K., Bauer, S., Vaughan, J.L., Hassan, H.K., Unnasch, T.R., Bell, J.A., 2012. Theoretical potential of passerine filariasis to enhance the enzootic transmission of West Nile virus. J. Med. Entomol. 49, 1430–1441.

Vellai, T., 2009. Autophagy genes and ageing. Cell Death Diff. 16, 94–102.

Wang, L., Karpac, J., Jasper, H., 2014. Promoting longevity by maintaining metabolic and proliferative homeostasis. J. Exp. Biol. 217, 109–118.

Wang, B., Pakpour, N., Napoli, E., Drexler, A., Glennon, E.K., Surachetpong, W., Cheung, K., Aguirre, A., Klyver, J.M., Lewis, E.E., 2015. *Anopheles stephensi* p38 MAPK signaling regulates innate immunity and bioenergetics during *Plasmodium falciparum* infection. Parasites and Vectors 8, 1–21.

Wang, Z., 1965. Knowledge on the control of silkworm disease in ancient China. Symp. Sci. Hist. 8, 15.

Wymann, M.P., Zvelebil, M., Laffargue, M., 2003. Phosphoinositide 3-kinase signalling–which way to target? Trends Pharmacol. Sci. 24, 366–376.

Yoon, Y., Lee, J., Hwang, S., Choi, K.S., Yoon, G., 2005. TGF β1 induces prolonged mitochondrial ROS generation through decreased complex IV activity with senescent arrest in Mv1Lu cells. Oncogene 24, 1895–1903.

Youle, R.J., van der Bliek, A.M., 2012. Mitochondrial fission, fusion, and stress. Science 337, 1062–1065.

Zhao, M., New, L., Kravchenko, V.V., Kato, Y., Gram, H., di Padova, F., Olson, E.N., Ulevitch, R.J., Han, J., 1999. Regulation of the MEF2 family of transcription factors by p38. Mol. Cell Biol. 19, 21–30.

Zhou, Y.Y., Li, Y., Jiang, W.Q., Zhou, L.F., 2015. MAPK/JNK signaling: a potential autophagy regulation pathway. Biosci. Rep. 35 (3), e00199.

CHAPTER

3

Wolbachia-Mediated Immunity Induction in Mosquito Vectors

Xiaoling Pan, Suzanne Thiem, Zhiyong Xi

Michigan State University, East Lansing, MI, United States

INTRODUCTION

Wolbachia pipientis is a maternally transmitted Gram-negative intracellular endosymbiotic bacterium that is estimated to infect over 65% of insect species (Werren et al., 2008). First discovered in *Culex pipiens* in 1924, *Wolbachia* have significant nonrandom distribution among different mosquito species (Hilgenboecker et al., 2008; Kittayapong et al., 2000). Approximately 42.1% of *Culex* species and 50% of *Mansonia* species carry *Wolbachia* in nature (Kittayapong et al., 2000). While *Aedes albopictus* is generally superinfected with two different *Wolbachia* strains, *Wolbachia* has never been found in *Aedes aegypti*. Furthermore, no native *Wolbachia* infection had been found in the 38 surveyed *Anopheles* species until a report identified a novel *Wolbachia* strain, related to but distinct from strains infecting other arthropods, in *Anopheles gambiae* (Baldini et al., 2014; Bourtzis et al., 2014; Kittayapong et al., 2000).

As alpha-proteobacteria closely related to Rickettsia, *Wolbachia* spp. have been divided into eight different supergroups based on 16S ribosomal RNA sequences and other sequence information (Casiraghi et al., 2005; Werren et al., 2008). With the exception of supergroups C and D that are restricted to filarial nematodes, the other supergroups are found primarily in arthropods, in which A and B are the most common (Werren et al., 2008). For example, the *Wolbachia* strain *w*Pip (Zele et al., 2012) of *Culex* species belongs to supergroup B while the two strains, *w*AlbA and *w*AlbB, of *Ae. albopictus* are in supergroup A and B, respectively. *Drosophila melanogaster* carries *w*Mel, a strain in supergroup A (Blagrove et al., 2012). These different strains interact with their hosts to maintain persistent infections and manipulate host reproduction, resulting in the expression of a variety of reproductive alterations (Werren et al., 2008). Currently studies of *Wolbachia*–mosquito interactions are mainly focused on four key elements of *Wolbachia* symbiosis: maternal transmission, cytoplasmic incompatibility, pathogen interference, and host fitness.

Maternal Transmission of *Wolbachia*

Although *Wolbachia* can move horizontally within and between species (Baldo et al., 2008;

Arthropod Vector: Controller of Disease Transmission, Volume 1
http://dx.doi.org/10.1016/B978-0-12-805350-8.00003-9

Raychoudhury et al., 2009; Toomey et al., 2013; Werren et al., 1995), maternal transmission is the main mode that is used by *Wolbachia* to facilitate their symbiotic association with their host. *Wolbachia* utilize their intrinsic factors and host factors to regulate their distribution and density during host oogenesis through inducing the tropism of ovarian stem cell niches (Fast et al., 2011; Serbus and Sullivan, 2007; Toomey et al., 2013). Consequently, *Wolbachia* localizes in the mature oocytes and reaches a high density in the germline, which warrants its vertical transmission to the next generation (Ferree et al., 2005; Serbus and Sullivan, 2007).

As the key feature of *Wolbachia* symbiosis, maternal transmission allows *Wolbachia* to form intimate associations with the hosts. Differing from a horizontally transmitted pathogen, presumably the level of *Wolbachia* infection should be finely controlled such that it would be sufficiently high to ensure fidelity of transovarial transmission but low enough not to cause overt host pathology. This process is determined by the strain of *Wolbachia*, host genetic background, the host microbiome, and environmental conditions (Hughes et al., 2014; McGraw et al., 2002). As observed in the other host–microbial symbiosis, the host immune system could play important role in establishment, promotion, and maintenance of this symbiotic relationship (Chu and Mazmanian, 2013; Hooper et al., 2012; Nyholm and Graf, 2012).

Although *Wolbachia* does not present naturally in *Ae. aegypti* and most of *Anopheles* mosquito species, *Wolbachia* can be transferred into these mosquito species to establish the transinfected mosquito lines via embryonic microinjection (Bian et al., 2013a; McMeniman et al., 2009; Walker et al., 2011; Xi et al., 2005). In these transinfected lines, *Wolbachia* can be maternally transmitted into the next generations with ~100% efficiency and this heritable infection can be stably maintained in both laboratory and field conditions. These transinfected lines are currently used to develop *Wolbachia*-based mosquito control strategies for disease control in the field (Hoffmann et al., 2011). They are different from the native *Wolbachia* infection in that it is only recently that *Wolbachia* forms symbiosis with the host, with the adaptation between *Wolbachia* and host just beginning. In addition, the transinfected lines differ from a transient somatic infection in that *Wolbachia* have passed though the germ line and thus present in the host before its immune system is developed. Because *Wolbachia* are not maternally transmitted in a system with transient somatic infection, a symbiotic relationship has not been formed between the *Wolbachia* and its host. Understanding the difference will aid in better interpretation of the results from studies using different *Wolbachia*/host associations.

Wolbachia-Mediated Cytoplasmic Incompatibility

One of the most common phenotypes induced by *Wolbachia* is cytoplasmic incompatibility (CI), in which embryonic death occurs when infected males mate with either uninfected females or infected females carrying different type of *Wolbachia* (Hoffmann and Turelli, 1997). First reported in *Culex* species in 1936, CI is the only *Wolbachia*-induced reproductive alteration that has been observed so far in mosquito hosts (Sinkins, 2004). Although the molecular mechanism of CI is still unknown, evidence supports a model in which when a sperm from an infected male fertilizes an egg that is either uninfected or infected with a different type of *Wolbachia*, improperly condensed paternal chromosomes are lost during early embryonic development. When the same type of *Wolbachia* that is carried by the male presents in the egg, this modification of sperm is rescued and the paternal material is restored functionality, resulting in the success of embryonic development (Poinsot et al., 2003).

The ability of *Wolbachia* to induce CI allows the development of two *Wolbachia*-based strategies to control mosquito-transmitted diseases. The first strategy is referred to as population suppression and entails the release of male mosquitoes infected with *Wolbachia*, resulting in sterile matings and a reduction in the mosquito population (Laven, 1967; O'Connor et al., 2012). The second strategy is population replacement, which would use *Wolbachia* to spread desired traits (such as an antipathogen phenotype) into the population via release of the *Wolbachia*-infected females (Bian et al., 2013a; Curtis and Sinkins, 1998; Hoffmann et al., 2011; Turelli and Hoffmann, 1999; Xi et al., 2005). Because the infected females can produce the infected offspring after they mate with either the infected or uninfected males, but the uninfected females can reproduce the offspring only if they mate with uninfected males, CI provides the infected females an advantage in reproduction, resulting in invasion of *Wolbachia* into the population.

Wolbachia-Mediated Pathogen Interference

A native *Wolbachia* infection was firstly discovered to protect flies from virus-induced mortality (Hedges et al., 2008). Subsequently, three independent studies showed that *Wolbachia* inhibit pathogens, including dengue, chikungunya and *Plasmodium* and filarial nematode in transinfected *Ae. aegypti* (Bian et al., 2010; Kambris et al., 2009; Moreira et al., 2009). The spectrum of pathogens that *Wolbachia* can inhibit continues to expand and now includes West Nile, yellow fever, Zika, and bluetongue virus (Table 3.1) (Dutra et al., 2016; Shaw et al., 2012). In addition, *Wolbachia* are also shown to protect *Ae. aegypti* from infection with both Gram-negative and Gram-positive bacteria (Ye et al., 2013) and increase survival rate of mosquitoes after challenge with fungi (unpublished data). Such a broad spectrum has generated significant interests in both understanding of the molecular mechanism underlying pathogen interference and implementation of *Wolbachia*-based approaches to modify mosquito population for disease control.

Wolbachia-mediated pathogen interference is mainly observed in three different systems: transinfected hosts, hosts with native *Wolbachia* infection, and hosts with a transient somatic infection. (1) In the transinfected mosquitoes (McMeniman et al., 2009; Walker et al., 2011; Xi et al., 2005), the three different *Wolbachia* strains, *w*AlbB, *w*Mel, and *w*MelPop, are able to significantly inhibit a variety of pathogens, including many positive single-stranded RNA viruses [(+) ssRNA], bacteria, and eukaryotic pathogens (Table 3.1) (Hussain et al., 2013; van den Hurk et al., 2012). In most cases, the strength of inhibition is positively correlated with *Wolbachia* density and *w*MelPop shows the strongest pathogen blocking while *w*AlbB and *w*Mel induces a similar level of resistance (Ferguson et al., 2015; Joubert et al., 2016). Furthermore, *w*AlbB induces pathogen interference in transinfected *Ae. aegypti*, *Aedes Polynesiensis,* and *Anopheles stephensi* (Bian et al., 2013a,b; Xi et al., 2005). There is an additive effect in viral inhibition in transinfected mosquitoes with superinfection of *w*AlbB and *w*Mel (Joubert et al., 2016). (2) In the mosquitoes or flies with native *Wolbachia* infection, either viral interference or host protection has been consistently observed. The antiviral spectrum includes flavivirus, alphanodavirus, cripavirus, norovirus, and orbivirus (Table 3.1). However, a lack of *Wolbachia*-induced interference effect on both viruses and bacteria has also been observed, presumably caused by either attenuation of *Wolbachia* titer or the type of pathogens (Baton et al., 2013; Micieli and Glaser, 2014; Panteleev et al., 2007; Tsai et al., 2006). For example, La Crosse virus, with negative single-stranded RNAs, and Invertebrate iridescent virus 6 (IIV-6), with double-stranded DNAs, are not affected by *Wolbachia* in *D. melanogaster* (Glaser and Meola, 2010; Teixeira et al.,

TABLE 3.1 *Wolbachia*-Mediated Pathogen Interference in Insects

Pathogen Interference	*Wolbachia* Strain	Infection Type of *Wolbachia*	Host	Pathogen (Group)	Pathogen (Genus)	Pathogen (Species)	Effect	References
Viral inhibition	*w*AlbB	Stable transinfection	*Aedes polynesiensis*	Group IV [(+) ssRNA]	*Flavivirus*	DENV	Reduced virus load and virus transmission	Bian et al. (2013b)
		Stable transinfection	*Aedes aegypti*	Group IV [(+) ssRNA]	*Alphavirus*	CHIKV	Reduced virus load and virus transmission	Unpublished data
					Flavivirus	DENV	Reduced infection rate, virus load, dissemination and transmission	Bian et al. (2010)
						ZIKV	Reduced virus load and virus transmission	Unpublished data
	*w*Mel	Stable transinfection	*Aedes albopictus*	Group IV [(+) ssRNA]	*Alphavirus*	CHIKV	Reduced transmission	Blagrove et al. (2013)
					Flavivirus	DENV	Reduced transmission	Blagrove et al. (2012)
		Stable transinfection	*Ae. aegypti*	Group IV [(+) ssRNA]	*Alphavirus*	CHIKV	Reduced virus load and dissemination	van den Hurk et al. (2012)
					Flavivirus	DENV	Reduced virus load and dissemination	Walker et al. (2011)
						WNV	Reduced transmission	Hussain et al. (2013)
						YFV	Reduced virus load	van den Hurk et al. (2012)
						ZIKV	Reduced virus load, dissemination and transmission	Dutra et al. (2016)
	*w*MelPop	Stable transinfection	*Ae. aegypti*	Group IV [(+) ssRNA]	*Alphavirus*	CHIKV	Reduced infection rate, virus load, and dissemination	Moreira et al. (2009)
					Flavivirus	DENV	Reduced infection rate, virus load, and dissemination	Moreira et al. (2009)
						WNV	Reduced infection rate, viral load, dissemination, and transmission	Hussain et al. (2013)
						YFV	Reduced infection rate and virus load	van den Hurk et al. (2012)
	*w*Pip	Natural infection	*Culex quinquefasciatus*	Group IV [(+) ssRNA]	*Flavivirus*	WNV	Reduced virus load, dissemination rate, and transmission rate	Glaser and Meola (2010)

	Natural infection	*Culex pipiens*	Group IV [(+) ssRNA]	*Flavivirus*	WNV	No effect	Micieli and Glaser (2014)
*w*AlbA and *w*AlbB	Natural infection	*Ades albopictus*	Group IV [(+) ssRNA]	*Alphavirus*	CHIKV	No effect	Mousson et al. (2010)
				Flavivirus	DENV	Reduced virus load in cell, no effect in mosquito	Lu et al. (2012)
					DENV	No effect on virus load; reduced virus transmission	Mousson et al. (2012)
	Introgressed	*Ae. albopictus*	Group IV [(+) ssRNA]	*Alphavirus*	CHIKV	No effect	Blagrove et al. (2013)
*w*PolA	Natural infection	*Ae. polynesiensis*	Group IV [(+) ssRNA]	*Flavivirus*	DENV	No effect?	Bian et al. (2013b)
Not subgrouped	Natural infection	*Armigeres subalbatus*	Group IV [(+) ssRNA]	*Flavivirus*	JEV	No effect in salivary gland cells	Tsai et al. (2006)
*w*MelCS	Natural infection	*Drosophila melanogaster*	Group I (dsDNA)	*Iridovirus*	IIV-6	Decreased life span of infected flies. No effect on virus load	Teixeira et al. (2008)
			Group IV [(+) ssRNA]	*Alphanodavirus*	FHV	Increased life span of infected flies; reduced virus load	Teixeira et al. (2008)
				Alphanodavirus	FHV	Increased life span of infected flies	Hedges et al. (2008)
				Cripavirus	DCV	Increased life span of infected flies; reduced virus load	Teixeira et al. (2008)
				Cripavirus	DCV	Increased life span of infected flies	Hedges et al. (2008)
				Cripavirus	CrPV	Increased life span of infected flies	Hedges et al. (2008)
				Norovirus	NoraV	Reduced virus load	Teixeira et al. (2008)
*w*MelPop	Natural infection	*D. melanogaster*	Group IV [(+) ssRNA]	*Alphanodavirus*	FHV	Increased life span of infected flies	Hedges et al. (2008)
				Cripavirus	DCV	Increased life span of infected flies; reduced virus load	Hedges et al. (2008)
				Cripavirus	CrPV	Increased life span of infected flies	Hedges et al. (2008)
*w*Mel	Natural infection	*D. melanogaster*	Group III (dsRNA)	*Orbivirus*	BTV	Reduced virus load	Shaw et al. (2012)

Continued

TABLE 3.1 *Wolbachia*-Mediated Pathogen Interference in Insects—cont'd

Pathogen Interference	*Wolbachia* Strain	Infection Type of *Wolbachia*	Host	Pathogen (Group)	Pathogen (Genus)	Pathogen (Species)	Effect	References
				Group V [(−) ssRNA]	*Orthobunyavirus*	LACV	No effect	Glaser and Meola (2010)
				Group IV [(+) ssRNA]	*Alphavirus*	CHIKV	No effect	Glaser and Meola (2010)
					Flavivirus	WNV	Reduced virus load	Glaser and Meola (2010)
					Flavivirus	DENV	Reduced infection rate and virus load	Rances et al. (2012)
		Stable transinfection	*Drosophila simulans*	Group IV [(+) ssRNA]	*Cripavirus*	DCV	Increased life span of infected flies; reduced virus load	Osborne et al. (2009)
	*w*Au	Natural infection	*D. simulans*	Group IV [(+) ssRNA]	*Alphanodavirus*	FHV	Increased life span of infected flies	Osborne et al. (2009)
					Cripavirus	DCV	Increased life span of infected flies; reduced virus load	Osborne et al. (2009)
	*w*Ri	Natural infection	*D. simulans*	Group IV [(+) ssRNA]	*Alphanodavirus*	FHV	Increased life span of infected flies	Osborne et al. (2009)
					Cripavirus	DCV	Increased life span of infected flies	Osborne et al. (2009)
	*w*Ha	Natural infection	*D. simulans*	Group IV [(+) ssRNA]	*Alphanodavirus*	FHV	No effect	Osborne et al. (2009)
					Cripavirus	DCV	No effect	Osborne et al. (2009)
	*w*No	Natural infection	*D. simulans*	Group IV [(+) ssRNA]	*Alphanodavirus*	FHV	No effect	Osborne et al. (2009)
					Cripavirus	DCV	No effect	Osborne et al. (2009)
	wExe1—supergroup B	Natural infection	*Spodoptera exempta* (African armyworm)	Group I (dsDNA)	*Alphabaculovirus*	SpexNPV	Positively associated with infection of SpexNPV in field Increased host mortality in laboratory bioassays	Graham et al. (2012)
	wExe2—supergroup B	Natural infection	*S. exempta* (African armyworm)	Group I (dsDNA)	*Alphabaculovirus*	SpexNPV	Positively associated with infection of SpexNPV in field	Graham et al. (2012)
	wExe3—supergroup A	Natural infection	*S. exempta* (African armyworm)	Group I (dsDNA)	*Alphabaculovirus*	SpexNPV	Positively associated with infection of SpexNPV in field	Graham et al. (2012)
Bacterium inhibition	*w*Mel	Stable transinfection	*Ae. aegypti*	Gram positive	*Mycobacterium*	*Mycobacterium marinum*	No effect	Ye et al. (2013)

			Gram negative	*Burkholderia*	*Burkholderia cepacia*	No effect on life span of infected mosquitoes, reduced bacterial density	Ye et al. (2013)
				Erwinia	*Erwinia carotovora*	Increased life span of infected mosquitoes, reduced bacterial density	Ye et al. (2013)
				Salmonella	*Salmonella typhimurium*	Increased life span of infected mosquitoes	Ye et al. (2013)
*w*MelPop	Stable transinfection	*Ae. aegypti*	Gram positive	*Micrococcus*	*Micrococcus luteus*	No effect	Kambris et al. (2009)
				Mycobacterium	*M. marinum*	Increased life span of infected mosquitoes, reduced bacterial density	Ye et al. (2013)
			Gram negative	*Burkholderia*	*Bu. cepacia*	Increased life span of infected mosquitoes, reduced bacterial density	Ye et al. (2013)
				Erwinia	*E. carotovora*	Increased life span of infected mosquitoes	Kambris et al. (2009)
				Erwinia	*E. carotovora*	Increased life span of infected mosquitoes, reduced bacterial density	Ye et al. (2013)
				Salmonella	*S. typhimurium*	Increased life span of infected mosquitoes, reduced bacterial density	Ye et al. (2013)
*w*AlbB	Stable transinfection	*Ae. aegypti*	Gram positive	*Micrococcus*	*M. luteus*	Increased life span of infected mosquitoes	Unpublished data
			Gram negative	*Enterobacter*	*Enterobacter cloacae*	Increased life span of infected mosquitoes	Unpublished data
*w*Au	Natural infection	*D. simulans*	Gram negative	*Erwinia*	*E. carotovora*	No effect	Wong et al. (2011)
				Pseudomonas	*Pseudomonas aeruginosa*	No effect	Wong et al. (2011)
				Serratia	*Serratia marcescens*	No effect	Wong et al. (2011)

Continued

TABLE 3.1 *Wolbachia*-Mediated Pathogen Interference in Insects—cont'd

Pathogen Interference	*Wolbachia* Strain	Infection Type of *Wolbachia*	Host	Pathogen (Group)	Pathogen (Genus)	Pathogen (Species)	Effect	References
	*w*Ri	Natural infection	*D. simulans*	Gram negative	*Erwinia*	*E. carotovora*	No effect	Wong et al. (2011)
					Pseudomonas	*P. aeruginosa*	No effect	Wong et al. (2011)
					Serratia	*S. marcescens*	No effect	Wong et al. (2011)
	*w*Mel	Natural infection	*D. simulans*	Gram negative	*Pseudomonas*	*P. aeruginosa*	No effect	Wong et al. (2011)
					Serratia	*S. marcescens*	No effect	Wong et al. (2011)
					Erwinia	*E. carotovora*	No effect	Wong et al. (2011)
	*w*Ha	Natural infection	*D. simulans*	Gram negative	*Erwinia*	*E. carotovora*	No effect	Wong et al. (2011)
					Pseudomonas	*P. aeruginosa*	No effect	Wong et al. (2011)
					Serratia	*S. marcescens*	No effect	Wong et al. (2011)
	*w*No	Natural infection	*D. simulans*	Gram negative	*Erwinia*	*E. carotovora*	No effect	Wong et al. (2011)
					Pseudomonas	*P. aeruginosa*	No effect	Wong et al. (2011)
					Serratia	*S. marcescens*	No effect	Wong et al. (2011)
	*w*MelCS	Natural infection	*D melanogaster*	Gram negative	*Erwinia*	*E. carotovora*	No effect	Wong et al. (2011)
					Pseudomonas	*P. aeruginosa*	No effect	Wong et al. (2011)
					Serratia	*S. marcescens*	No effect	Wong et al. (2011)
	*w*Mel	Natural infection	*D. melanogaster*	Gram positive	*Listeria*	*Listeria monocytogenes*	No effect	Rottschaefer and Lazzaro (2012)
					Mycobacterium	*M. marinum*	No effect	Ye et al. (2013)
				Gram negative	*Burkholderia*	*Bu. cepacia*	No effect	Ye et al. (2013)
					Erwinia	*E. carotovora*	No effect	Ye et al. (2013)
					Providencia	*Providencia rettgeri*	No effect	Rottschaefer and Lazzaro (2012)
					Salmonella	*S. typhimurium*	No effect	Rottschaefer and Lazzaro (2012)
					Salmonella	*S. typhimurium*	No effect	Ye et al. (2013)

	*w*MelPop	Natural infection	*D. melanogaster*	Gram positive	*Mycobacterium*	*M. marinum*	No effect	Ye et al. (2013)
				Gram negative	*Erwinia*	*Erwinia carotovora*	No effect	Ye et al. (2013)
					Burkholderia	*Bu. cepacia*	No effect	Ye et al. (2013)
					Salmonella	*S. typhimurium*	No effect	Ye et al. (2013)
Eukaryotic pathogen inhibition	*w*Mel	Stable transinfection	*Ae. aegypti*	Filarial nematode	*Brugia*	*Brugia pahangi*	Increased life span of infected mosquitoes, reduced worm load	Kambris et al. (2009)
	*w*MelPop	Stable transinfection	*Ae. aegypti*	Protozoan parasite	*Plasmodium*	*Plasmodium gallinaceum*	Reduced infection rate and parasite load	Moreira et al. (2009)
						Plasmodium falciparum	Reduced parasite load	Hughes et al. (2011)
	*w*AlbB	Stable transinfection	*Ae. aegypti*	Fungi	*Beauveria*	*Beauveria bassiana*	Increased life span of infected mosquitoes	Unpublished data
			Ae. polynesiensis	Filarial nematode	*Brugia*	*Br. pahangi*	Increased life span of infected mosquitoes, reduced worm load	Andrews et al. (2012)
		Stable transinfection	*Anopheles stephensi*	Protozoan parasite	*Plasmodium*	*Plasmodium berghei*	Reduced parasite load and transmission	Unpublished data
						P. falciparum	Reduced parasite load and transmission	Bian et al. (2013a)
	*w*Flu	Natural infection	*Aedes fluviatilis*	Protozoan parasite	*Plasmodium*	*P. gallinaceum*	Increased parasite loads	Baton et al. (2013)
	*w*Pip	Natural infection	*Culex pipiens quinquefasciatus*	Protozoan parasite	*Plasmodium*	*Plasmodium relictum*	Increased life span of infected mosquitoes	Zele et al. (2012)
	Not subgrouped	Natural infection	*D. melanogaster*	Fungi	*Beauveria*	*Be. bassiana*	Enhanced the viability of infected *drosophila*	Panteleev et al. (2007)

BTV, bluetongue virus; *CHIKV*, chikungunya virus; *CrPV*, cricket paralysis virus; *DCV*, *Drosophila* C virus; *DENV*, dengue virus; *FHV*, Flock House virus; *IIV-6*, invertebrate iridescent virus 6, also named Chilo iridescent virus (CIV); *JEV*, Japanese encephalitis virus; *LACV*, La Crosse virus; *NoraV*, noravirus; *SpexNPV*, *Spodoptera exempta* nucleopolyhedrovirus; *WNV*, West Nile virus; *YFV*, yellow fever virus; *ZIKV*, Zika virus. Reduced virus transmission means the reduction of virus load in the mosquito saliva; Reduced virus transmission rate means the reduction in the number of individuals with infection in saliva. Reduced virus dissemination means the reduction of virus load in the mosquito head or leg. Reduced virus dissemination rate means the reduction in the number of individuals with infection in head or leg. Reduced virus load means the reduction in either viral gene copies by real-time PCR or virus titer by plaque assay or TCID50. Reduced virus infection rate means the decrease in the number of individuals infected with virus. Reduced bacterial density means the reduction in either bacterial gene copies or bacterial titer. Reduced parasite infection rate means the decrease in the number of individuals infected with plasmodium. Reduced parasite loads means reduction of oocysts in midgut from infected mosquitoes. Increased parasite loads means increasing of oocysts in midgut from infected mosquitoes. Reduced parasite transmission means the reduction of sporozoite load in the mosquito saliva gland. Reduced worm load means reduction in number of infective stage L3s from infected mosquitoes.

2008). Furthermore, flies infected with both IIV-6 and *Wolbachia* have a reduced life span, indicating an increase in fitness cost due to co-infection (Teixeira et al., 2008). (3) Due to its ease of manipulation, mosquito with a transient somatic *Wolbachia* infection is used as a model to study its impact on pathogens. Different with pathogen inhibition observed consistently in a stable infected system, this transiently infected model results in both pathogen interference and enhancement (Hughes et al., 2011, 2012; Kambris et al., 2010). For example, *w*AlbB was reported to increase West Nile infection in *Culex Tarsalis* (Dodson et al., 2014), while an opposite effect was observed in *Ae. aegypti*. Although a stable *w*AlbB infection inhibits *Plasmodium berghei* in *An. stephensi*, a transient *w*AlbB infection was reported to enhance *P. berghei* in *An. gambiae* (Hughes et al., 2012). Further studies are needed to determine whether the disparity is an artifact or is due to host difference.

Wolbachia-Associated Fitness

As symbiotic bacteria and reproductive parasites, *Wolbachia* also impact host fitness, including longevity, fecundity, fertility, male mating competitiveness, and immature survivorship and development (Blagrove et al., 2013; Dobson et al., 2002; Joshi et al., 2014; McMeniman et al., 2009; McMeniman and O'Neill, 2010). Depending on the *Wolbachia* strain, host background, sex, and environmental condition, *Wolbachia* may provide a fitness advantage, cost, or neutral effect to hosts. For example, *Wolbachia* *w*AlbA and *w*AlbB increase mosquito life spans and fecundity in *Ae. albopictus* (Dobson et al., 2002). After transferring into *Ae. aegypti* and *An. stephensi*, *w*AlbB increases the life span of the transinfected mosquitoes (Bian et al., 2010; Joshi et al., 2014). A similar increase in the life span is also observed in *w*Mel-transinfected *Ae. aegypti* males but not females (Blagrove et al., 2013). However, *w*MelPop significantly reduces host longevity in both the original host *D. melanogaster* and transinfected *Ae. aegypti* (McMeniman et al., 2009; Min and Benzer, 1997). The *w*MelPop infection also negatively affects *Ae. aegypti* larval and pupal development time, adult body size, and wing size (McMeniman and O'Neill, 2010; Yeap et al., 2011). *w*MelPop- and *w*Mel-transinfected *Ae. aegypti* females are less fecund than are wild-type uninfected females (McMeniman et al., 2011). In contrast, *Wolbachia* infection can confer a positive fecundity benefit for *D. melanogaster* reared on iron-restricted or iron-overloaded diets (Brownlie et al., 2009). *D. mauritiana* infected with a native *Wolbachia* *w*Mau strain produces about four times more eggs than the noninfected counterpart due to an increase in the mitotic activity of germline stem cells, as well as a decrease in programmed cell death in the germarium of the infected flies (Fast et al., 2011).

WOLBACHIA-MEDIATED IMMUNE INDUCTIONS

Insects such as fruit flies and mosquitoes have developed highly effective immune systems to defend against invading microorganisms. As a symbiotic intracellular bacterium, *Wolbachia* have formed intimate association with its host, raising the question of how this guest microbe interacts with its host immune system. Compared to most other commensal bacteria within mosquitoes, *Wolbachia* may interact with the host immune system in a very different way due to its transmission mode, cellular location, and evolutionary relationship with its hosts. Because of its vertical transmission, *Wolbachia* are present in its host before host immune system starts to develop, providing the opportunity for *Wolbachia* to shape the development of host immunity. As an intracellular bacterium residing within a vacuole, *Wolbachia* may not be sensitive to attack by many immune effectors, such as secreted antimicrobial peptides.

In some *Wolbachia*–host associations that have experienced long evolutionary histories, both *Wolbachia* and host immunity may have coadapted in such a manner that *Wolbachia* are no longer recognized as foreign organisms by host pattern recognition receptors (PRRs). From this perspective, a recent *Wolbachia*/host association would be a better system to study the cross talk between *Wolbachia* and host immunity. This is because these new associations represent a time window when coadaptation between *Wolbachia* and host immune responses is initiated, but not masked by a deep evolutionary history. Thus, it is not surprising that most of our knowledge on the interaction of *Wolbachia* with host immunity has been elucidated from transinfected mosquitoes (Fig. 3.1), in which hosts have maintained the maternal heritable novel *Wolbachia* infection for less than 10 years.

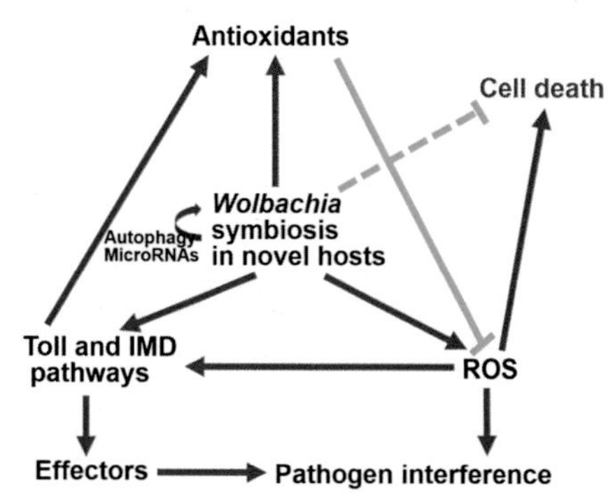

FIGURE 3.1 ***Wolbachia*-mediated immunity induction in mosquito vectors.** When *Wolbachia* are introduced into and forced to form symbiosis with a novel mosquito host, *Wolbachia* induce oxidative stress and production of ROS, resulting in alteration of the intracellular redox balance. If loss of control occurs, it will result in cell death such as apoptosis. To maintain symbiosis, *Wolbachia* induces the production of antioxidants either directly or through activation of the Toll pathway. By removing free radicals and inhibiting chain reactions and oxidation, antioxidants prevent their damage to the cells. *Wolbachia* may also directly inhibit cell apoptosis as reported in several studies (Bazzocchi et al., 2007; Fast et al., 2011; Pannebakker et al., 2007). In the transinfected mosquitoes, *Wolbachia* may activate the Toll and Imd pathways through either ROS as a signal molecule, an intracellular PRR such as PGRPLE, or both. When *Wolbachia* form an association with a new host, activation of host basal immunity may function to foster and protect the formation of symbiosis, while microRNAs and autophagy play roles in regulating *Wolbachia* density. As a by-product of this symbiosis process, production of ROS and expression of Toll/Imd-regulating effectors, together with associated metabolic changes, result in pathogen interference.

Wolbachia Recognition by Hosts

Host immune system may be able to distinguish beneficial bacteria from harmful ones, resulting in either symbiosis or clearance of microbes, respectively. Evidence reveals that invertebrate innate immunity can be finely adjusted to promote the establishment and maintenance of beneficial bacterium symbiosis, which may be accomplished through long-term evolution (Nyholm and Graf, 2012). In *Drosophila* with a native *Wolbachia* infection, host immunity is weakly boosted by *Wolbachia* (Rances et al., 2012; Wong et al., 2011). However, in *Ae. aegypti* lines transinfected with *w*AlbB, *w*Mel, and *w*MelPop, *Wolbachia* is consistently observed to activate host immunity and result in strong induction of expression for immune effector molecules (Moreira et al., 2009; Pan et al., 2012; Rances et al., 2012). As in the case with other symbiotic bacteria (Chu and Mazmanian, 2013; Nyholm and Graf, 2012), microorganism-associated molecular patterns (MAMPs) of *Wolbachia* may be recognized by host immune system via PRR.

Based on the annotation of *Wolbachia* genome, it is unlikely that the cell wall of *Wolbachia* contains LPS-like molecules (Foster et al., 2005; Wu et al., 2004). Evidence also shows that the *Wolbachia* cell wall does not have the same peptidoglycan as extracellular Gram-negative bacteria (Foster et al., 2005; Wu et al., 2004). However, it has been shown that *Wolbachia* are able to synthesize de novo meso-diaminopimelic acid (*m*-DAP) and lipid II (Foster et al., 2005). In *Drosophila*, DAP-type peptidoglycan is detected by the receptors PGRP-LC and PGRP-LE, which result in the activation of the

Imd pathway (Kaneko et al., 2006). The major surface protein and soluble endotoxin-like products of *Wolbachia* endosymbionts of filarial nematodes are implicated in the induction of the immune response in patients with filarial diseases (Cross et al., 2001; Taylor et al., 2001). In mosquito cells, the *Wolbachia* surface protein (WSP) has been reported to induce innate immune responses (Pinto et al., 2012). Thus, *Wolbachia* contain MAMPs that can be detected by the host immune system.

Currently we have no conclusive evidence to show which PRRs can detect the intracellular *Wolbachia* infection. PGRP-LE can be an intracellular sensor of Gram-negative bacteria in *Drosophila* (Kaneko et al., 2006), raising the possibility that it may be able to detect *Wolbachia*. In addition to inducing the Imd pathway, recognition of intracellular bacteria by PGRP-LE can induce autophagy in *Drosophila* (Yano et al., 2008). In both *Drosophila* and the filarial nematode, the presence of *Wolbachia* initiates and activates autophagy, resulting in inhibition of *Wolbachia* populations (Voronin et al., 2012). Further evidence is needed to conclude whether PGRP-LE recognizes *Wolbachia* to induce autophagy or not. Interestingly, PGRP-LE is not induced by *w*Mel or *w*MelPop in the transinfected *Ae. aegypti* lines (Pan et al., 2012; Rances et al., 2012), with the exception of slight induction by *w*AlbB in mosquito midgut, indicating that *Wolbachia* may be able to avoid host recognition and thus prevent activation of autophagy.

Wolbachia Regulates Host Toll and Imd Signaling Pathways

Wolbachia infection significantly induces the Toll signaling pathway in both blood-fed and blood-unfed transinfected mosquitoes (Pan et al., 2012). This activation is unexpected considering the fact that *Wolbachia* is a Gram-negative bacterium and this pathway is normally induced by Gram-positive bacteria. Evidence indicates that *Wolbachia* induces production of reactive oxygen species (ROS) in the mosquito tissues in the transinfected *Ae. aegypti* mosquitoes carrying stable *w*AlbB infection (Pan et al., 2012). As observed in the mammalian systems, high level of ROS can serve as the signal molecule to activate the Toll pathway in mosquitoes (Pan et al., 2012).

Our knowledge on the interactions of *Wolbachia* with the Imd pathway is relatively sparse compared to the Toll pathway. In 5- to 7-day-old females before a blood meal, the expression of the Imd pathway-related genes are induced by both *w*AlbB and *w*Melpop in transinfected *Ae. aegypti* (Moreira et al., 2009; Pan et al., 2012). But a similar induction has not been observed in the *w*AlbB-infected mosquitoes 12 days after feeding on DENV-infected blood. Considering that Toll and Imd pathways synergistically activate innate immune responses in insects (Zou et al., 2011), mild induction of Imd pathway, together with the activated Toll pathway, may strongly boost the expression of the downstream effector molecules. Consistently, very high levels of antimicrobial peptides are induced by *Wolbachia* in the transinfected mosquitoes. For example, in transinfected *Ae. Aegypti*, 17-fold and up to 100-fold increase in the expression of antimicrobial peptides were induced by *Wolbachia w*AlbB and *w*Melpop, respectively (Moreira et al., 2009; Pan et al., 2012).

Wolbachia Induces Production of Reactive Oxygen Species

Evidence shows that *Wolbachia* induce oxidative stress and ROS production in both mosquito tissues and cells (Andrews et al., 2012; Brennan et al., 2008; Pan et al., 2012). The association of *Wolbachia* and high level of ROS production in the host was first reported in *Ae. albopictus* cell line with a natural *Wolbachia* infection (Brennan et al., 2008). Subsequently, a *w*AlbB-induced increase in ROS levels was observed in *Ae. aegypti* (Pan et al., 2012),

Ae. polynesiensis (Andrews et al., 2012), and *An. stephensi* (Bian et al., 2013a). In the transinfected *Ae. aegypti*, *w*AlbB induced the expression of the two enzymes involved in production of ROS, NADPH oxidase (NOXM), and dual oxidase (DUOX2), which subsequently triggered the activation of Toll pathway and production of AMPs and antioxidants (Pan et al., 2012). In addition to ROS produced through these enzymes, over 90% of cellular ROS can be traced back to a mitochondrial origin in the cell. As a parasite, *Wolbachia* may hijack the host's energy metabolism to exploit its energy resource and metabolites for its own growth. This leads to a hypothesis that the increased rate of energy metabolism in host cells as a result of *Wolbachia* infection may lead to the overproduction of ROS through the mitochondrial respiratory chain.

This elevated level of ROS may also result in pathogen interference and/or expression of CI. In addition to activation of the host immune pathway as signaling molecules, ROS can also serve as effector molecules to directly inhibit pathogens. The ROS-based host immune defense against bacteria and fungi is widely present among various insect species, including the cockroach *Blaberus discoidalis* (Whitten and Ratcliffe, 1999), the silkworm *Bombyx mori* (Ishii et al., 2008), the scale insect *Dactylopius coccus* (Garcia-Gil De Munoz et al., 2007), the moths *Galleria mellonella* (Bergin et al., 2005), and *Parasemia plantaginis* (Mikonranta et al., 2014), and the sand fly *Lutzomyia longipalpis* (Diaz-Albiter et al., 2012). *In An. gambiae*, elevated ROS levels are necessary to induce an effective immune response against *Plasmodium* and bacteria (Kumar et al., 2003; Molina-Cruz et al., 2008). The level of ROS increases significantly in *Drosophila simulans* naturally infected with *Wolbachia* as compared with uninfected flies (Wong et al., 2015), and the high ROS levels are associated with an increase in DNA strand breaks in meiotic spermatocytes, which could be related to sperm modification in *Wolbachia*-infected males (Brennan et al., 2012).

The Interaction of *Wolbachia* With MicroRNAs

MicroRNAs (miRNAs) are endogenous small noncoding RNA molecules (snRNA) that function in RNA silencing and posttranscriptional regulation of target mRNAs expression. In mosquitoes, miRNAs play important roles in regulating blood feeding, female reproduction, and antipathogen immunity (Bryant et al., 2010; Lucas et al., 2015; Slonchak et al., 2014; Zhu et al., 2014). *Wolbachia* has been observed to alter the expression profile of miRNAs in mosquitoes (Hussain et al., 2011; Mayoral et al., 2014a; Zhang et al., 2013), which may facilitate symbiosis formation by *Wolbachia* and induction of pathogen interference. In transinfected *Ae. aegypti*, *w*MelPop-CLA induces expression of the miRNA aae-miR-2940, which suppresses the DNA (cytosine-5) methyltransferase gene *AaDnmt2*, resulting in inhibition of DENV replication. aae-miR-2940 also targets the metalloprotease coding *m41 ftsh* in mosquitoes, and either the inhibition of aae-miR-2940 or downregulation of the metalloprotease gene leads to a decrease in *Wolbachia* density (Hussain et al., 2011). aae-miR-12 is another miRNA induced by *w*MelPop-CLA in *Ae. aegypti* Aag-2 cells. It affects two host genes coding for DNA replication licensing factor (MCM6) and monocarboxylate transporter (MCT1) and plays essential roles in the replication and maintenance of *Wolbachia* in the host cells (Osei-Amo et al., 2012).

In addition to host miRNAs, two small noncoding *Wolbachia* RNAs (WsnRNAs), WsnRNA-46 and WsnRNA-49, are expressed by *Wolbachia* in the transinfected *Ae. aegypti* and naturally infected *D. melanogaster* and *D. simulans* (Mayoral et al., 2014b). These *Wolbachia*-encoded WsnRNAs may potentially regulate expression of host and *Wolbachia* genes, supporting the possibility that the production of functional snRNAs by *Wolbachia* play roles in cross-kingdom communication between the endosymbiont and its host.

THE ROLE OF *WOLBACHIA*-INDUCED IMMUNITY IN PATHOGEN INTERFERENCE

Although the molecular mechanism of *Wolbachia*-mediated pathogen interference is still unclear, immune priming is one of the major models that can explain many observations in different *Wolbachia*/host associations. In this model, *Wolbachia* act like a vaccine to boost host immune system, resulting in resistance of host to pathogens, including those transmitted to humans such as arboviruses and *Plasmodia* and those continuously exposed to mosquitoes in nature such as environmental bacteria and fungi (Fig. 3.1).

In order to determine the role of immune priming in *Wolbachia*-mediated pathogen interference, it is important to confront the model with available experimental facts. Below are the facts that support the role of *Wolbachia*-induced immunity in pathogen interference.

1. Mosquito innate immunity is strongly induced in the *w*AlbB (Bian et al., 2010; Pan et al., 2012), *w*MelPop, and *w*Mel-transinfected mosquitoes (Moreira et al., 2009), all of which are resistant to pathogens. These activated immune responses include the Toll pathway and ROS, which are known to inhibit viruses, Gram-positive bacteria, fungi, and *Plasmodia*.
2. Silencing the antimicrobial peptides genes, which are induced by *Wolbachia*, increases dengue viral infection in *w*AlbB-transinfected *Ae. aegypti*. Overexpression of these antimicrobial peptide genes results in inhibition of dengue virus in *Ae. aegypti* in the absence of *Wolbachia* (Pan et al., 2012).
3. There is a positive correlation between *Wolbachia* density and viral interference (Lu et al., 2012; Osborne et al., 2009, 2012). A similar positive correlation is also present between *Wolbachia* density and the expression level of antimicrobial peptides (Lu et al., 2012).
4. Some of the activated immune effectors have a broad antibiotic spectrum, targeting both prokaryotic and eukaryotic pathogens. ROS appears to be such an effector that can both inhibit a variety of pathogens and activate the Toll pathway to enhance the antipathogen effect (Bian et al., 2013a; Pan et al., 2012). The presence of ROS both intracellularly and extracellularly supports its inhibitory role in virus dissemination and replication in mosquitoes. Furthermore, the local effect of ROS also matches the lack of pathogen interference noted in some naturally infected insects with *Wolbachia* that concentrate mainly in the germline cells (Lu et al., 2012). In addition to its role in immune defense, ROS can also impact host cell metabolism (Kremer et al., 2009), resulting in viral interference, because as intracellular parasites viruses depend on the biosynthetic mechanisms of their host cells.

Additional evidence is required for the immune priming model to better explain the observed pathogen interference. For example, the fact that *Wolbachia* do not protect their host from infections with DNA viruses (Graham et al., 2012; Teixeira et al., 2008) indicates that immune effectors induced by *Wolbachia* may be present and function in the host cell cytoplasm and not in the nucleus. The ability of both baculoviruses and iridoviruses to replicate in the nucleus may allow them to avoid the inhibitory effects of *Wolbachia*. It is also worthy to note that *Wolbachia* can enhance the pathogenesis of these two viral infections in their hosts although there is no evidence for an increase in viral load (Graham et al., 2012; Teixeira et al., 2008).

Although there is a strong correlation between *Wolbachia*-induced immunity and pathogen interference in transinfected mosquitoes, such linkages appear to be weak in insects naturally infected with *Wolbachia*. *D. melanogaster* can be artificially infected with dengue virus and the native *Wolbachia* is able to inhibit viral replication in flies (Rances et al., 2012). The *Drosophila* homologs of the mosquito immune genes that are induced by *Wolbachia* in the transinfected *Ae. aegypti*, show either no or minor regulation by these native *Wolbachia* symbionts in flies (Bourtzis et al., 2000; Rances et al., 2012; Wong et al., 2011). In *Drosophila* lines deficient for Toll and Imd pathways, *Wolbachia* retain the ability to inhibit dengue virus, indicating that neither pathway is required for the expression of the dengue virus blocking phenotype in the *Drosophila* host with a natural *Wolbachia* infection (Rances et al., 2013). However, caution is needed to interpret these results in mosquito. Because *Drosophila* is not a natural host, flies were injected intrathoracically with dengue virus and the whole insect body was used for analyses in these studies. It is known that the important physiological barriers, including midgut/salivary gland infection and escape barriers, determine mosquito vector competence. In order to determine the role of *Wolbachia*-induced immunity in pathogen interference in mosquito, it is essential to study the *Wolbachia*-induced local immune response in the key tissues where dengue virus replicates and migrates.

Both *Ae. albopictus* and *Ae. polynesiensis* are naturally infected with *Wolbachia* but can serve as dengue mosquito vectors. Evidence indicates that the long-term association of *Wolbachia* with these mosquito species results in a significant reduction in *Wolbachia* titer, in particular in the somatic tissues where dengue virus replicates and passes through before successful transmission to vertebrate hosts (Lu et al., 2012). Consistently, native *Wolbachia* have not been observed to boost immunity in *Ae. albopictus* (Blagrove et al., 2013; Lu et al., 2012; Mousson et al., 2010, 2012). However, native *Wolbachia* can clear dengue virus in *Ae. albopictus* cells (Lu et al., 2012). Replacing the native *Wolbachia* with a novel strain can cause pathogen interference in both mosquito species (Bian et al., 2013b; Blagrove et al., 2012). These novel infections result in either an increase in ROS production or induction of immunity, although not as robust as seen in the transinfected *Ae. aegypti* (Bian et al., 2013a; Blagrove et al., 2012; Pan et al., 2012). It would be interesting to know whether the level of *Wolbachia*-induced pathogen interference in the transinfected *Ae. albopictus* is lower than that observed in *Ae. aegypti*, and whether previous exposure of the host immune system to a *Wolbachia* strain will facilitate its adaptation to another strain such that there is accelerated waning of *Wolbachia*-induced immune protection.

THE ROLE OF *WOLBACHIA*-INDUCED IMMUNITY IN SYMBIOSIS FORMATION

It is still unknown how *Wolbachia* interact with their host to maintain persistent infection and how the density of *Wolbachia* is regulated by their host. The fact that *Wolbachia* are maintained in the transinfected lines with highly boosted immunity indicates that *Wolbachia* may not be sensitive to the host immune effectors. It is likely that maternal transmission mechanism may provide endosymbionts the opportunity to shape host immunity for their establishment and maintenance. Consistent with this predication, another parasitic endosymbiont, *Spiroplasma*, which can protect *Drosophila* from a parasitic nematode, is not susceptible to Toll and Imd responses in flies (Herren and Lemaitre, 2011). In addition, activation of humoral immune responses increases endosymbiont load in *D. melanogaster* (Herren and Lemaitre, 2011). Further studies are needed to determine whether *Wolbachia* also utilizes the host immune system to promote the establishment of a symbiotic relationship with its new host.

Evidence supports the idea that the ability to maintain the redox homeostasis is essential for *Wolbachia* to maintain symbiosis with its hosts (Fig. 3.1). In both transinfected mosquitoes and naturally infected mosquito cells, expression of antioxidants is significantly induced by *Wolbachia* (Brennan et al., 2008; Pan et al., 2012), presumably to balance the high ROS and oxidative stress caused by *Wolbachia* infection. In both *Drosophila* and mosquitoes, treatment with antioxidants results in an increase in *Wolbachia* density, indicating that the *Wolbachia* density is sensitive to redox balance (Brennan et al., 2012; Monnin et al., 2016). The ability of *Wolbachia* to protect its host against oxidative stress is also supported by the observation that *Wolbachia*-infected flies produced more eggs than uninfected flies when they reared on diets containing high amounts of iron (Brownlie et al., 2009). Furthermore, the expression of eight antioxidants in the *wAlbB*-transinfected *Ae. aegypti* is regulated by the activated Toll pathway (Pan et al., 2012), supporting the role of the Toll pathway in promotion of the symbiosis formation between *Wolbachia* and a new host.

As a conserved intracellular immune barrier and a regulator of cell homeostasis, autophagy is essential for cell survival during infections and under nutrient limitation. Moreover, as a membrane-traffic housekeeper, autophagy plays a crucial role in the inhibition of intracellular bacteria (Choy and Roy, 2013). In both *Drosophila* and filarial nematodes, *Wolbachia* induce the expression of autophagy-based immune genes. Recognition and activation of host autophagy is particularly apparent in rapidly replicating strains of *Wolbachia* found in somatic tissues of *Drosophila* and filarial nematodes (Voronin et al., 2012). Furthermore, activation of host autophagy reduces *Wolbachia* loads while its inhibition results in an increase in the bacterial titer. In the woodlouse *Armadillidium vulgare* and the wasp *Asobara tabida*, *Wolbachia* reduce the expression of autophagy-associated genes (Chevalier et al., 2012; Kremer et al., 2012), indicating that regulation of autophagy by *Wolbachia* is required for bacterial survival and symbiosis maintenance.

The low level of *Wolbachia* density in the naturally infected insects is associated with a low level of host immunity while a high level of *Wolbachia* present in the transinfected mosquitoes correlates with an elevated level of host immunity. To better understand the cross talk between *Wolbachia* and host immune system, additional experiments are needed to understand whether *Wolbachia* is sensitive to the induced immune effectors, or immune inductions will promote symbiosis formation. Presumably autophagy will inhibit *Wolbachia* and thus needs to be finely regulated such that it can prevent *Wolbachia* from overgrowth, but will not eliminate *Wolbachia*. *Wolbachia* may not be sensitive to antimicrobial peptides. Activation of the Toll pathway might function to not only remove those sensitive microbes that compete with *Wolbachia* for a special niche, but also induced antioxidant expression to balance the oxidative stress caused by *Wolbachia* infection.

THE IMPACT OF *WOLBACHIA*-INDUCED IMMUNITY ON MICROBIOTA

If *Wolbachia* induce expression of immune effectors to which themselves are not sensitive, the boosted immunity may facilitate *Wolbachia* to dominate the microbial community within insects by suppressing the susceptible taxa which may otherwise compete for the same niche. Consistently, *Wolbachia* are the dominant bacterial species in *Cx. pipiens* accounting for 91% of total microbiota, while the dominant bacterial species, *Sphingomonas*, in *Wolbachia*-free *Culex restuans* accounts for 31% of the total microbiota (Muturi et al., 2016). Activation of the Toll and Imd pathways or production of ROS in response to *Wolbachia* is expected to suppress the sensitive Gram-positive and Gram-negative bacteria in mosquito. In both *Ae. aegypti* and *An. stephensi*, *Wolbachia* were found to inhibit *Asaia*, an acetic acid symbiont located in the gut, salivary glands, and reproductive organs of adult mosquitoes although a negative impact of *Asaia* on maternal transmission of *Wolbachia* was also observed (Rossi et al., 2015). Both *w*Mel and *w*MelPop inhibit the extracellular bacteria *Erwinia carotovora* and *Burkholderia cepacia* in *Ae. aegypti*, while only *w*MelPop infection results in reduced densities of the two intracellular bacteria *Salmonella typhimurium* and *Mycobacterium marinum* (Ye et al., 2013). Suppression of these bacteria was also found to correlate with an increase in mosquito survival (Ye et al., 2013). However, the impact of *Wolbachia* on microbiota may occur in a tissue-specific manner and depend on the specific mosquito–microbe associations. A study showed that persistent infection by *Wolbachia* *w*AlbB has no effect on composition of the gut microbiota in adult female *An. stephensi* (Chen et al., 2016)

Evidence further supports that the impact of *Wolbachia* on the other microbes is mediated by host immunity. In *D. simulans* and *D. melanogaster* with native *Wolbachia* infections, the expression of antibacterial immune genes was not induced (Rances et al., 2012; Wong et al., 2011). As expected, no antibacterial protection was observed in the insects with or without *Wolbachia* after challenge with the pathogenic bacteria *Pseudomonas aeruginosa* PA01, *Serratia marcescens,* or *Erwinia carotovora* (Wong et al., 2011). In addition, there is no evidence that *Wolbachia* alter their host's ability to suppress proliferation of any of intracellular bacteria *Listeria monocytogenes* and *Salmonella typhimurium* in *D. melanogaster* (Rottschaefer and Lazzaro, 2012).

EVOLUTION OF *WOLBACHIA*-MEDIATED IMMUNE INDUCTIONS AND ITS IMPACT ON DISEASE CONTROL

Immune induction is costly to the host, thus selection may be in favor of a reduction in immune response and attenuation of *Wolbachia* titer. Consistently, *Wolbachia*-mediated immune activation was mostly observed in the transinfected mosquito lines while low or absence of immune induction was detected in insects carrying native *Wolbachia*. A positive correlation between *Wolbachia* density and expression of immune effectors was observed in mosquito cells (Lu et al., 2012), supporting attenuation of *Wolbachia* titer as a direct cause of waning of *Wolbachia*-induced immune response during the evolution of the symbiotic relationship. As a vertically transmitted symbiont, it is essential for *Wolbachia* to maintain a persistent infection in the germline tissue of a host. In contrast, *Wolbachia* in somatic tissues including midgut, fat body, and salivary gland may be viewed as spillover from reproductive tissues, which will not be passed onto offspring. The attenuation of *Wolbachia* titer may eventually result in the disappearance of *Wolbachia* and loss of local immune induction in those somatic tissues. During this process, some tissues may lose *Wolbachia* earlier than others.

It is difficult to predict how the evolution of *Wolbachia*-mediated immune induction will influence mosquito–pathogen interactions and the implementation of *Wolbachia* for disease control. First, it is unknown how long it will take for the *Wolbachia*-induced immune response to wane below a threshold which is required for effective pathogen blocking. Evidence indicates that it will be a slow process because strong viral blocking has been observed in the *Ae. aegypti* strain with *w*AlbB introduced more than 10 years ago (Pan et al., 2012; Xi et al., 2005). No change in viral interference was detected in *w*Mel-infected *Ae. aegypti,* which was collected from the field site 2 years after establishment (Hoffmann et al., 2011). Evidence also indicates that mosquitoes may still be resistant to pathogens even without immune priming because other mechanisms, such as competition for host factors between *Wolbachia* and pathogens, may function. As observed in *Ae. albopictus*, however, loss of pathogen interference is expected to occur if *Wolbachia* no longer reside in those tissues where pathogens will migrate to, replicate, or absent outside of the germline. Second, it would be interesting to know whether pathogens can develop resistance to host immune effects induced by *Wolbachia*. Considering the rapid evolution of arbovirus genomes and leakiness of the blocking effect for virus infection, it would not be impossible to see that viruses develop mechanisms to evade the blocking mechanism through selection.

TRANSLATIONAL OPPORTUNITIES FOR DISEASE CONTROL AND PREVENTION

The ability of *Wolbachia* to boost host innate immunity makes this symbiotic bacterium act like a vaccine for mosquito vectors to defend against the human pathogens they transmit. Currently, a population replacement strategy is being tested in field trials to spread *Wolbachia* *w*Mel into *Ae. aegypti* for dengue and Zika virus disease control (Aliota et al., 2016). Its impact on disease control will depend on the efficacy of *Wolbachia*-mediated pathogen blocking. A strong blocking requires a high density of *Wolbachia*, while overgrowth of a *Wolbachia* strain, such as *w*MelPop, will lead to an increase in fitness cost to host and failure in population replacement (Iturbe-Ormaetxe et al., 2011; Turelli, 2010). Thus, an ideal *Wolbachia* strain or multiple strain combination would be one that can develop a level of infection which is high enough to block pathogen but low enough not to prevent population replacement.

Mosquitoes, such as *Ae. polynesiensis* and *Ae. albopictus*, carry a low level of native *Wolbachia* infection and serve as vectors of human pathogens. Efforts have been developed to repair the protection effect of *Wolbachia* on arboviruses such as dengue virus. Replacing a native *Wolbachia* with *w*AlbB from *Ae. albopictus* resulted in restoring resistance to dengue virus in *Ae. polynesiensis* (Bian et al., 2013b). In order to develop *Ae. albopictus* that is resistant to dengue and Zika virus, *Wolbachia* *w*Pip from *Cx. pipiens f. molestus* was added into *Ae. albopictus* to generate a mosquito strain carrying triple *Wolbachia* infection (Zhang et al., 2016; Zhang et al., 2015a,b). These strategies will also be useful if attenuation of *Wolbachia* titer results in loss of *Wolbachia*-mediated immune protection and pathogen interference in *Ae. aegypti* in the future.

Better understanding the mechanism of *Wolbachia*-induced immunity and its role in pathogen interference may result in novel translational opportunities for disease control and prevention. A *Wolbachia* strain may be selected or designed to induce strong immune priming even at low titer to avoid its density-associated fitness cost. The other maternally inheritable microbes may be combined with *Wolbachia* to improve priming. Furthermore, *Wolbachia* itself, instead of mosquito, can be used as a target to develop novel insecticides. For example, by manipulating the mosquito immune system, novel genetic approaches may be developed to boost the level of native *Wolbachia* infection in mosquito vectors for blocking disease transmission.

FUTURE RESEARCH DIRECTIONS

The discovery of *Wolbachia*-mediated immune induction has opened a number of exciting directions for the field. The cross talk between *Wolbachia* and host immunity in symbiosis will be one of the most exciting areas. The questions to be answered include: (1) what the key MAMPs of *Wolbachia* involved in determining the strength of immune priming are, (2) how *Wolbachia* are sensed by the host immune system, (3) whether PGRP-LE plays a role in recognizing *Wolbachia* as observed in detecting intracellular bacteria in *Drosophila*, and (4) how the expression of immune effectors facilitates or regulates *Wolbachia* growth to form symbiosis with a novel host. It will be helpful to address most of these fundamental questions if a technology can be developed to genetically modify *Wolbachia*. Furthermore, an increase in production of both extracellular and intracellular ROS is a hallmark of *Wolbachia*–host interactions, in particular when *Wolbachia* has recently formed symbiosis with its mosquito host. It would be important to know what the main source of the elevated ROS is in response to *Wolbachia* infection.

Further studies on the evolution of *Wolbachia*-induced immunity and the mechanisms determining *Wolbachia* host range will facilitate moving the field forward. Although technology was developed to establish *Wolbachia* symbiosis in different insect species, there have been challenges in introducing *Wolbachia* into other species, including the *An. gambiae* complex. In those challenging systems, a common observation is that a high level of *Wolbachia* infection results in either death or sterility of the host, while *Wolbachia* at a low titer will show low fidelity of maternal transmission and be gradually lost in the subsequent generations. Further studies are required to determine what roles host immunity or microbiota plays in the process to form a novel symbiosis with *Wolbachia* and how the host immune system is shaped by *Wolbachia* during the initiation of symbiosis.

Another important direction is to determine the relative contribution of immune induction on *Wolbachia*-mediated pathogen interference. Evidence indicates that native *Wolbachia* can induce resistance to virus in *Drosophila* even without an apparent immune activation, raising the possibility that multiple mechanisms contribute to *Wolbachia*-mediated pathogen interference. However, precise dissection of the mechanisms will require a better understanding of the profile of *Wolbachia*-induced immune effectors that play roles in pathogen blocking. Some effectors, like ROS, which are involved in both metabolism and immunity, are likely to present challenges to this study. Our improved knowledge on *Wolbachia*-mediated pathogen interference will facilitate development of mosquito strains with complete and stable blocking effects and high fitness, which can greatly aid in the current global effort to develop *Wolbachia* for control of mosquito-borne diseases, including malaria, dengue, and Zika.

References

Aliota, M.T., Peinado, S.A., Velez, I.D., Osorio, J.E., 2016. The wMel strain of *Wolbachia* reduces transmission of zika virus by *Aedes aegypti*. Sci. Rep. 6, 28792.

Andrews, E.S., Crain, P.R., Fu, Y., Howe, D.K., Dobson, S.L., 2012. Reactive oxygen species production and *Brugia pahangi* survivorship in *Aedes polynesiensis* with artificial *Wolbachia* infection types. PLoS Pathog. 8, e1003075.

Baldini, F., Segata, N., Pompon, J., Marcenac, P., Robert Shaw, W., Dabire, R.K., Diabate, A., Levashina, E.A., Catteruccia, F., 2014. Evidence of natural *Wolbachia* infections in field populations of *Anopheles gambiae*. Nat. Commun. 5, 3985.

Baldo, L., Ayoub, N.A., Hayashi, C.Y., Russell, J.A., Stahlhut, J.K., Werren, J.H., 2008. Insight into the routes of *Wolbachia* invasion: high levels of horizontal transfer in the spider genus *Agelenopsis* revealed by *Wolbachia* strain and mitochondrial DNA diversity. Mol. Ecol. 17, 557–569.

Baton, L.A., Pacidonio, E.C., Goncalves, D.S., Moreira, L.A., 2013. wFlu: characterization and evaluation of a native *Wolbachia* from the mosquito *Aedes fluviatilis* as a potential vector control agent. PLoS One 8, e59619.

Bazzocchi, C., Comazzi, S., Santoni, R., Bandi, C., Genchi, C., Mortarino, M., 2007. *Wolbachia* surface protein (WSP) inhibits apoptosis in human neutrophils. Parasite Immunol. 29, 73–79.

Bergin, D., Reeves, E.P., Renwick, J., Wientjes, F.B., Kavanagh, K., 2005. Superoxide production in *Galleria mellonella* hemocytes: identification of proteins homologous to the NADPH oxidase complex of human neutrophils. Infect. Immun. 73, 4161–4170.

Bian, G., Xu, Y., Lu, P., Xie, Y., Xi, Z., 2010. The endosymbiotic bacterium *Wolbachia* induces resistance to dengue virus in *Aedes aegypti*. PLoS Pathog. 6, e1000833.

Bian, G., Joshi, D., Dong, Y., Lu, P., Zhou, G., Pan, X., Xu, Y., Dimopoulos, G., Xi, Z., 2013a. *Wolbachia* invades *Anopheles stephensi* populations and induces refractoriness to *Plasmodium* infection. Science 340, 748–751.

Bian, G., Zhou, G., Lu, P., Xi, Z., 2013b. Replacing a native *Wolbachia* with a novel strain results in an increase in endosymbiont load and resistance to dengue virus in a mosquito vector. PLoS Negl. Trop. Dis. 7, e2250.

Blagrove, M.S., Arias-Goeta, C., Failloux, A.B., Sinkins, S.P., 2012. *Wolbachia* strain wMel induces cytoplasmic incompatibility and blocks dengue transmission in *Aedes albopictus*. Proc. Natl. Acad. Sci. U.S.A. 109, 255–260.

Blagrove, M.S., Arias-Goeta, C., Di Genua, C., Failloux, A.B., Sinkins, S.P., 2013. A *Wolbachia* wMel transinfection in *Aedes albopictus* is not detrimental to host fitness and inhibits Chikungunya virus. PLoS Negl. Trop. Dis. 7, e2152.

Bourtzis, K., Pettigrew, M.M., O'Neill, S.L., 2000. *Wolbachia* neither induces nor suppresses transcripts encoding antimicrobial peptides. Insect Mol. Biol. 9, 635–639.

Bourtzis, K., Dobson, S.L., Xi, Z., Rasgon, J.L., Calvitti, M., Moreira, L.A., Bossin, H.C., Moretti, R., Baton, L.A., Hughes, G.L., Mavingui, P., Gilles, J.R., 2014. Harnessing mosquito-*Wolbachia* symbiosis for vector and disease control. Acta Trop. 132 (Suppl.), S150–S163.

Brennan, L.J., Keddie, B.A., Braig, H.R., Harris, H.L., 2008. The endosymbiont *Wolbachia pipientis* induces the expression of host antioxidant proteins in an *Aedes albopictus* cell line. PLoS One 3, e2083.

Brennan, L.J., Haukedal, J.A., Earle, J.C., Keddie, B., Harris, H.L., 2012. Disruption of redox homeostasis leads to oxidative DNA damage in spermatocytes of *Wolbachia*-infected *Drosophila simulans*. Insect Mol. Biol. 21, 510–520.

Brownlie, J.C., Cass, B.N., Riegler, M., Witsenburg, J.J., Iturbe-Ormaetxe, I., McGraw, E.A., O'Neill, S.L., 2009. Evidence for metabolic provisioning by a common invertebrate endosymbiont, *Wolbachia pipientis*, during periods of nutritional stress. PLoS Pathog. 5, e1000368.

Bryant, B., Macdonald, W., Raikhel, A.S., 2010. microRNA miR-275 is indispensable for blood digestion and egg development in the mosquito *Aedes aegypti*. Proc. Natl. Acad. Sci. U.S.A. 107, 22391–22398.

Casiraghi, M., Bordenstein, S.R., Baldo, L., Lo, N., Beninati, T., Wernegreen, J.J., Werren, J.H., Bandi, C., 2005. Phylogeny of *Wolbachia pipientis* based on gltA, groEL and ftsZ gene sequences: clustering of arthropod and nematode symbionts in the F supergroup, and evidence for further diversity in the *Wolbachia* tree. Microbiology 151, 4015–4022.

Chen, S., Zhao, J., Joshi, D., Xi, Z., Norman, B., Walker, E., 2016. Persistent infection by *Wolbachia* wAlbB has no effect on composition of the gut microbiota in adult female *Anopheles stephensi*. Front. Microbiol. 7. http://dx.doi.org/10.3389/fmicb.2016.01485.

Chevalier, F., Herbiniere-Gaboreau, J., Charif, D., Mitta, G., Gavory, F., Wincker, P., Greve, P., Braquart-Varnier, C., Bouchon, D., 2012. Feminizing *Wolbachia*: a transcriptomics approach with insights on the immune response genes in *Armadillidium vulgare*. BMC Microbiol. 12 (Suppl. 1), S1.

Choy, A., Roy, C.R., 2013. Autophagy and bacterial infection: an evolving arms race. Trends Microbiol. 21, 451–456.

Chu, H., Mazmanian, S.K., 2013. Innate immune recognition of the microbiota promotes host-microbial symbiosis. Nat. Immunol. 14, 668–675.

Cross, H.F., Haarbrink, M., Egerton, G., Yazdanbakhsh, M., Taylor, M.J., 2001. Severe reactions to filarial chemotherapy and release of *Wolbachia* endosymbionts into blood. Lancet 358, 1873–1875.

Curtis, C.F., Sinkins, S.P., 1998. *Wolbachia* as a possible means of driving genes into populations. Parasitology 116 (Suppl.), S111–S115.

Diaz-Albiter, H., Sant'Anna, M.R., Genta, F.A., Dillon, R.J., 2012. Reactive oxygen species-mediated immunity against *Leishmania mexicana* and *Serratia marcescens* in the sand phlebotomine fly *Lutzomyia longipalpis*. J. Biol. Chem. 287, 23995–24003.

Dobson, S.L., Marsland, E.J., Rattanadechakul, W., 2002. Mutualistic *Wolbachia* infection in *Aedes albopictus*: accelerating cytoplasmic drive. Genetics 160, 1087–1094.

Dodson, B.L., Hughes, G.L., Paul, O., Matacchiero, A.C., Kramer, L.D., Rasgon, J.L., 2014. *Wolbachia* enhances West Nile virus (WNV) infection in the mosquito *Culex tarsalis*. PLoS Negl. Trop. Dis. 8, e2965.

Dutra, H.L., Rocha, M.N., Dias, F.B., Mansur, S.B., Caragata, E.P., Moreira, L.A., 2016. *Wolbachia* blocks currently circulating zika virus isolates in Brazilian *Aedes aegypti* mosquitoes. Cell Host Microbe 19, 771–774.

Fast, E.M., Toomey, M.E., Panaram, K., Desjardins, D., Kolaczyk, E.D., Frydman, H.M., 2011. *Wolbachia* enhance *Drosophila* stem cell proliferation and target the germline stem cell niche. Science 334, 990–992.

Ferguson, N.M., Kien, D.T., Clapham, H., Aguas, R., Trung, V.T., Chau, T.N., Popovici, J., Ryan, P.A., O'Neill, S.L., McGraw, E.A., Long, V.T., Dui le, T., Nguyen, H.L., Chau, N.V., Wills, B., Simmons, C.P., 2015. Modeling the impact on virus transmission of *Wolbachia*-mediated blocking of dengue virus infection of *Aedes aegypti*. Sci. Transl. Med. 7, 279ra237.

Ferree, P.M., Frydman, H.M., Li, J.M., Cao, J., Wieschaus, E., Sullivan, W., 2005. *Wolbachia* utilizes host microtubules and Dynein for anterior localization in the *Drosophila* oocyte. PLoS Pathog. 1, e14.

Foster, J., Ganatra, M., Kamal, I., Ware, J., Makarova, K., Ivanova, N., Bhattacharyya, A., Kapatral, V., Kumar, S., Posfai, J., Vincze, T., Ingram, J., Moran, L., Lapidus, A., Omelchenko, M., Kyrpides, N., Ghedin, E., Wang, S., Goltsman, E., Joukov, V., Ostrovskaya, O., Tsukerman, K., Mazur, M., Comb, D., Koonin, E., Slatko, B., 2005. The *Wolbachia* genome of *Brugia malayi*: endosymbiont evolution within a human pathogenic nematode. PLoS Biol. 3, e121.

Garcia-Gil De Munoz, F., Lanz-Mendoza, H., Hernandez-Hernandez, F.C., 2007. Free radical generation during the activation of hemolymph prepared from the homopteran *Dactylopius coccus*. Arch. Insect Biochem. Physiol. 65, 20–28.

Glaser, R.L., Meola, M.A., 2010. The native *Wolbachia* endosymbionts of *Drosophila melanogaster* and *Culex quinquefasciatus* increase host resistance to West Nile virus infection. PLoS One 5, e11977.

Graham, R.I., Grzywacz, D., Mushobozi, W.L., Wilson, K., 2012. *Wolbachia* in a major African crop pest increases susceptibility to viral disease rather than protects. Ecol. Lett. 15, 993–1000.

Hedges, L.M., Brownlie, J.C., O'Neill, S.L., Johnson, K.N., 2008. *Wolbachia* and virus protection in insects. Science 322, 702.

Herren, J.K., Lemaitre, B., 2011. Spiroplasma and host immunity: activation of humoral immune responses increases endosymbiont load and susceptibility to certain Gram-negative bacterial pathogens in *Drosophila melanogaster*. Cell Microbiol. 13, 1385–1396.

Hilgenboecker, K., Hammerstein, P., Schlattmann, P., Telschow, A., Werren, J.H., 2008. How many species are infected with *Wolbachia*?–A statistical analysis of current data. FEMS Microbiol. Lett. 281, 215–220.

Hoffmann, A.A., Turelli, M., 1997. Cytoplasmic incompatibility in insects. In: O'Neill, S.L., Hoffmann, A.A., Werren, J.H. (Eds.), Influential Passengers: Inherited Microorganisms and Arthropod Reproduction. Oxford University Press, Oxford, pp. 42–80.

Hoffmann, A.A., Montgomery, B.L., Popovici, J., Iturbe-Ormaetxe, I., Johnson, P.H., Muzzi, F., Greenfield, M., Durkan, M., Leong, Y.S., Dong, Y., Cook, H., Axford, J., Callahan, A.G., Kenny, N., Omodei, C., McGraw, E.A., Ryan, P.A., Ritchie, S.A., Turelli, M., O'Neill, S.L., 2011. Successful establishment of *Wolbachia* in *Aedes* populations to suppress dengue transmission. Nature 476, 454–457.

Hooper, L.V., Littman, D.R., Macpherson, A.J., 2012. Interactions between the microbiota and the immune system. Science 336, 1268–1273.

Hughes, G.L., Koga, R., Xue, P., Fukatsu, T., Rasgon, J.L., 2011. *Wolbachia* infections are virulent and inhibit the human malaria parasite *Plasmodium falciparum* in anopheles gambiae. PLoS Pathog. 7, e1002043.

Hughes, G.L., Vega-Rodriguez, J., Xue, P., Rasgon, J.L., 2012. *Wolbachia* strain wAlbB enhances infection by the rodent malaria parasite *Plasmodium berghei* in *Anopheles gambiae* mosquitoes. Appl. Environ. Microbiol. 78, 1491–1495.

Hughes, G.L., Dodson, B.L., Johnson, R.M., Murdock, C.C., Tsujimoto, H., Suzuki, Y., Patt, A.A., Cui, L., Nossa, C.W., Barry, R.M., Sakamoto, J.M., Hornett, E.A., Rasgon, J.L., 2014. Native microbiome impedes vertical transmission of *Wolbachia* in *Anopheles* mosquitoes. Proc. Natl. Acad. Sci. U.S.A. 111, 12498–12503.

Hussain, M., Frentiu, F.D., Moreira, L.A., O'Neill, S.L., Asgari, S., 2011. *Wolbachia* uses host microRNAs to manipulate host gene expression and facilitate colonization of the dengue vector *Aedes aegypti*. Proc. Natl. Acad. Sci. U.S.A. 108, 9250–9255.

Hussain, M., Lu, G., Torres, S., Edmonds, J.H., Kay, B.H., Khromykh, A.A., Asgari, S., 2013. Effect of *Wolbachia* on replication of West Nile virus in a mosquito cell line and adult mosquitoes. J. Virol. 87, 851–858.

Ishii, K., Hamamoto, H., Kamimura, M., Sekimizu, K., 2008. Activation of the silkworm cytokine by bacterial and fungal cell wall components via a reactive oxygen species-triggered mechanism. J. Biol. Chem. 283, 2185–2191.

Iturbe-Ormaetxe, I., Walker, T., O'Neill, S.L., 2011. *Wolbachia* and the biological control of mosquito-borne disease. EMBO Rep. 12, 508–518.

Joshi, D., McFadden, M.J., Bevins, D., Zhang, F., Xi, Z., 2014. *Wolbachia* strain wAlbB confers both fitness costs and benefit on *Anopheles stephensi*. Parasit. Vectors 7, 336.

Joubert, D.A., Walker, T., Carrington, L.B., De Bruyne, J.T., Kien, D.H., Hoang Nle, T., Chau, N.V., Iturbe-Ormaetxe, I., Simmons, C.P., O'Neill, S.L., 2016. Establishment of a *Wolbachia* superinfection in *Aedes aegypti* mosquitoes as a potential approach for future resistance management. PLoS Pathog. 12, e1005434.

Kambris, Z., Cook, P.E., Phuc, H.K., Sinkins, S.P., 2009. Immune activation by life-shortening *Wolbachia* and reduced filarial competence in mosquitoes. Science 326, 134–136.

Kambris, Z., Blagborough, A.M., Pinto, S.B., Blagrove, M.S., Godfray, H.C., Sinden, R.E., Sinkins, S.P., 2010. *Wolbachia* stimulates immune gene expression and inhibits plasmodium development in *Anopheles gambiae*. PLoS Pathog. 6.

Kaneko, T., Yano, T., Aggarwal, K., Lim, J.H., Ueda, K., Oshima, Y., Peach, C., Erturk-Hasdemir, D., Goldman, W.E., Oh, B.H., Kurata, S., Silverman, N., 2006. PGRP-LC and PGRP-LE have essential yet distinct functions in the drosophila immune response to monomeric DAP-type peptidoglycan. Nat. Immunol. 7, 715–723.

Kittayapong, P., Baisley, K.J., Baimai, V., O'Neill, S.L., 2000. Distribution and diversity of *Wolbachia* infections in Southeast Asian mosquitoes (Diptera: Culicidae). J. Med. Entomol. 37, 340–345.

Kremer, N., Voronin, D., Charif, D., Mavingui, P., Mollereau, B., Vavre, F., 2009. *Wolbachia* interferes with ferritin expression and iron metabolism in insects. PLoS Pathog. 5, e1000630.

Kremer, N., Charif, D., Henri, H., Gavory, F., Wincker, P., Mavingui, P., Vavre, F., 2012. Influence of *Wolbachia* on host gene expression in an obligatory symbiosis. BMC Microbiol. 12 (Suppl. 1), S7.

Kumar, S., Christophides, G.K., Cantera, R., Charles, B., Han, Y.S., Meister, S., Dimopoulos, G., Kafatos, F.C., Barillas-Mury, C., 2003. The role of reactive oxygen species on *Plasmodium* melanotic encapsulation in *Anopheles gambiae*. Proc. Natl. Acad. Sci. U.S.A. 100, 14139–14144.

Laven, H., 1967. Eradication of *Culex pipiens fatigans* through cytoplasmic incompatibility. Nature 216, 383–384.

Lu, P., Bian, G., Pan, X., Xi, Z., 2012. *Wolbachia* induces density-dependent inhibition to dengue virus in mosquito cells. PLoS Negl. Trop. Dis. 6, e1754.

Lucas, K.J., Roy, S., Ha, J., Gervaise, A.L., Kokoza, V.A., Raikhel, A.S., 2015. MicroRNA-8 targets the Wingless signaling pathway in the female mosquito fat body to regulate reproductive processes. Proc. Natl. Acad. Sci. U.S.A. 112, 1440–1445.

Mayoral, J.G., Etebari, K., Hussain, M., Khromykh, A.A., Asgari, S., 2014a. *Wolbachia* infection modifies the profile, shuttling and structure of microRNAs in a mosquito cell line. PLoS One 9, e96107.

Mayoral, J.G., Hussain, M., Joubert, D.A., Iturbe-Ormaetxe, I., O'Neill, S.L., Asgari, S., 2014b. *Wolbachia* small noncoding RNAs and their role in cross-kingdom communications. Proc. Natl. Acad. Sci. U.S.A. 111, 18721–18726.

McGraw, E.A., Merritt, D.J., Droller, J.N., O'Neill, S.L., 2002. *Wolbachia* density and virulence attenuation after transfer into a novel host. Proc. Natl. Acad. Sci. U.S.A. 99, 2918–2923.

McMeniman, C.J., O'Neill, S.L., 2010. A virulent *Wolbachia* infection decreases the viability of the dengue vector *Aedes aegypti* during periods of embryonic quiescence. PLoS Negl. Trop. Dis. 4, e748.

McMeniman, C.J., Lane, R.V., Cass, B.N., Fong, A.W., Sidhu, M., Wang, Y.F., O'Neill, S.L., 2009. Stable introduction of a life-shortening *Wolbachia* infection into the mosquito *Aedes aegypti*. Science 323, 141–144.

McMeniman, C.J., Hughes, G.L., O'Neill, S.L., 2011. A *Wolbachia* symbiont in *Aedes aegypti* disrupts mosquito egg development to a greater extent when mosquitoes feed on nonhuman versus human blood. J. Med. Entomol. 48, 76–84.

Micieli, M.V., Glaser, R.L., 2014. Somatic *Wolbachia* (Rickettsiales: Rickettsiaceae) levels in *Culex quinquefasciatus* and *Culex pipiens* (Diptera: Culicidae) and resistance to West Nile virus infection. J. Med. Entomol. 51, 189–199.

Mikonranta, L., Mappes, J., Kaukoniitty, M., Freitak, D., 2014. Insect immunity: oral exposure to a bacterial pathogen elicits free radical response and protects from a recurring infection. Front. Zool. 11, 23.

Min, K.T., Benzer, S., 1997. *Wolbachia*, normally a symbiont of *Drosophila*, can be virulent, causing degeneration and early death. Proc. Natl. Acad. Sci. U.S.A. 94, 10792–10796.

Molina-Cruz, A., DeJong, R.J., Charles, B., Gupta, L., Kumar, S., Jaramillo-Gutierrez, G., Barillas-Mury, C., 2008. Reactive oxygen species modulate *Anopheles gambiae* immunity against bacteria and *Plasmodium*. J. Biol. Chem. 283, 3217–3223.

Monnin, D., Kremer, N., Berny, C., Henri, H., Dumet, A., Voituron, Y., Desouhant, E., Vavre, F., 2016. Influence of oxidative homeostasis on bacterial density and cost of infection in *Drosophila-Wolbachia* symbioses. J. Evol. Biol. 29, 1211–1222.

Moreira, L.A., Iturbe-Ormaetxe, I., Jeffery, J.A., Lu, G., Pyke, A.T., Hedges, L.M., Rocha, B.C., Hall-Mendelin, S., Day, A., Riegler, M., Hugo, L.E., Johnson, K.N., Kay, B.H., McGraw, E.A., van den Hurk, A.F., Ryan, P.A., O'Neill, S.L., 2009. A *Wolbachia* symbiont in *Aedes aegypti* limits infection with dengue, Chikungunya, and *Plasmodium*. Cell 139, 1268–1278.

Mousson, L., Martin, E., Zouache, K., Madec, Y., Mavingui, P., Failloux, A.B., 2010. *Wolbachia* modulates Chikungunya replication in *Aedes albopictus*. Mol. Ecol. 19, 1953–1964.

Mousson, L., Zouache, K., Arias-Goeta, C., Raquin, V., Mavingui, P., Failloux, A.B., 2012. The native *Wolbachia* symbionts limit transmission of dengue virus in *Aedes albopictus*. PLoS Negl. Trop. Dis. 6, e1989.

Muturi, E.J., Kim, C.H., Bara, J., Bach, E.M., Siddappaji, M.H., 2016. *Culex pipiens* and *Culex restuans* mosquitoes harbor distinct microbiota dominated by few bacterial taxa. Parasit. Vectors 9, 18.

Nyholm, S.V., Graf, J., 2012. Knowing your friends: invertebrate innate immunity fosters beneficial bacterial symbioses. Nat. Rev. Microbiol. 10, 815–827.

O'Connor, L., Plichart, C., Sang, A.C., Brelsfoard, C.L., Bossin, H.C., Dobson, S.L., 2012. Open release of male mosquitoes infected with a *Wolbachia* biopesticide: field performance and infection containment. PLoS Negl. Trop. Dis. 6, e1797.

Osborne, S.E., Leong, Y.S., O'Neill, S.L., Johnson, K.N., 2009. Variation in antiviral protection mediated by different *Wolbachia* strains in *Drosophila simulans*. PLoS Pathog. 5, e1000656.

Osborne, S.E., Iturbe-Ormaetxe, I., Brownlie, J.C., O'Neill, S.L., Johnson, K.N., 2012. Antiviral protection and the importance of *Wolbachia* density and tissue tropism in *Drosophila simulans*. Appl. Environ. Microbiol. 78, 6922–6929.

Osei-Amo, S., Hussain, M., O'Neill, S.L., Asgari, S., 2012. *Wolbachia*-induced aae-miR-12 miRNA negatively regulates the expression of MCT1 and MCM6 genes in *Wolbachia*-infected mosquito cell line. PLoS One 7, e50049.

Pan, X., Zhou, G., Wu, J., Bian, G., Lu, P., Raikhel, A.S., Xi, Z., 2012. *Wolbachia* induces reactive oxygen species (ROS)-dependent activation of the Toll pathway to control dengue virus in the mosquito *Aedes aegypti*. Proc. Natl. Acad. Sci. U.S.A. 109, E23–E31.

Pannebakker, B.A., Loppin, B., Elemans, C.P., Humblot, L., Vavre, F., 2007. Parasitic inhibition of cell death facilitates symbiosis. Proc. Natl. Acad. Sci. U.S.A. 104, 213–215.

Panteleev, D., Goriacheva, I.I., Andrianov, B.V., Reznik, N.L., Lazebnyi, O.E., Kulikov, A.M., 2007. The endosymbiotic bacterium *Wolbachia* enhances the nonspecific resistance to insect pathogens and alters behavior of *Drosophila melanogaster*. Genetika 43, 1277–1280.

Pinto, S.B., Mariconti, M., Bazzocchi, C., Bandi, C., Sinkins, S.P., 2012. *Wolbachia* surface protein induces innate immune responses in mosquito cells. BMC Microbiol. 12 (Suppl. 1), S11.

Poinsot, D., Charlat, S., Mercot, H., 2003. On the mechanism of *Wolbachia*-induced cytoplasmic incompatibility: confronting the models with the facts. Bioessays 25, 259–265.

Rances, E., Ye, Y.H., Woolfit, M., McGraw, E.A., O'Neill, S.L., 2012. The relative importance of innate immune priming in *Wolbachia*-mediated dengue interference. PLoS Pathog. 8, e1002548.

Rances, E., Johnson, T.K., Popovici, J., Iturbe-Ormaetxe, I., Zakir, T., Warr, C.G., O'Neill, S.L., 2013. The toll and Imd pathways are not required for *Wolbachia*-mediated dengue virus interference. J. Virol. 87, 11945–11949.

Raychoudhury, R., Baldo, L., Oliveira, D.C., Werren, J.H., 2009. Modes of acquisition of *Wolbachia*: horizontal transfer, hybrid introgression, and codivergence in the *Nasonia* species complex. Evolution 63, 165–183.

Rossi, P., Ricci, I., Cappelli, A., Damiani, C., Ulissi, U., Mancini, M.V., Valzano, M., Capone, A., Epis, S., Crotti, E., Chouaia, B., Scuppa, P., Joshi, D., Xi, Z., Mandrioli, M., Sacchi, L., O'Neill, S.L., Favia, G., 2015. Mutual exclusion of *Asaia* and *Wolbachia* in the reproductive organs of mosquito vectors. Parasit. Vectors 8, 278.

Rottschaefer, S.M., Lazzaro, B.P., 2012. No effect of *Wolbachia* on resistance to intracellular infection by pathogenic bacteria in *Drosophila melanogaster*. PLoS One 7, e40500.

Serbus, L.R., Sullivan, W., 2007. A cellular basis for *Wolbachia* recruitment to the host germline. PLoS Pathog. 3, e190.

Shaw, A.E., Veronesi, E., Maurin, G., Ftaich, N., Guiguen, F., Rixon, F., Ratinier, M., Mertens, P., Carpenter, S., Palmarini, M., Terzian, C., Arnaud, F., 2012. *Drosophila melanogaster* as a model organism for bluetongue virus replication and tropism. J. Virol. 86, 9015–9024.

Sinkins, S.P., 2004. *Wolbachia* and cytoplasmic incompatibility in mosquitoes. Insect Biochem. Mol. Biol. 34, 723–729.

Slonchak, A., Hussain, M., Torres, S., Asgari, S., Khromykh, A.A., 2014. Expression of mosquito microRNA Aae-miR-2940-5p is downregulated in response to West Nile virus infection to restrict viral replication. J. Virol. 88, 8457–8467.

Taylor, M.J., Cross, H.F., Ford, L., Makunde, W.H., Prasad, G.B., Bilo, K., 2001. *Wolbachia* bacteria in filarial immunity and disease. Parasite Immunol. 23, 401–409.

Teixeira, L., Ferreira, A., Ashburner, M., 2008. The bacterial symbiont *Wolbachia* induces resistance to RNA viral infections in *Drosophila melanogaster*. PLoS Biol. 6, e2.

Toomey, M.E., Panaram, K., Fast, E.M., Beatty, C., Frydman, H.M., 2013. Evolutionarily conserved *Wolbachia*-encoded factors control pattern of stem-cell niche tropism in *Drosophila* ovaries and favor infection. Proc. Natl. Acad. Sci. U.S.A. 110, 10788–10793.

Tsai, K.H., Huang, C.G., Wu, W.J., Chuang, C.K., Lin, C.C., Chen, W.J., 2006. Parallel infection of Japanese encephalitis virus and *Wolbachia* within cells of mosquito salivary glands. J. Med. Entomol. 43, 752–756.

Turelli, M., Hoffmann, A.A., 1999. Microbe-induced cytoplasmic incompatibility as a mechanism for introducing transgenes into arthropod populations. Insect Mol. Biol. 8, 243–255.

Turelli, M., 2010. Cytoplasmic incompatibility in populations with overlapping generations. Evolution 64, 232–241.

van den Hurk, A.F., Hall-Mendelin, S., Pyke, A.T., Frentiu, F.D., McElroy, K., Day, A., Higgs, S., O'Neill, S.L., 2012. Impact of *Wolbachia* on infection with chikungunya and yellow fever viruses in the mosquito vector *Aedes aegypti*. PLoS Negl. Trop. Dis. 6, e1892.

Voronin, D., Cook, D.A., Steven, A., Taylor, M.J., 2012. Autophagy regulates *Wolbachia* populations across diverse symbiotic associations. Proc. Natl. Acad. Sci. U.S.A. 109, E1638–E1646.

Walker, T., Johnson, P.H., Moreira, L.A., Iturbe-Ormaetxe, I., Frentiu, F.D., McMeniman, C.J., Leong, Y.S., Dong, Y., Axford, J., Kriesner, P., Lloyd, A.L., Ritchie, S.A., O'Neill, S.L., Hoffmann, A.A., 2011. The wMel *Wolbachia* strain blocks dengue and invades caged *Aedes aegypti* populations. Nature 476, 450–453.

Werren, J.H., Zhang, W., Guo, L.R., 1995. Evolution and phylogeny of *Wolbachia*: reproductive parasites of arthropods. Proc. Royal Soc. London. Ser. B: Biol. Sci. 261, 55–63.

Werren, J.H., Baldo, L., Clark, M.E., 2008. *Wolbachia*: master manipulators of invertebrate biology. Nat. Rev. Microbiol. 6, 741–751.

Whitten, M.M., Ratcliffe, N.A., 1999. In vitro superoxide activity in the haemolymph of the West Indian leaf cockroach, *Blaberus discoidalis*. J. Insect Physiol. 45, 667–675.

Wong, Z.S., Hedges, L.M., Brownlie, J.C., Johnson, K.N., 2011. *Wolbachia*-mediated antibacterial protection and immune gene regulation in *Drosophila*. PLoS One 6, e25430.

Wong, Z.S., Brownlie, J.C., Johnson, K.N., 2015. Oxidative stress correlates with *Wolbachia*-mediated antiviral protection in *Wolbachia-Drosophila* associations. Appl. Environ. Microbiol. 81, 3001–3005.

Wu, M., Sun, L.V., Vamathevan, J., Riegler, M., Deboy, R., Brownlie, J.C., McGraw, E.A., Martin, W., Esser, C., Ahmadinejad, N., Wiegand, C., Madupu, R., Beanan, M.J., Brinkac, L.M., Daugherty, S.C., Durkin, A.S., Kolonay, J.F., Nelson, W.C., Mohamoud, Y., Lee, P., Berry, K., Young, M.B., Utterback, T., Weidman, J., Nierman, W.C., Paulsen, I.T., Nelson, K.E., Tettelin, H., O'Neill, S.L., Eisen, J.A., 2004. Phylogenomics of the reproductive parasite *Wolbachia pipientis* *w*Mel: a streamlined genome overrun by mobile genetic elements. PLoS Biol. 2, E69.

Xi, Z., Khoo, C.C., Dobson, S.L., 2005. *Wolbachia* establishment and invasion in an *Aedes aegypti* laboratory population. Science 310, 326–328.

Yano, T., Mita, S., Ohmori, H., Oshima, Y., Fujimoto, Y., Ueda, R., Takada, H., Goldman, W.E., Fukase, K., Silverman, N., Yoshimori, T., Kurata, S., 2008. Autophagic control of listeria through intracellular innate immune recognition in drosophila. Nat. Immunol. 9, 908–916.

Ye, Y.H., Woolfit, M., Rances, E., O'Neill, S.L., McGraw, E.A., 2013. *Wolbachia*-associated bacterial protection in the mosquito *Aedes aegypti*. PLoS Negl. Trop. Dis. 7, e2362.

Yeap, H.L., Mee, P., Walker, T., Weeks, A.R., O'Neill, S.L., Johnson, P., Ritchie, S.A., Richardson, K.M., Doig, C., Endersby, N.M., Hoffmann, A.A., 2011. Dynamics of the "popcorn" *Wolbachia* infection in outbred *Aedes aegypti* informs prospects for mosquito vector control. Genetics 187, 583–595.

Zele, F., Nicot, A., Duron, O., Rivero, A., 2012. Infection with *Wolbachia* protects mosquitoes against *Plasmodium*-induced mortality in a natural system. J. Evol. Biol. 25, 1243–1252.

Zhang, G., Hussain, M., O'Neill, S.L., Asgari, S., 2013. *Wolbachia* uses a host microRNA to regulate transcripts of a methyltransferase, contributing to dengue virus inhibition in *Aedes aegypti*. Proc. Natl. Acad. Sci. U.S.A. 110, 10276–10281.

Zhang, D., Lees, R.S., Xi, Z., Gilles, J.R., Bourtzis, K., 2015a. Combining the sterile insect technique with *Wolbachia*-based approaches: II–A safer approach to *Aedes albopictus* population suppression programmes, designed to minimize the consequences of inadvertent female release. PLoS One 10, e0135194.

Zhang, D., Zheng, X., Xi, Z., Bourtzis, K., Gilles, J.R., 2015b. Combining the sterile insect technique with the incompatible insect technique: I-impact of *Wolbachia* infection on the fitness of triple- and double-infected strains of *Aedes albopictus*. PLoS One 10, e0121126.

Zhang, D., Lees, R.S., Xi, Z., Bourtzis, K., Gilles, J.R., 2016. Combining the sterile insect technique with the incompatible insect technique: III-Robust mating competitiveness of irradiated triple *Wolbachia*-infected *Aedes albopictus* males under semi-field conditions. PLoS One 11, e0151864.

Zhu, X., He, Z., Hu, Y., Wen, W., Lin, C., Yu, J., Pan, J., Li, R., Deng, H., Liao, S., Yuan, J., Wu, J., Li, J., Li, M., 2014. MicroRNA-30e* suppresses dengue virus replication by promoting NF-kappaB-dependent IFN production. PLoS Negl. Trop. Dis. 8, e3088.

Zou, Z., Souza-Neto, J., Xi, Z., Kokoza, V., Shin, S.W., Dimopoulos, G., Raikhel, A., 2011. Transcriptome analysis of *Aedes aegypti* transgenic mosquitoes with altered immunity. PLoS Pathog. 7, e1002394.

CHAPTER

4

Modulation of Mosquito Immune Defenses as a Control Strategy

Victoria L.M. Rhodes, Kristin Michel

Kansas State University, Manhattan, KS, United States

INTRODUCTION

Human malaria, caused by five species of apicomplexan protozoans in the genus *Plasmodium*, requires the parasite to successfully infect and survive within its mosquito vector and mammalian host. Successful completion of the *Plasmodium* life cycle depends on a number of nongenetic determinants, as well the interactions between the genotypes of the parasite, host, and vector. The ability of mosquitoes to acquire, maintain, and transmit human malaria parasite species, also referred to as vector competence, is rare in nature. The majority of mosquito species are unable to sustain the development of human malaria parasites and are refractory to infection. Indeed, even in competent vector species the majority of parasites are blocked or killed at various stages in their development. Mechanistic insights into these processes carry the promise that naturally susceptible mosquito populations could be rendered refractory as an intervention strategy for malaria.

Susceptible mosquitoes allow *Plasmodium* spp. to undergo sporogony (Fig. 4.1, reviewed in Baton and Ranford-Cartwright, 2005; Beier, 1998), and serve as the definitive host, because sexual reproduction of the parasite occurs strictly within the insect's midgut. The mosquito becomes infected when taking a blood meal from a human containing male and female gametocytes. These undergo gametogenesis within minutes of reaching the midgut lumen. Fertilization takes place within the next hour, and, in the following 2h, the zygote undergoes meiosis. Within the first day post blood feeding, the zygote transforms into the motile ookinete, which then traverses the peritrophic matrix surrounding the blood bolus and enters the midgut epithelium. Upon traversal, the ookinete emerges on the basal side of the midgut epithelium, rounds up, and forms the oocyst. In the subsequent 4–5days, the oocyst undergoes mitosis, ultimately leading to the formation of several thousand sporozoites (Rosenberg and Rungsiwongse, 1991). Sporozoites are released into the hemocoel and distributed passively by hemolymph flow throughout the open circulatory system (Hillyer et al., 2007). Between 10 and 14 days after the initial gametocyte-containing blood meal, the mosquito becomes infective to the

Arthropod Vector: Controller of Disease Transmission, Volume 1
http://dx.doi.org/10.1016/B978-0-12-805350-8.00004-0

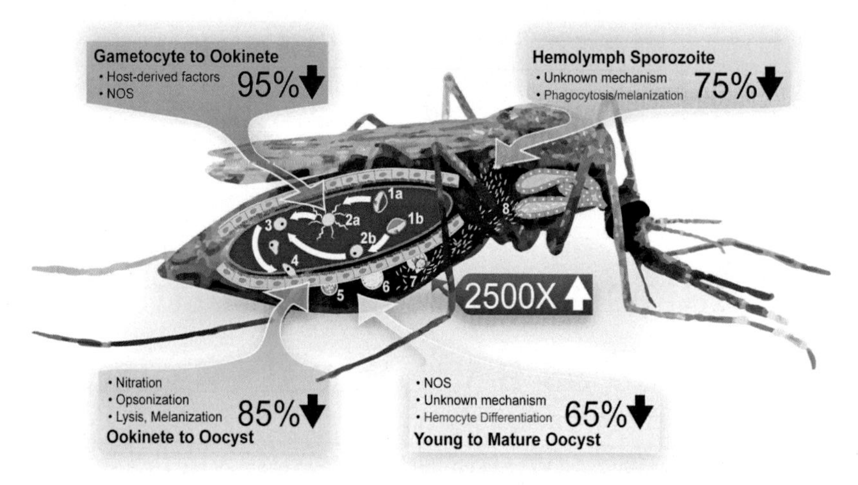

FIGURE 4.1 **The *Plasmodium* spp. life stages affected by mosquito immune reactions.** The sporogonic cycle of *Plasmodium* parasites begins with the ingestion of male microgametocytes and female macrogametocytes (1a/1b). Exflagellated microgametes (2a) fertilize female macrogametocytes (2b) to produce diploid zygotes (3). Zygotes develop into mobile ookinetes (4), which penetrate through the midgut epithelium and form early oocyst beneath the basal lamina (5). Mature oocysts (6) rupture and release thousands of sporozoites (7) into the hemolymph, and the hemolymph flow carries them to the salivary gland (8). The colored boxes depict major parasite life stage transitions, the immune mechanisms that act upon them, and the percent increase or decrease in parasite numbers that occur at each of these transitions. *The figure was adapted from the iconic life cycle depiction by Baton, L.A., Ranford-Cartwright, L.C., 2005. Spreading the seeds of million-murdering death: metamorphoses of malaria in the mosquito. Trends Parasitol. 21, 573–580.*

next human host, as sporozoites cross the salivary gland epithelium, and reach the mosquito saliva. Malaria parasite losses occur at multiple stages of the sporogonic life cycle (Fig. 4.1). The major bottleneck occurs during the first 2–5 days of infection, during the transition of gametocyte to oocyst, leading to elimination more than 99% of parasites (Gouagna et al., 2004). In addition, less than 25% of sporozoites released into the hemocoel eventually reach the salivary gland lumen (Rosenberg and Rungsiwongse, 1991).

The molecular mechanisms that control early parasite development within the mosquito vector and the molecules that are required for parasite passage throughout the mosquito have been reviewed elsewhere, and will not be repeated here (Bennink et al., 2016; Mueller et al., 2010; Whitten et al., 2006). Instead, this chapter will present a synthesis of how our understanding of the mechanisms underlying parasite losses has led to proof of principle studies that render competent vector species refractory to malaria parasites. We provide a brief summary of the data that demonstrated the genetic basis of vector competence and their link to mosquito immunity. Sections Current Knowledge of Antiparasite Immune Reactions in the Mosquito Vector and The Regulation of Anti-Parasite Immunity by Canonical Signal Transduction Pathways detail our understanding of mosquito immunity as it relates to parasite killing. The section Creating Malaria-Refractory Mosquitoes in the Laboratory: The Proof of Principle then summarizes experimental approaches to manipulate the mosquito's immune system to kill and ultimately eliminate malaria parasites within its vector. Finally, we discuss these approaches in the context of ongoing vector control strategies and their potential place in future integrated malaria elimination agendas.

THE GENETIC BASIS OF VECTOR COMPETENCE AND ITS LINK TO MOSQUITO IMMUNITY

The experiments performed in the late 1800s by Ross, Bignami, Grassi, Dionisi, and Bastianelli not only demonstrated that malaria parasites indeed are transmitted by mosquitoes, but that vector competence for malaria varies widely among mosquito species (for an excellent early summary of this work see Lyon, 1900). During his time in Secunderabad, India, Ronald Ross discovered and visualized the pigmented cells (oocysts) on the midguts of "dapple-winged" mosquitoes, now known as *Anopheles* spp., after the ingestion of gametocytes from a malaria-infected patient (Ross and Smyth, 1897). However, experiments conducted during the previous 2 years with "gray" mosquitoes, most likely mosquitoes in the genus *Aedes*, fed on malaria-infected humans never produced oocysts. Similarly, experiments performed by Bignami at the same time in Italy, also demonstrated that mosquitoes outside the genus *Anopheles* were not susceptible to human malaria infection (Bastianelli and Bignami, 1900; Ross, 1899). Variation in vector competence is not only seen across but also within species. Huff observed early that *Culex pipiens* shows variation in natural susceptibility to avian malaria parasites (Huff, 1929, 1927). *C. pipiens*, allowed to take repeated blood meals from birds infected with *Plasmodium cathemerium*, *Plasmodium relictum*, or *Plasmodium elongatum* displayed similar infection rates between first and second feedings with the same *Plasmodium* species. However, the susceptibility of *C. pipiens* was parasite species specific, suggesting that vector competence for a given *Plasmodium* spp. is fixed and thus hereditary (Huff, 1930). These early results have been confirmed subsequently across several mosquito species, which range widely in their ability to support development of distinct *Plasmodium* spp., depending on the vector and parasite species combination (Alavi et al., 2003; Huff, 1929, 1927; Vaughan et al., 1994). Similar observations were made in field-derived populations of *Anopheles gambiae* and *Plasmodium falciparum* in different locations of human malaria transmission in sub-Saharan Africa (Lambrechts et al., 2005; Niaré et al., 2002).

The genetic basis of vector competence was elegantly demonstrated in selection experiments performed by Huff. He produced *C. pipiens* lines by selective mating that were either susceptible or resistant to the avian malaria parasite, *P. cathemerium*. Crosses of these lines yielded F1 progeny that exhibited Mendelian ratios of susceptibility (Huff, 1935, 1931). The heritability of vector competence for bird malaria was confirmed subsequently in a number of *Plasmodium* spp. and mosquito species combinations, demonstrating increased susceptibility after several generations of selective breeding (Micks, 1949; Trager, 1942). Similarly, selection experiments were used to generate two independent refractory lines of *An. gambiae*, a species usually highly susceptible to human malaria parasite infection (Collins et al., 1986; Vernick et al., 1995). For one of these lines, L35, the genetic basis of refractoriness was mapped to one major and two minor quantitative trait loci (Gorman et al., 1997; Zheng et al., 1997).

As early as 1927, Huff had proposed that the immune system of the mosquito kills malaria parasites and thus is an important determinant of vector competence (Huff, 1927). In fact, the earliest description of a mosquito humoral immune response against malaria parasites had been made by Ross, when he documented degenerated and melanized oocysts as "black spores" (Daniels, 1898; Knowles and Basu, 1933; Sinden and Garnham, 1973). Ultimately, the selection experiments performed by Collins and Vernick confirmed Huff's early proposal. The L35 line kills rodent malaria parasites and allopatric human *P. falciparum* ookinetes through an immune reaction called melanization (Collins et al., 1986). In the SUAF2 line, ookinetes of the bird malaria parasite *Plasmodium gallinaceum* are killed by an immune effector mechanism called lysis (Vernick et al., 1995). These selection experiments hold the promise that targeted manipulation of the mosquito immune system can render epidemiologically important vector species refractory and prevent malaria transmission. Over the last three decades, this field has made significant advances in our understanding of the mosquito immune system, including the processes that kill different parasite stages within the mosquito vector.

CURRENT KNOWLEDGE OF ANTIPARASITE IMMUNE REACTIONS IN THE MOSQUITO VECTOR

Antimalarial immune reactions, resulting in parasite killing throughout its life cycle within the mosquito, can be divided into distinct phases based on (1) the parasite life stages that are targeted, and (2) the compartments in which they occur: (a) Parasites are killed during gametogenesis and zygote formation inside the midgut lumen. (b) Ookinetes and young oocysts are killed either while traversing the midgut epithelium or during and after emergence into the subepithelial space. (c) Oocysts are killed during maturation while attached to the basal portion of the midgut epithelium. (d) Sporozoites are killed after being released from oocysts while circulating in the hemolymph. (e) Sporozoites are targeted within the salivary glands. The following section will summarize our current knowledge of these immune reactions and their effect on parasite survival.

Antimalarial Immunity in the Midgut Lumen

After taking an infectious blood meal, the first line of mosquito defense is not truly their own but belongs to the host they are feeding on. While inside the mosquito gut, host-derived white blood cells phagocytize *Plasmodium* spp. gametes (Lensen et al., 1997; Ross, 1898). In addition, host plasma components reduce the infectivity of the mosquito vector by the parasite (Gouagna et al., 2004). This effect is at least partially mediated by vertebrate complement, which, within the first hour after blood ingestion targets *Plasmodium* spp. during gametogenesis and young zygote stages (Margos et al., 2001; Simon et al., 2013). This complement-induced lysis is not effective for later developmental stages of the parasite, as *Plasmodium* spp. zygotes bind vertebrate complement inhibitor factor H, thereby coopting this protective protein to evade human complement factor C3b-mediated lysis within the midgut (Simon et al., 2013).

At the same time, mosquito physiology switches to digestion of the blood meal, as evidenced by significant changes of one-third of the mosquito transcriptome within 3 h post blood ingestion (Marinotti et al., 2006). One of the genes that is upregulated early and specifically upon an infectious blood meal is *nitric oxide synthase* (*NOS*, Luckhart et al., 2003). This upregulation is likely controlled by host blood-derived TGFβ-1, which persists in the mosquito midgut and is recognized by mosquito cells inducing *NOS* expression (Luckhart et al., 2003). Early NOS induction in the midgut of *Anopheles stephensi* most likely drives the increased nitric oxide (NO) levels observed 12.5 h post blood meal, coinciding with zygote to ookinete transition within the midgut lumen (Luckhart et al., 2003). Experimentally enhancing NOS activity through provision of the NOS substrate, L-arginine, in mosquito diets significantly decreases *Plasmodium* oocyst counts. Conversely, decreasing NOS activity by including the inhibitor L-NAME within an infective blood meal has the opposite effect, showing conclusively that NOS activity negatively impacts *Plasmodium* spp. parasites during the early stages of development, most likely even before reaching the midgut epithelium (Luckhart et al., 1998).

Plasmodium spp. parasites, which escape attack by the human immune system, survive increased NO levels, and transform from the zygote to the motile ookinete stage, must leave the blood bolus in order to avoid excretion. To escape the blood bolus and reach the midgut epithelium, *Plasmodium* parasites must successfully transverse the mosquito's first physical defense, the peritrophic matrix (PM) (Richards and Richards, 1977). The PM is composed of an organized matrix of chitin, proteins, and glycans that present a dense proteinaceous barrier. This barrier contains pores or channels that allow the passage of digestive enzymes to the blood bolus while blocking pathogen access to midgut epithelia (Edwards and Jacobs-Lorena, 2000; Hegedus et al., 2009). In response to this chitinous physical barrier, *Plasmodium* spp. ookinetes secrete chitinases to enzymatically degrade the PM (Huber et al., 1991; Tsai et al., 2001). In addition, the process requires binding of the ookinete to the fibrinogen-related protein 1 (FREP1), which is part of the mosquito PM (Zhang et al., 2015). A successful breach of the PM then allows the first interaction between the parasite and the mosquito epithelium.

Antimalarial Immunity Against Parasites Traversing the Midgut Epithelium

Our current understanding of mosquito epithelial immunity centers on responses to *Plasmodium* ookinete midgut invasion, since the transmigration of ookinetes across the mosquito midgut coincides with the most critical population bottleneck during the parasite's life cycle (Fig. 4.1). Experiments were most often performed in the laboratory using the rodent malaria parasite model *Plasmodium berghei*, and to a lesser degree, the most pathogenic human malaria parasite, *P. falciparum*. This aspect of mosquito immunity has been the subject of numerous reviews (Barillas-Mury and Kumar, 2005; Crompton et al., 2014; Meister et al., 2004; Osta et al., 2004b; Saraiva et al., 2016; Smith et al., 2014; Smith and Barillas-Mury, 2016; Whitten et al., 2006). We will briefly summarize the major findings here, in as far as they pertain to the potential of manipulating these immune reactions to eliminate parasite transmission.

The response to parasite invasion into the midgut epithelium is drastic. Invaded cells undergo apoptosis, as observed by nuclear swelling and membrane blebbing, and are extruded from the epithelial layer into the midgut lumen (Gupta et al., 2005; Han et al., 2000; Kumar et al., 2004; Zieler and Dvorak, 2000). As the ookinete attempts to transverse the midgut, the epithelial cells respond by inducing *NOS*, *heme peroxidase 2* (*HPX2*), and *NADPH oxidase 5* (*NOX5*) expression. Increased abundance of these enzymes leads to elevated levels of reactive nitrogen compounds (NO and nitrogen dioxide) that mediate the nitration of proteins, which causes damage to the parasite surface (Kumar et al., 2010, 2004; Luckhart et al., 1998). This damage is considered the first step in the process that ultimately kills ookinetes upon egression from the midgut epithelium.

The second step in this process is opsonization of the ookinete through the activation and deposition of the thioester-containing protein 1 (TEP1; Fig. 4.2A), which is also referred to as the complement-like pathway in mosquitoes (Blandin et al., 2008). Nitration seems to be a prerequisite for this process, as knockdown of *HPX2* or *NOX5* levels significantly decrease protein nitration in the midgut epithelium and TEP1 binding to *P. berghei* (Oliveira et al., 2012). TEP1 is a complement-like opsonin that shares striking similarities with vertebrate C3 complement factor (Levashina et al., 2001). It contains a thioester binding motif, which allows it to covalently bind target substrates (Baxter et al., 2007). TEP1 initially circulates as a monomer in the hemolymph. Upon initial activation by proteolytic cleavage, cleaved TEP1 is stabilized in the hemolymph in a heterotrimer bound to leucine-rich immune molecule 1 (LRIM1) and *Anopheles Plasmodium*-responsive leucine-rich

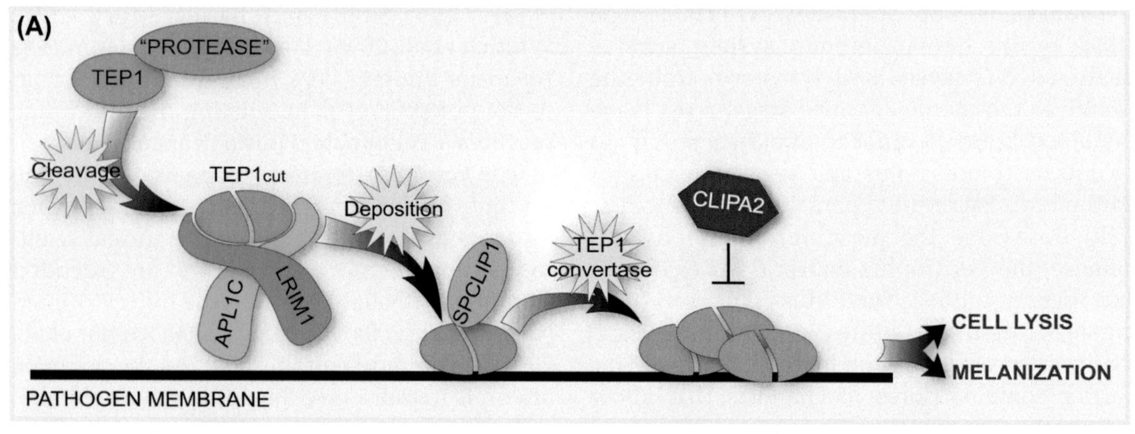

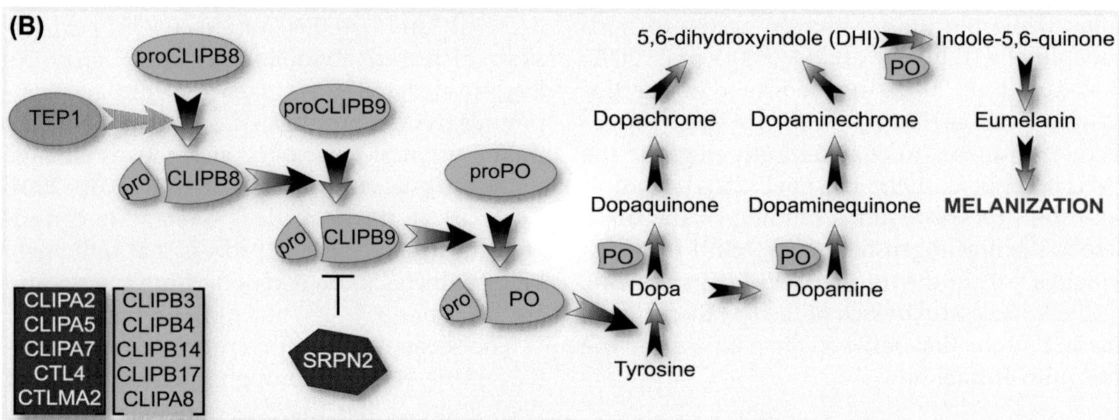

FIGURE 4.2 **Models of TEP1 complement-like pathway and melanization cascade.** (A) Opsonization of pathogen surfaces by TEP1 begins with cleavage of TEP1 by an unknown protease(s), which is stabilized in the hemolymph by the LRR proteins APL1C and LRIM1. During infection, TEP1 is deposited on the surface of pathogens, at which time a putative TEP1 convertase results in the rapid deposition of more cleaved TEP1. SPCLIP1 enhances this process, while CLIPA2 inhibits this deposition. (B) The melanization cascade is characterized by a clip-domain serine protease (CLIP) cascade leading to the activation of proPO to PO by proteolytic cleavage. Active PO then initiates the conversion of tyrosine into eumelanin, leading to melanization of microbial surfaces. ProPO pathway activators (green) and inhibitors (red) that thus far cannot be placed in any particular order within the cascade are shown at the bottom left. Red hexagons, major inhibitors of each pathway. *(A) Figure adapted from Povelones, M., Bhagavatula, L., Yassine, H., Tan, L.A., Upton, L.M., Osta, M.A., Christophides, G.K., 2013. The CLIP-domain serine protease homolog SPCLIP1 regulates complement recruitment to microbial surfaces in the malaria mosquito* Anopheles gambiae. *PLoS Pathog. 9, e1003623.*

repeat protein 1 (APL1C) (Fraiture et al., 2009; Povelones et al., 2009). An unknown mechanism leads to the deposition of cleaved TEP1 to microbial surfaces. Initial deposition of cleaved TEP1 is currently thought to result in the formation of a putative TEP1 convertase, similar to the C3 convertase in vertebrates on the microbial surface (Povelones et al., 2013). The putative TEP1 convertase is thought to then rapidly opsonize the ookinete by recruiting additional full-length TEP1 molecules from the hemolymph and converting it to the cut, active form on the ookinete surface. This process is regulated by two nonfunctional proteases, the clip-domain serine protease homologs SPCLIP1 and CLIPA2, which are required for and inhibit TEP1 accumulation, respectively (Povelones et al., 2013; Yassine et al., 2014). Opsonization of ookinetes by the complement-like pathway are critical to the developmental bottleneck observed during the transition from ookinete to oocyst, as depletion of any of its factors, TEP1, LRIM1, APL1C, or SPCLIP1, increases the number of *P. berghei* oocysts by two- to fivefold (Blandin et al., 2004; Povelones et al., 2013; Riehle et al., 2008; Yassine et al., 2012).

In addition to TEP1 deposition, the fibrinogen-like protein, FBN9 also binds to the surface of ookinetes (Dong and Dimopoulos, 2009). Similar to TEP1, LRIM1, APL1C, or SPCLIP1 depletion, *FBN9* knockdown also increases the number of developing oocysts (Dong and Dimopoulos, 2009; Garver et al., 2013). Similarly, different isoforms of the Down syndrome cell adhesion molecule (Dscam), which mediates axon guidance (Wojtowicz et al., 2004) and functions as a hypervariable pattern recognition receptor of the immune system (Dong et al., 2006), bind to the surface of *Plasmodium* spp. ookinetes (Dong et al., 2012a). Knockdown of *Dscam* increases the number of oocysts of *Plasmodium* spp. parasites, and thus phenocopies LRIM1, APL1C, TEP1, and FBN9 depletion (Dong et al., 2012a, 2006). The same phenotype can also be achieved by knockdown of two infection-responsive immunoglobulin domain proteins, *IRID4* and *6* (Garver et al., 2008). Taken together, these studies demonstrate that proteins containing leucine-rich repeat, fibronectin, or Ig domains bind to the ookinete surface. It is however currently unclear whether FBN9, Dscam, IRID4, or IRID6 are mechanistically linked to the complement-like pathway.

Opsonization ultimately leads to killing of the ookinete stage by lysis and/or melanization. Ookinete lysis is characterized by extensive membrane blebbing, followed by parasite fragmentation, and the observation of free hemozoin granules (Blandin et al., 2004; Vernick et al., 1995). However, while TEP1 binding is required to initiate lysis, the specific molecules required to carry out lysis are unknown. Ookinete lysis is the major process by which parasites are killed in both susceptible and refractory mosquitoes. In addition, ookinetes can be killed by melanization, which is characterized by the deposition of eumelanin on the parasite surface. This response is however more rarely observed in malaria-susceptible mosquitoes (Riehle et al., 2006; Schwartz and Koella, 2002). Melanization has received considerable attention, as it is a selectable phenotype for refractoriness in the laboratory (Collins, 1986; Hurd et al., 2005). Like all insects, mosquitoes utilize melanin deposition to seal wounds and physically surround pathogens in a thick, dense melanin coat (Gorman et al., 2007; Lai et al., 2002). The melanization pathway is triggered by activation of a clip-domain serine protease (CLIP) cascade leading to the activation of prophenoloxidase (proPO) by proteolytic cleavage to PO (Fig. 4.2B). PO is a key enzyme in the biochemical pathway that converts tyrosine into eumelanin (reviewed in Christensen et al., 2005). Opsonization by TEP1 and LRIM1 is required to initiate melanization on foreign surfaces including Sephadex beads, as well as *P. berghei* ookinetes (Habtewold et al., 2008; Warr et al., 2006). The protease cascade that regulates melanization, by controlling cleavage of proPO and potentially TEP1 includes CLIPB3, 4, 8, 14, and 17, as knockdown of each *CLIP* results in decreased melanization of *P. berghei* ookinetes and partially also Sephadex beads (Paskewitz et al., 2006; Volz et al., 2006, 2005). The serine protease homolog CLIPA8 is indispensable for melanization, as knockdown of this gene dramatically decreases melanization of *P. berghei* ookinetes, injected bacteria, and injections of conidia of the entomopathogenic fungus, *Beauveria bassiana* (Schnitger et al., 2007; Volz et al., 2006; Yassine et al., 2012). Currently, CLIPB9 is the only biochemically confirmed CLIP protease in mosquitoes that directly cleaves proPO (An et al., 2011). Protease activity of the proPO activation cascade is tightly controlled through direct inhibition of CLIPBs by serine protease inhibitors, including the serine protease inhibitors (SRPN) 1 and 2 (An et al., 2011; Michel et al., 2006, 2005). In addition, melanization of *P. berghei* is inhibited by the serine protease homologs CLIPA2 and 5 (Volz et al., 2006) and the C-type lectins CTL4 and CTLMA2 (Osta et al., 2004a).

Antimalarial Immunity Against Developing Malaria Oocysts

Those ookinetes that egress the midgut epithelium and manage to escape lysis or melanization round up to become oocysts. TEP1 deposition on young oocysts is initially low as compared to ookinetes, but increases significantly over time. TEP1 deposition is highest in mature oocysts, but does not appear to kill the parasite (Blandin et al., 2004). Nevertheless, oocyst numbers decrease continuously by up to 50%–75% (Awono-Ambene and Robert, 1998; Goulielmaki et al., 2014; Gupta et al., 2009), suggesting that at least young oocysts are recognized and killed by the mosquito immune system. TEP1 binding is higher in older oocysts compared to younger ones, suggesting that this parasite stage continues to be recognized by the mosquito immune system (Blandin et al., 2004). Knockdown of *TEP1* prior to blood feeding only increases oocyst formation, but does not seemingly affect oocyst survival (Gupta et al., 2009; Smith et al., 2015). These data are difficult to interpret, as it is unclear if RNAi-mediated knockdown of *TEP1* persists to this time point. The knockout *TEP1* mosquito lines will be helpful in assessing conclusively TEP1's role in oocyst killing (Smidler et al., 2013). Oocyst killing may also be affected by NO, as feeding the NOS inhibitor L-NAME at the time of transition from ookinete to oocyst increases mature oocyst numbers (Gupta et al., 2009). In addition, hemocyte differentiation has been linked to antioocyst immunity (Smith et al., 2015). While the mechanisms of oocyst killing remain to be elucidated, these data emphasize that the oocyst life stage remains susceptible to attack by the mosquito's immune system.

Antimalarial Immunity Against Sporozoites in Hemolymph and Salivary Glands

Sporozoite escape from the oocyst marks the beginning of the interaction of this parasite life stage with the mosquito hemocoel. Sporozoites are transported by hemolymph flow and have been observed in all body parts of the mosquito including appendages (Hillyer et al., 2007). This interaction is transient, as sporozoites injected into the hemocoel rapidly decline in numbers from this compartment (Hillyer et al., 2007), and ultimately only 10%–20% of sporozoites that were released into the hemocoel reach the salivary glands (Barreau et al., 1995; Hillyer et al., 2007; Rosenberg and Rungsiwongse, 1991). Within the first 8 h after injection, roughly 20% of sporozoites have reached the salivary gland, while already more than 40% of sporozoites have died. The remainder of the hemolymph sporozoites do not make it to the salivary glands but are killed in the hemolymph rapidly within the next 16 h, and have entirely disappeared from this compartment by 7 days (Hillyer et al., 2007). Phagocytosis and melanization of sporozoites in the hemolymph have been recorded (Hernández-Martínez et al., 2002; Hillyer et al., 2007, 2003). However, the rate of either is not sufficient to explain the large losses of roughly 80%–90% of sporozoites that are released into the hemocoel. Instead, sporozoites are rapidly killed and degraded in the hemocoel by a yet to be described mechanism (Hillyer et al., 2007). About 10% of the hemocyte transcriptome undergoes significant changes at the time of sporozoite migration through the hemolymph (Baton et al., 2009). Some of the genes that are upregulated during this time are opsonins and members of the protease network that controls melanization. It is therefore tempting to speculate that sporozoites are killed by the same immune mechanisms that kill ookinetes.

While the sporozoite population undergoes a major bottleneck during migration through the hemocoel, migration across the salivary gland epithelium does not seem to negatively affect sporozoites. Salivary glands of adult sugar-fed female mosquitoes encode antimicrobial peptides, lysozyme, and pathogen pattern recognition polypeptides (Arca et al., 2005; Calvo et al., 2010). However, in contrast to the mosquito midgut, which provides multiple immunological defenses against the ookinete stage, salivary glands appear to exhibit a subtle response to sporozoite invasion. Although hundreds of malaria sporozoites may enter a single salivary gland epithelial cell, the characteristic NOS upregulation observed during ookinete invasion of the midgut is absent in salivary gland epithelia, and apoptosis is not induced in invaded cells (Dimopoulos et al., 1998; Pinto et al., 2008; Rosinski-Chupin et al., 2007). Transcriptomic analyses of salivary glands showed transcript responses to sporozoite invasion, with 37 immune-related genes upregulated in infected versus control glands, including antimicrobial peptides *defensin1* and *cecropin2* (Rosinski-Chupin et al., 2007). Thus far, only four genes are known to be upregulated in response to parasite invasion of salivary glands and midgut epithelia (Pinto et al., 2008; Rodrigues et al., 2012; Rosinski-Chupin et al., 2007). Knockdown of the protease inhibitor *SRPN6* leads to higher numbers of parasites in each tissue (Abraham et al., 2005; Pinto et al., 2008), while depletion of the protease *ESP* reduces parasite numbers (Rodrigues et al., 2012). However, these proteins seem to interfere with the ability of parasites to invade the epithelia, rather than partaking in an antiplasmodial immune response that kills the parasite.

Together, the innate immune responses of the mosquito form a strong attack against every life stage of the *Plasmodium* parasite. These attacks in the end however are insufficient to kill the entire parasite population, especially after the approximately 2500-fold increase in number of parasites due to sporogony (Rosenberg and Rungsiwongse, 1991). Understanding immune system regulation may offer clues on how this sophisticated system strikes the balance between parasite killing and protecting itself from the harmful consequences of an overly active immune system.

THE REGULATION OF ANTI-PARASITE IMMUNITY BY CANONICAL SIGNAL TRANSDUCTION PATHWAYS

Insect innate immunity is largely regulated through canonical signal transduction pathways that regulate transcription of the majority of innate immune factors. These include the nuclear factor (NF)-κB-dependent Toll and Imd pathways and the Janus kinase (JAK)-signal transducer and activator of transcription (STAT) pathway. In addition, mitogen-activated protein kinase (MAPK) signaling also regulates mosquito innate immunity through three pathways, including signaling through ERK, Jun-N-terminal kinase (JNK), and p38 MAP kinase. The basis of our understanding of the molecular makeup and function of these pathways largely derives from studies in the model organism *Drosophila melanogaster*. Comparative genomic analyses have shown that these pathways are highly conserved across all sequenced mosquito species (Bartholomay et al., 2010; Christophides et al., 2002; Neafsey et al., 2015; Waterhouse et al., 2007). The following section will provide a brief overview of these pathways, focusing on their roles in regulating antimalarial immunity in mosquitoes.

The Toll Pathway Controls Immune Reactions Targeting Broad Classes of Pathogens

The Toll pathway was initially described in *D. melanogaster* as critical to establish the dorsoventral axis in the early embryo (Anderson and

Nüsslein-Volhard, 1984), and is named after its transmembrane receptor, Toll (Hashimoto et al., 1988). The role of the Toll pathway in immunity was initially hypothesized based on the sequence similarity between the cytoplasmic domains of the *D. melanogaster* Toll and human interleukin-1 (IL-1) receptor (Gay and Keith, 1991), and later experimentally confirmed to control antimicrobial peptide expression in *D. melanogaster* (Lemaitre et al., 1996). This pathway is a major immune signaling pathway in the insect fat body. In mosquitoes, the Toll pathway has been linked to hemocyte activation (Ramirez et al., 2014), mediates antibacterial and antifungal immune responses (Bian et al., 2005; Dong et al., 2012b; Shin et al., 2006), and regulates malaria parasite killing (Frolet et al., 2006).

The Toll pathway consists of an extracellular protease cascade and an intracellular signal transduction pathway, culminating in nuclear translocation of the NF-κB transcription factors, Dorsal and the Dorsal-related immunity factor (Dif) (Fig. 4.3; reviewed in Valanne et al., 2011). The Toll pathway is activated through recognition of conserved pathogen-associated molecular patterns by extracellular pattern recognition receptors such as Gram-negative binding proteins and peptidoglycan recognition proteins (PGRPs) (Bischoff et al., 2004; Gobert et al., 2003; Gottar et al., 2006; Michel et al., 2001). Alternatively, Toll activation occurs by sensing "danger" signals that include fungal or bacterial virulence factors, such as digestive proteases, as well as damage-associated molecular patterns that arise from stressed or damaged cells during infection (El Chamy et al., 2008; Gottar et al., 2006; Ming et al., 2014). Detection of either signal culminates in the proteolytic cleavage of the Toll ligand Spätzle, which, upon binding to the Toll receptor, triggers an intracellular signaling cascade, characterized by the death domain proteins Myd88, Tube, and Pelle that phosphorylate the inhibitor of (I)κB, Cactus. Cactus, in its unphosphorylated state, is bound to Dif/Dorsal, preventing it from entering the nucleus. Upon phosphorylation, Cactus releases Dif/Dorsal, resulting in the translocation of these NF-κB transcription factors to the nucleus to activate gene transcription.

The Toll pathway is tightly controlled by inhibitors at several different intervention points. The extracellular protease cascade is inhibited by the serpin Necrotic (Spn43Ac, Nec), preventing Toll signaling in the absence of fungal infection (Levashina et al., 1999). Furthermore, serpin-1 (Spn1) potentially regulates Spätzle activation, possibly by inhibition of the modular serine protease ModSP (Fullaondo et al., 2011). In addition to Cactus, the intracellular signal transduction pathway downstream of Toll is downregulated by ubiquitination. The ubiquitin E3 ligase Pellino associates with MyD88, thereby targeting MyD88 for degradation and, in turn, inhibiting Toll signaling (Ji et al., 2014). Finally, Toll pathway activity is not only controlled through increased Cactus phosphorylation, but also through decreased *Cactus* transcription controlled by the Hippo signal transduction pathway (Liu et al., 2016).

Components of the intracellular signal transduction pathway are conserved by orthology across insects, and orthologs of each protein have been identified in all sequenced mosquito genomes (Bartholomay et al., 2010; Christophides et al., 2002; Neafsey et al., 2015; Waterhouse et al., 2007). In contrast, no members of the extracellular protease cascade that regulate Spätzle cleavage have been identified, largely because these enzymes are members of protein families with numerous clade-specific expansions and losses that defy orthology assignments (An et al., 2011; Christophides et al., 2002). The impact of the Toll pathway on *Plasmodium* spp. development has been analyzed in a number of mosquito species. Knockdown of *Cactus* increases the number of *Plasmodium* spp. oocysts in multiple parasite–vector species combinations (Garver et al., 2009; Mitri et al., 2009; Riehle et al., 2008;

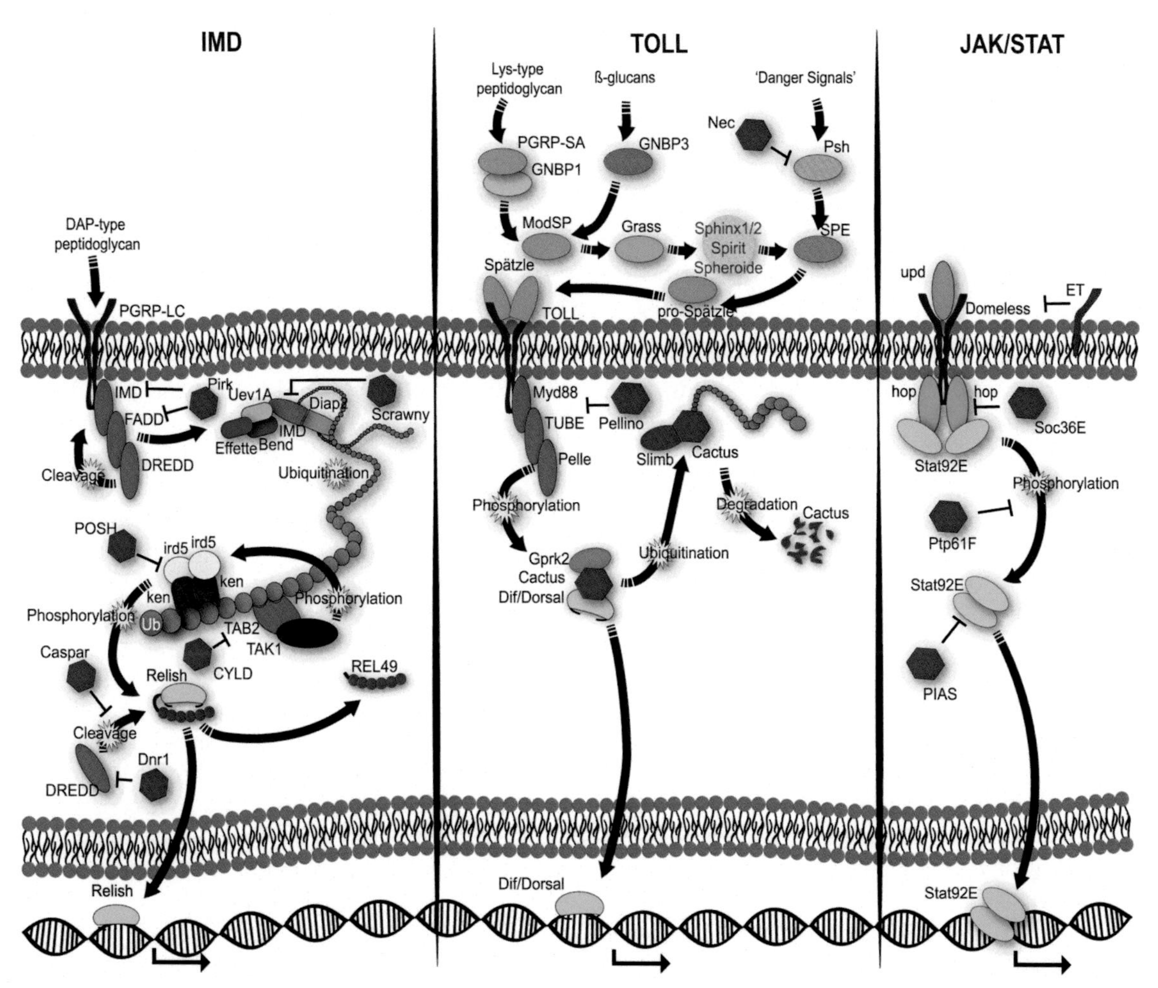

FIGURE 4.3 **Models of Imd, Toll, and JAK/STAT signaling pathways.** The Imd, Toll, and Jak/STAT pathways are major regulators of the immunotranscriptome of insects. Toll and JAK/STAT signaling is initiated through the cytokines spätzle and unpaired (upd), to the Toll and Domeless receptors, respectively. In contrast, Imd is activated through direct interaction of its receptor, PGRP-LC with Gram-negative microbial cell walls. The NF-κB signaling pathways, Imd and Toll are characterized by a substantial number of pathway-specific proteins. Their interactions through phosphorylation, ubiquitination, and protease cleavage culminate in the activation of NF-kB transcription factors Relish (REL2 in mosquitoes) and Dif/Dorsal (REL1 in mosquitoes). In contrast, the JAK/STAT pathway is much simpler, consisting of 10 components, culminating in the phosphorylation of its transcription factor STAT. Red hexagons, major inhibitors of each pathway, Light blue, transcription factors. Further details and citations are provided in The Toll Pathway Controls Immune Reactions Targeting Broad Classes of Pathogens, The Immunodeficiency (Imd) Pathway Is a Major Regulator of Gut Immunity, and The JAK/STAT Pathway Regulates the Antiviral Response and Cellular and Gut Immunity sections.

Zou et al., 2011). Several *Plasmodium* spp. killing mechanisms are controlled by the Toll pathway, including antimicrobial peptides (AMPs), transcriptional upregulation of components of the complement-like pathway, as well as a negative regulator of the melanization cascade (Frolet et al., 2006). Overexpression of *REL1*, the ortholog of *Drosophila Dif/Dorsal*, or knockdown of *Cactus* affects the expression of 264 and 1850 genes (Zou et al., 2011), respectively, and thus likely elicits pleiotropic effects beyond canonical immune effector mechanisms that affect parasite survival.

The Immunodeficiency (Imd) Pathway Is a Major Regulator of Gut Immunity

The immunodeficiency (Imd) pathway is a critical humoral immune pathway responsible for the regulation of AMPs in insects (Fig. 4.3, reviewed in Kleino and Silverman, 2014; Myllymäki et al., 2014). This pathway is the major immune signaling pathway in the gut (reviewed in Buchon et al., 2014, 2013), contributes to the maintenance of a healthy gut microbiota, and regulates malaria parasite killing in mosquitoes across a wide range of vector species–malaria parasite species combinations (Antonova et al., 2009; Chen et al., 2012; Garver et al., 2012, 2009; Meister et al., 2009, 2005; Dong et al., 2011).

Research using *D. melanogaster* shows this pathway is responsible for inducing the expression of the majority of AMPs following recognition of bacterial DAP-type peptidoglycan (PGN) present in the cell walls of some Gram-positive bacilli and all Gram-negative bacteria by PGRPs (Leulier et al., 2003). The membrane-bound receptor PGRP-LC and the intra- and extracellular forms of PGRP-LE are the primary and secondary receptor responsible for Imd pathway activation (Choe et al., 2002; Gottar et al., 2002; Kaneko et al., 2006). Binding of PGN to PGRP-LC, PGRP-LE, or PGRP-LC/PGRP-LE dimers recruits and assembles a death domain protein signaling complex consisting of Imd (Lemaitre et al., 1995), the adaptor protein dFadd (Leulier et al., 2002), and the Death-related ced-3/Nedd2-like (DREDD) caspase (Leulier et al., 2000). Formation of this complex leads to posttranslational modification of DREDD by the E3-ligase inhibitor of apoptosis 2 (Iap2) (Meinander et al., 2012). Iap2, along with the E2-ubiquitin-conjugating enzymes, UEV1a, Bendless, and Effete ubiquitinate and activate DREDD (Paquette et al., 2010; Zhou et al., 2005). Active DREDD then cleaves Imd and exposes an Iap2 binding site, leading to the subsequent ubiquitination of Imd (Paquette et al., 2010). Cleaved and ubiquitinated Imd recruits and activates transforming growth factor-activated kinase 1 (Tak1) and Tak1-associated binding protein 2 (Tab2) (Kleino et al., 2005). The Tak1/Tab2 complex directly phosphorylates the Imd pathway IκB kinase (IKK) complex, composed of the regulatory subunit kenny and the kinase ird5 (Lu et al., 2001; Silverman et al., 2000). This IKK complex then phosphorylates the NF-κB transcription factor, Relish (Ertürk-Hasdemir et al., 2009). Interestingly, this phosphorylation does not lead to the nuclear translocation of Relish, but enhances the ability of Relish to recruit RNA polymerase II to promote and upregulate the transcription of target genes, such as AMPs, once Relish enters the nucleus (Ertürk-Hasdemir et al., 2009). Nuclear translocation of Relish is induced through proteolytic cleavage by DREDD, which separates the translocation-inhibiting ankyrin repeat domains in the C-terminus from the N-terminal Rel homology DNA-binding domain (Ertürk-Hasdemir et al., 2009; Stoven et al., 2003).

Negative regulation of the Imd pathway occurs at multiple points (Lee and Ferrandon, 2011). The activating signal is downregulated by PGRP amidases that degrade PGN (Mellroth et al., 2003). The PGRP-LF receptor competes with PGRP-LC for dimerization, thus producing inactive heterodimers. The poor Imd response upon knock-in (PIRK) intracellular regulator disrupts interactions with the cytoplasmic tail of PGRP-LC, and high expression of *PIRK* leads to intracellular sequestration of PGRP-LC, inhibiting its ability to activate Imd signaling at the cell membrane (Lhocine et al., 2008). Multiple proteins, including SKPA/SLMB/DCUL1, dUSP36, CYLD, POSH, DNR-1, and Caspar (Kim et al., 2006) alter the ubiquitination status of Imd pathway components. An additional level of regulation is provided in the posterior midgut by the Caudal transcription factor that blocks the transcription of AMP genes, but not of other IMD targets (Ryu et al., 2008). Expression of PGRP amidases, PGRP-LF, and PIRK is regulated by Imd pathway, providing a negative feedback loop.

The Imd pathway is remarkably conserved between insects and mammals (Bartholomay et al., 2010; Christophides et al., 2002; Viljakainen, 2015; Waterhouse et al., 2007). A difference of note is that *An. gambiae Relish* (Rel2) possesses two isoforms arising from splice variation. The full-length form (REL2-F) shares the same domain structure as its *Drosophila* counterpart, while the shortened form (REL2-S) lacks the complex-forming Death domain and the inhibitory ankyrin repeats that normally keeps Relish from translocating to the nucleus (Meister et al., 2005). Rel2-F appears to share functional orthology with *Drosophila Relish*, while REL2-S phenocopies *Drosophila Dif* mutants, exhibiting impaired survival against Gram-positive bacteria (Meister et al., 2005). *Ae. aegypti Relish* possesses three splice isoforms, coding either the full-length protein, only the DNA-binding Rel homology domain, or only the ankyrin repeat domain, which are all upregulated in response to bacterial challenge (Shin et al., 2002). In addition, no orthologs of PIRK could thus far be identified outside of the genus *Drosophila*. Parasite killing within the mosquito vector is controlled majorly by the Imd pathway. Knockdown of various signaling components, including *PGRP-LC, IMD, FADD, CaspL1*, or *REL2-L*, increase the number of rodent and human malaria parasite oocysts (Antonova et al., 2009; Garver et al., 2012; Meister et al., 2009, 2005; Mitri et al., 2009). Consistent with these findings, overexpression of *REL2* decreases the number of bird malaria oocysts in *Ae. aegypti* and human malaria ookinetes in the midgut lumen of *An. gambiae* (Antonova et al., 2009; Dong et al., 2011). In addition, knockdown of *Caspar* and *Caudal* decreases the number of human malaria parasites in *An. gambiae* (Garver et al., 2012, 2009; Clayton et al., 2013). The Imd pathway regulates the antimalarial immunity against parasites traversing the midgut epithelium, as well as developing oocysts (Garver et al., 2012). It controls the expression of several known antimalarial factors including TEP1 and LRIM1, and serine proteases required for proPO activation (Fig. 4.2; see section Antimalarial Immunity Against Parasites Traversing the Midgut Epithelium). REL2 also controls the alternative splice repertoire of Dscam, through transcriptional control of the splice factors caper and IRSF1, which confers antiparasite immunity (Dong et al., 2012a).

The JAK/STAT Pathway Regulates the Antiviral Response and Cellular and Gut Immunity

The JAK-STAT pathway regulates fundamental biological functions in insects, including embryonic patterning, eye and wing formation, stem cell niche maintenance, and immunity (reviewed in Kingsolver et al., 2013; Myllymäki and Rämet, 2014). In mosquitoes, this pathway affects antiviral immunity (Carissimo et al., 2015; Paradkar et al., 2012; Souza-Neto et al., 2009) and also contributes to malaria parasite killing (Gupta et al., 2009).

The JAK/STAT signaling core in *D. melanogaster* consists of four principal components: the unpaired family ligands upd, upd2, upd3, the receptor Domeless (Dome), the JAK Hopscotch (hop), and the transcription factor Stat92E (reviewed in Zeidler and Bausek, 2013). JAK/STAT activation and signaling follows a canonical model conserved between insects and mammals (reviewed in Liongue and Ward, 2013). Upd ligand binding to the Dome receptor induces conformational changes in the ligand that activates the intracellularly associated hop. These phosphorylate the receptor complex that recruits the transcription factor Stat92E, which in turn is phosphorylated, dimerizes, and enters the nucleus to facilitate transcription of its target genes.

JAK/STAT signaling is inhibited principally by three types of negative regulators (reviewed in Liongue and Ward, 2013; Valentino and Pierre, 2006). Phosphotyrosine phosphatases (SHPs, CD45, PTP1B/TC-PTP) dephosphorylate activated JAK, STAT, or cytokine receptors. The protein inhibitor of activated STAT (PIAS) proteins bind to activated STAT dimers, blocking their capacity to activate transcription. In addition, PIAS acts as E3-type ligase to sumoylate and modify the function of other proteins (reviewed in Morales et al., 2010). Suppressor of cytokine signaling (SOCS) proteins interact with activated JAK or with the phosphorylated receptors, inhibiting the recruitment of STAT, the activation of the JAK enzymatic activity, or inducing the proteasome-dependent degradation of activated JAK or receptors. While the first two inhibitor classes are constitutively expressed, SOCS expression is induced by cytokines, providing a negative feedback loop for the JAK/STAT pathway. In *D. melanogaster*, the phosphotyrosine phosphatase Ptp61F is also transcriptionally regulated by the pathway, thus serving as an additional negative feedback regulator (Baeg et al., 2005).

JAK/STAT core components (Dome, Hop, and Stat92E), as well as their inhibitors (PTP61F, three SOCS, and PIAS), are conserved in mosquitoes (Souza-Neto et al., 2009; Waterhouse et al., 2007), with two copies of the transcription factor STAT, STATA and B existing in a subclade of anophelines (Neafsey et al., 2015). Silencing *STATB* by RNAi led to decreased expression of *STATA* (and not vice versa), indicating that STATB regulates the mRNA levels of *STATA* within *An. gambiae* (Gupta et al., 2009). The observation that *STATA* is expressed in the fat body and enters the nucleus following bacterial challenge was the first link of this pathway to immunity in mosquitoes (Barillas-Mury et al., 1999). Knockdown of the *Anopheles aquasalis Stat92E* ortholog increased oocyst numbers following *Plasmodium vivax* infection, indicating that JAK/STAT plays a functional role in inhibiting transition of ookinetes to oocysts in the early stages of infection (Bahia et al., 2011). In *An. gambiae*, silencing of *STATA* increased oocyst numbers of rodent and human malaria parasites, while silencing of *SOCS* had the opposite effect (Gupta et al., 2009). This phenotype is at least in part explained by the role of the JAK/STAT pathway in regulating the expression of antiplasmodial immune factors. Silencing of *STATA* lowered baseline levels of *NOS* expression and abolished *NOS* and *TEP1* upregulation upon *P. berghei* ookinetes invasion (Gupta et al., 2009). In addition, the JAK/STAT pathway is required for hemocyte differentiation, which may contribute to antimalarial immunity against developing malaria oocysts (Gupta et al., 2009; Smith et al., 2015).

Mitogen-Activated Protein Kinase Signaling Affects Mosquito Midgut Homeostasis and Hemocyte Proliferation

MAPK signaling regulates many physiologies ranging from stress responses to cell proliferation and differentiation, apoptosis, and inflammation. In the last few years, MAPK signaling has emerged as a major regulator of mosquito immunity against malaria parasites (Bryant and Michel, 2014; Drexler et al., 2013, 2014; Garver et al., 2013). MAPK signaling regulates mosquito innate immunity through three pathways (Fig. 4.4), including signaling through ERK, JNK, and p38 MAP kinase (p38). Upon ligand binding, the receptor tyrosine kinases (RTKs) recruit adaptor proteins that signal through a core of kinases in the MAP3K/MAP2K/MAPK sequence. Upon phosphorylation, the terminal kinase in each

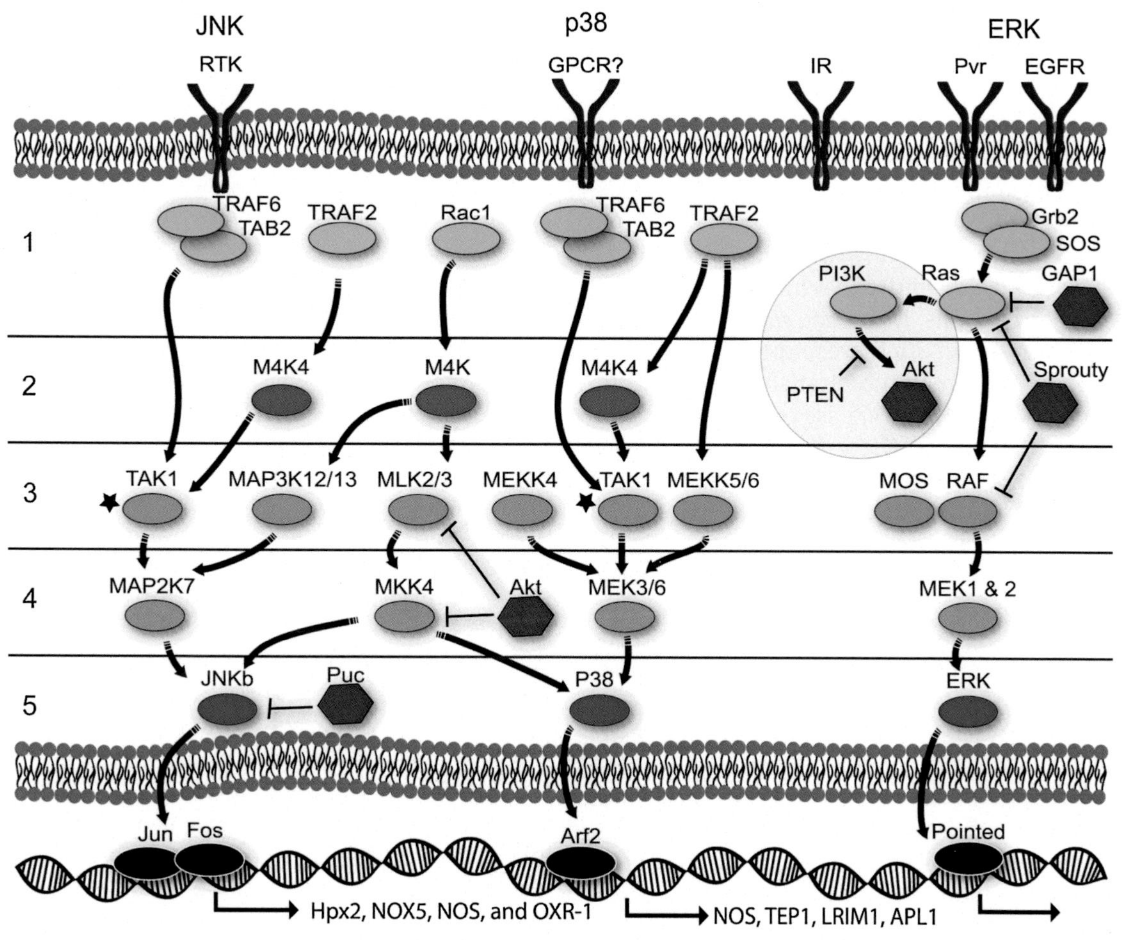

FIGURE 4.4 The predicted MAPK signaling pathways in mosquitoes. MAPK signaling pathways follow a generalized sequence of activation by a series of protein interaction and/or phosphorylation steps, indicated by arrows: 1. Upon ligand binding, the transmembrane receptor tyrosine kinases (RTKs) recruit adaptor proteins (green). 2. The adaptor proteins activate in some cases MAP4Ks (dark blue). 3. MAP3Ks (light blue) are activated by phosphorylation through either adaptor proteins or MAP4Ks. 4. pMAP3Ks phosphorylate MAP2Ks (pink), which 5. phosphorylate the terminal MAPKs, ERK, p38, or JNK (purple), which then enter the nucleus and activate their respective transcription factors (black). Each pathway is regulated by inhibitors at one or more intervention points (*red hexagons*). Cross talk to the Imd pathway is mediated by TAK1 (star, also see Fig. 4.3). In addition, Ras also signals through the PI3K/Akt pathway (*yellow circle*), which in turn inhibits MKK4 upstream of JNK and p38. Pathway members were identified by Horton et al. (2011) through phylogenetic analysis, and connections were drawn based on the human MAPK pathway provided by the KEGG pathway database (Kanehisa et al., 2010). Further details and citations are provided in the Mitogen-Activated Protein Kinase Signaling Affects Mosquito Midgut Homeostasis and Hemocyte Proliferation section.

of these pathways, ERK, JNK, and p38, enters the nucleus and phosphorylates transcription factors that mediate transcriptional regulation of pathway targets. In *An. gambiae*, the MAP kinome has been annotated, and based on orthology the likely molecular makeup of the three MAPK signaling pathways has been described (Horton et al., 2011).

ERK Signaling

The ERK signaling pathway is initiated with the recruitment of Grb2 to the RTK that then recruits the guanine nucleotide exchange factor son of sevenless, which in turn activates the GTPase Ras. Ras then activates Raf, which activates MEK1/2, which activates ERK. One transcription factor downstream of ERK in *D. melanogaster* is Pointed, which is conserved across arthropods, including mosquitoes. ERK phosphorylation increases upon addition of either human insulin or transforming growth factor (TGF)β to *An. gambiae* cells in vitro, suggesting that the RTKs of these two mitogens may function as receptors for this signaling pathway (Horton et al., 2011). Based on data from *D. melanogaster*, RTKs upstream of ERK are the epidermal growth factor receptor, and the PDGF/VEGF Receptor (Pvr), both of which have orthologs in mosquitoes. ERK signaling is inhibited at two points in the cascade, with GAP1 and Sprouty inhibiting Ras, and Sprouty also inhibiting Raf (Casci et al., 1999; Mason et al., 2006).

The role of ERK signaling in mosquito immunity is linked to proliferation of hemocytes upon blood feeding (Bryant and Michel, 2014). Whether the concomitant activation of these hemocytes, including the increased expression of TEP1, also results from ERK signaling is currently unclear (Bryant and Michel, 2016). In addition, ERK signaling is required for the expression of insulin-like peptides (Pietri et al., 2015), which in turn downregulate the innate immune response in response to parasite infection.

Ras Also Signals Through the PI3K/Akt Pathway

Besides its role in ERK signaling, RTK/Ras also activates other signaling pathways, including phosphatidylinositol 3-kinase (PI3K)/Akt signaling (reviewed in Castellano and Downward, 2011; Hay, 2011; Steelman et al., 2011). When active, PI3K converts phosphatidylinositol (4,5)-bisphosphate (PIP_2) into phosphatidylinositol (3,4,5)-trisphosphate (PIP_3). PIP_3, in turn, stimulates the kinase activity of Akt, which results in the inhibition of the forkhead box O (FOXO) transcription factor and activation of target of rapamycin (TOR) complex 1 (TORC1). The phosphatase and tensin homolog deleted on chromosome 10 (PTEN) tumor suppressor protein is the major inhibitor of the PI3K/Akt pathway. The PI3K/Akt pathway also negatively regulates p38 and JNK signaling, as Akt inhibits MKK4. The PI3K/Akt signaling pathway has a conserved role in downregulating immune responses. A PI3K mutant in the *Caenorhabditis elegans* worm model (*age-1*) extends life span and increased resistance to bacterial pathogens (Garsin et al., 2003). Similarly, overexpression of PTEN or a myristoylated constitutively active form of Akt (myrAkt) increases life span and reduces human malaria parasite oocyst numbers (Corby-Harris et al., 2010; Hauck et al., 2013; see also The Challenges and Opportunities for Boosting Mosquito Immunity in the Field section). The phenotype of parasite killing is linked to increased mitophagy and mitochondrial dysfunction, emphasizing the role of this pathway in midgut homeostasis (Luckhart et al., 2013).

Jun-N-Terminal Kinase/p38 Signaling

RTKs also signal through the JNK and p38 MAPKs (reviewed in Hotamisligil and Davis, 2016; Martín-Blanco, 2000; Zarubin and Han, 2005; Zeke et al., 2016). JNK and p38 signaling pathways use a larger repertoire of MAPKs along the MAP4K/MAP3K/MAP2K/MAPK axis as compared to ERK activation, multiple of which are conserved in *An. gambiae* (Fig. 4.4, Horton et al., 2011). The majority of MAPKs can signal through JNK and p38, including (1) the adaptor proteins Rac1, TRAF2, 6, and TAB2, (2) the MAP4Ks, M4K and M4K4, (3) the MAP3Ks, TAK1 and MLK2/3, and (4) the MAP2K,

MKK4. Both pathways in mosquitoes can be stimulated by the phosphoantigen HMBPP [(E)-4-hydroxy-3-methyl-but-2-enyl pyrophosphate], a microbial metabolite produced by most eubacteria, as well as *Plasmodium* spp. (Lindberg et al., 2013). In addition, p38 is phosphorylated upon stimulation with insulin, hydrogen peroxide, and LPS, while JNK is also activated upon LPS stimulation (Horton et al., 2011; Wang et al., 2015). As mentioned earlier, both pathways are negatively regulated by PI3K/Akt signaling through inhibition of MKK4 by Akt (Zhao et al., 2015). JNK signaling is further inhibited by the JNK phosphatase puckered (*Puc*) (Martín-Blanco et al., 1998; McEwen and Peifer, 2005).

Multiple studies have implicated JNK signaling in antimalarial immunity in mosquitoes. P38 signaling is induced in *An. stephensi* midguts upon infection, and pharmacological inhibition of p38 increased the parasite-inducible expression of *NOS*, *TEP1*, *LRIM1*, and *APL1* in *An. stephensi* midguts (Wang et al., 2015). Consistent with this observation, pharmacological inhibition of p38 decreased *P. falciparum* development in the mosquito midgut.

In *An. gambiae*, JNK apparently regulates the nitration response to parasites (Garver et al., 2013). Knockdown of multiple members of the JNK signaling pathway, including *JNK* and its downstream transcription factors *Jun* and *Fos*, decreases the expression of enzymes required for nitration, including *Hpx2*, *NOX5*, *OXR-1* (Garver et al., 2013; Jaramillo-Gutierrez et al., 2010). In addition, knockdown of *Jun* and *Fos* reduces the baseline expression of *TEP1* and *FBN9*, factors known to bind to the ookinetes surface and limit parasite survival. Knockdown of *Puc* in turn increased *TEP1* and *FBN* expression, suggesting that the JNK signaling pathway also regulates the complement-like pathway and opsonization (Garver et al., 2013). Knockdown of *MEK7* (*Hep*), *JNK*, and *Fos* increased rodent *P. berghei* oocyst numbers, while knockdown of *Puc* caused a decrease. Furthermore, knockdown of *JNK* during invasion of malaria parasites, significantly reduced caspase activity induced by parasite invasion, providing experimental evidence that JNK signaling is required for apoptosis in mosquitoes at least under certain physiological conditions (Ramphul et al., 2015). Finally, JNK signaling in the midgut is induced by low levels of human insulin growth factor (IGF)1, taken up by the mosquito in the blood meal (Drexler et al., 2014). Low levels of IGF1 in the blood meal induced *NOS* expression, increasing ROS. Low levels of IGF1 decreased *P. falciparum* oocyst numbers and extended the life span of *An. stephensi*, phenocopying overexpression of PTEN and myrAkt (Corby-Harris et al., 2010; Hauck et al., 2013).

The research summarized in this section highlights the complexity of immune system regulation. Dysregulation of any of these pathways has major consequences not only for the parasite and its survival, but also for the physiology of the mosquito. The substantial cross talk between these signaling cascades makes it difficult to tease apart not only their respective relative contribution to parasite killing, but also to determine the precise molecular mechanisms that underlie their ability to kill parasites. In addition, these pathways often provide responses in rapid spurts and function within and across many tissues. Exploring their temporal and spatial intricacies requires the use of new tools beyond RNAi by dsRNA injection. This especially holds true for signaling pathways that majorly rely on rapid activation by proteolytic cleavage or phosphorylation.

CREATING MALARIA-REFRACTORY MOSQUITOES IN THE LABORATORY: THE PROOF OF PRINCIPLE

The previous sections have highlighted how our knowledge of the mechanisms leading to malaria parasite killing has increased greatly over the last decade. Several proof of principle studies have since demonstrated that genetic manipulation of the mosquito's immune system can render susceptible mosquito species refractory to malaria parasite infection. The following section will provide a brief overview of the studies that through knockdown of negative regulators of immune reactions or by knocking in parasite antagonists generated such refractory mosquitoes. Emphasis is given on the putative mechanism of their action and their specificity with regards to the malaria species that can be targeted.

Transient Inhibition of *Plasmodium* spp. Development by RNAi

Within the array of studies utilizing RNAi to characterize specific components of the immune system in *An. gambiae* and *Ae. aegypti*, several mosquito agonists of *Plasmodium* spp. development within their vectors were discovered. Among this group of proteins is a class of putative negative immune system regulators. Their knockdown leads to increased mosquito defenses against the parasite, lowering surviving parasite numbers, and effectively halting the *Plasmodium* life cycle. Summarized in this section are studies that highlight key immune genes involved in melanization, and AMP production, which, after RNAi knockdown, can boost basal mosquito immunity to reduce parasite load and prevalence.

Knockdown of several members of the complement and melanization pathways prove to have powerful effects on *Plasmodium* spp. development. The majority of these studies were performed using rodent malaria parasites. Clip-domain serine proteases CLIPA2, CLIPA5, and CLIPA7 were shown to be agonists of rodent *P. berghei* development. Knockdown of these genes lowered ookinete and oocyst numbers through increased melanization but did not alter the 95% infection prevalence of *Plasmodium* within *An. gambiae* (Volz et al., 2006). It remains to be seen whether this effect holds true against *P. falciparum*. In addition, knockdown of C-type lectin *CTL4* resulted in 97% of ookinetes being melanized and knockdown of the lectin *CTLMA2* led to less melanized ookinetes (53%) (Osta et al., 2004a). Furthermore, this study observed that knockdown of either *CTL4* or *CTLMA2* resulted in complete melanization of ookinetes, with 33% and 8% of mosquitoes respectively to be completely refractory to infection. However, this phenotype was not observed with field isolates of *P. falciparum* (Cohuet et al., 2006), reemphasizing that the outcome of infection depends on the genetic makeup of both vector and parasite species. Knockdown of another melanization pathway regulator, *SRPN2*, resulted in a dysregulated melanization response and greatly enhanced ability to melanize *P. berghei* ookinetes, reducing infection prevalence by 50% or more (Michel et al., 2005). In addition, these mosquitoes exhibited decreased survivorship compared to controls, demonstrating the potential fitness cost of an immune system gone rogue. As observed in the case with CTL4, knockdown of *SRPN2* prior to infection with the human malaria parasite *P. falciparum* did not exhibit the same phenotype (Michel et al., 2006). Nevertheless, these studies provided the first formal proof that the immune system of the mosquito can be boosted to completely eliminate the malaria parasite. They also demonstrated that going forward, studies for the purpose of discovering putative targets needed to be performed in species combinations critical to human disease transmission.

Knockdown of signal transduction pathway inhibitors that control the transcription of large segments of the immune effector repertoire can also affect parasite development by enhancing the expression of anti-*Plasmodium* effector molecules. All signal transduction pathways currently known to contribute to antiplasmodial immunity include one or more inhibitors that control shutoff of these pathways (Figs. 4.3 and 4.4). For instance, knockdown of Toll pathway inhibitor, *Cactus*, boosts the expression of

472 genes, while downregulating 116 (Garver et al., 2009), including key immune genes such as *TEP1* and *CTL4* (Frolet et al., 2006). *Cactus* knockdown completely aborts rodent *P. berghei* development within mosquitoes, resulting in no live ookinetes present 10 days after infection. Dissected midguts also revealed that ookinetes were heavily melanized compared to controls (Frolet et al., 2006). Knockdown of *Ae. aegypti Cactus* produced similar results for *P. gallinaceum* (Zou et al., 2011). In comparison, the effect of *Cactus* knockdown on *P. falciparum* development is subtle, and in high infections reduces parasite load with a *P* value of 0.08 and no apparent effect on prevalence (Garver et al., 2009).

Knockdown of the major Imd pathway inhibitor, *Caspar*, has a smaller impact on the mosquito transcriptome, inducing the expression of 61 genes and repressing 55 genes (Garver et al., 2009). *Caspar* knockdown decreased *P. falciparum* oocyst numbers in *An. gambiae*, *An. albimanus*, and *An. stephensi*. This decrease can be drastic, as in the case of *An. gambiae*, where *Caspar* knockdown leads to a nearly fourfold reduction in prevalence from ds*GFP*-injected controls (Garver et al., 2009). Surprisingly, knockdown of *Caspar* in *Ae. aegypti* had no effect on the number of oocysts of the bird malaria species *P. gallinaceum*, providing another example that RNAi approaches can have drastically different outcomes in different host/parasite systems (Zou et al., 2011).

The impact of boosting JAK/STAT signaling thus far has been less explored. Knockdown of *PIAS* in *Ae. aegypti* induced the expression of 63 genes, while reducing 66, and increased the immunity against *P. gallinaceum* in a similar fashion to *Cactus* (Zou et al., 2011). To our knowledge, the impact of *PIAS* knockdown has not been explored in any other parasite–vector species combination. Given the impact of JAK/STAT signaling on parasite survival of *P. falciparum* and *P. vivax* (discussed in The JAK/STAT Pathway Regulates the Antiviral Response and Cellular and Gut Immunity section above), similar results to those obtained with *P. gallinaceum* are to be expected.

Taken together, these studies revealed a small number of potential genes that when knocked down can be used to boost immunity to malaria parasite species that affect human health. However, in contrast to transient knockdown achieved by RNAi, complete knockout of any of these genes is likely to cause pleiotropic effects that are deleterious to the mosquito in one or more of its life stages. Importantly, these studies have identified a few pathways that when boosted in a temporally controlled fashion may feasibly result in malaria refractory mosquitoes.

Inherited Boosting of Antiparasite Immunity in Mosquitoes

Several paths can be used to introduce inheritable traits in mosquitoes, including generation of knockout lines using CRISPR-Cas, introduction of novel symbionts (e.g., *Wolbachia*, see Chapters 1 and 7), paratransgenesis (Chapter 13), and the generation of transgenic mosquitoes using different knock-in technologies. This section will discuss the latter approach to boosting innate immunity. The reader is referred to Chapter 3 for the impact of *Wolbachia* endosymbionts on immune inductions.

The generation of transgenic mosquito lines remains far from a standard laboratory tool. This holds true especially for anopheline species, and as a consequence only a limited number of transgenic lines have been established targeting malaria parasite development. Studies utilizing mosquito transgenic approaches primarily place the gene of interest (transgene) under the expressional control of either the midgut-specific *carboxypeptidase A* (*Cp*) promoter or the fat body-specific *vitellogenin* (*Vg*) promoter, limiting expression to females during the first 2 days following blood meal-ingestion, respectively. This sex-specific and temporal control of transgene expression has the particular advantage of expression coinciding with the major bottleneck of

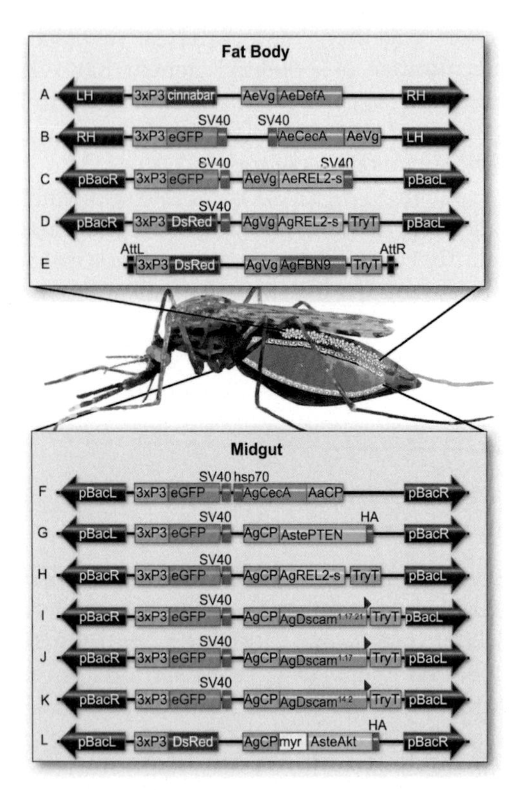

FIGURE 4.5 Transgenic mosquito lines with increased immunity against malaria parasites. This figure provides an overview of the constructs that have bene used to genetically engineer the immune responses in *Aedes. aegypti* (A–C), *Anopheles. gambiae* (D–F, H–K), and *An. stephensi* (G, L). All resulting transgenic mosquito lines exhibited decreased infection loads of one or more *Plasmodium* species. Phenotypes and effector genes are detailed in the Creating Malaria-Refractory Mosquitoes in the Laboratory: The Proof of Principle section in this chapter. In all cases, effector gene expression is female sex-specific and blood meal-induced, either driven in the fat body by the *vitellogenin* promoter (*Vg*, A–E, *yellow box*) or in the midgut by the *carboxypeptidase* promoter (*Cp*, F–L, *blue box*). All lines utilize the *3xP3* promoter to drive eye-specific expression of the marker genes *EGFP* (green), *DsRed* (red), and *cinnabar* (red), respectively. Transcriptional termination signals from SV40, the *Drosophila melanogaster heat shock protein (hsp)70*, *hemagglutinin (HA)*, and *An. gambiae trypsin* are marked where applicable. Most lines were generated using the type II transposable elements *Hermes* (A, B; Warren et al., 1994), or *piggyBac* (pBAC; C, D, F–L; Fraser et al., 1996), which require inverted terminal repeats for integration into the host genome (*black arrows*). Construct E was integrated by site-specific recombination using the phiC31 integrase system (Groth et al., 2004), which generates the attL and R sites upon recombination. *Red triangle*, FLAG tag.

ookinete to oocyst transition (Fig. 4.1). At the same time, it limits the putative deleterious effects resulting from overexpression of these immune factors.

Four general classes of effector molecules have been utilized to boost basal immunity, including AMPs, transcription factors of immune pathways, modulators of such pathways, and pathogen opsonins (Fig. 4.5). The resulting transgenic lines and their phenotypes with regards to *Plasmodium* transmission are discussed later.

Overexpression of Antimicrobial Peptides

The first transgenic mosquito lines had boosted basal immunity by overexpression of endogenous AMPs (Fig. 4.5A, B, and F). Different AMPs had been shown previously to negatively impact parasite survival, lending credence to this approach (Gwadz et al., 1989; Shahabuddin et al., 1998). Expression of *Cecropin A* (*CecA*) under the control of the *Cp* promoter in two independent transgenic *An. gambiae* (Kim et al., 2004) resulted in *CecA* transcripts in the posterior midgut 24h after blood feeding, coinciding with both the site and the timing of ookinete invasion. These transgenics displayed an average 61% lower *P. berghei* oocyst load 2 weeks after infection compared to nontransgenic control insects, while the prevalence of infection did not change between treatment groups (Kim et al., 2004). Similarly, expression of *CecA* and *Defensin (Def)A*, under the control of the *Vg* promoter effected *P. gallinaceum* development (Kokoza et al., 2000, 2010). The transgenic *Ae. aegypti* lines, expressing either *CecA* or *DefA*, and the offspring of their cross, resulted in a significant reduction in oocyst number 7days post infection as compared to the parental strain. Additionally, the offspring of the crossed lines contain no sporozoites in salivary glands 14days post infectious blood meal, if transgene expression had been reactivated by a noninfectious blood meal 1 week after the infectious blood meal (Kokoza et al., 2000, 2010).

Overexpression of Immune Pathway Transcription Factors

Four immune-enhanced mosquito transgenic lines overexpressing the IMD transcription factor *REL2*, either under the control of the *Vg* or *Cp* promoter have been developed. Two transgenic *Ae. aegypti* strains, designated REL2-A and REL2-B, overexpressing *REL2* under the control of the *Cp* promoter significantly decreased *P. gallinaceum* oocyst number (Antonova et al., 2009). Position effects resulting from the random integration of the transgene into the target genome was evident, as there was a significant difference in the magnitude of *P. gallinaceum* inhibition with REL2-A and REL2-B strains exhibiting a 89%–94% and 46%–51% reduction in oocyst counts, respectively (Antonova et al., 2009). Additionally, sporozoite counts 14 days post infection (with a second naïve blood meal provided 7 days post infection to reactivate the transgene) were 53%–61% lower in the REL2-A line (Antonova et al., 2009).

Similar experiments performed in *An. stephensi* also led to the suppression of *P. falciparum* numbers within midgut and salivary glands (48.6% and 33.1% from controls, respectively), and affected *Plasmodium* prevalence (93.9% and 87.5% from controls, respectively). Even better reduction was achieved when the *REL2* transgene was expressed under the control of the *Cp* promoter. *P. falciparum* oocyst and sporozoite loads were decreased by 77.1% and 74.7%, respectively, and overall parasite prevalence by 84.6% and 69.2%, respectively (Dong et al., 2011). *REL2* overexpression in the midgut reduced ookinete numbers within the midgut lumen, as well the transition from ookinete to oocyst, suggesting that the Imd pathway affects multiple parasite killing mechanisms (Dong et al., 2011). The *Cp-REL2* transgenic line showed no significant difference in life span, the number of eggs laid per female, or the hatch rate as compared to wild type (Dong et al., 2011). However, experimental dysregulation of the Imd pathway by either overexpression of REL2 or knockdown of *Caspar* changes the expression profiles of more than 100 genes well beyond canonical immunity genes, including those associated with digestion, redox and stress, and cell structure (Pike et al., 2014; Zou et al., 2011). Therefore, potential impact on mosquito fitness may be revealed only upon more varied or less optimal rearing conditions.

Overexpression of Immune Pathway Modulators

Transgenic *An. stephensi* lines overexpressing components of the PI3K/Akt signaling cascade (reviewed in Fig. 4.4, and the section Ras Also Signals Through the PI3K/Akt Pathway) show enhanced immunity to *P. falciparum* (Corby-Harris et al., 2010). An *An. stephensi* line overexpressing the active, myristoylated form of the signaling protein Akt under control of the *Cp* promoter completely blocked *P. falciparum* development in homozygous females with extended life span to boost. In heterozygotes, *Plasmodium* prevalence was decreased from 58.5% (36%–86%) in controls to 10.5% (2%–14%). Overall oocyst number decreased from an average of 3.9 oocysts/midgut in controls to 0.18 oocyst/midgut in transgenics (Corby-Harris et al., 2010). Transgenics exhibited decreased levels of pERK, p38, and pJNK, highlighting the inhibitory nature of Akt in MAPK signaling. Transgenic mosquitoes exhibited morphological changes in midgut epithelial integrity, with stalled autophagosomes and mitochondrial autophagy evident in adult females (Luckhart et al., 2013). The effects of Akt overactivation are complex and were discussed in Chapter 2. Overall, misregulation of mitochondrial activity, and overproduction of NO are seemingly responsible for the anti-*Plasmodium* effect the transgenic line possesses.

A transgenic *An. stephensi* line overexpressing the PI3K/Akt signaling pathway antagonist phosphatase and tensin homolog (PTEN) increased mosquito life span and inhibited *P. falciparum* development within *An. stephensi*. The reduction in parasite development, however, does not seem to stem from innate immune overreaction (such as increased *NOS* expression within the gut, as discussed previously), but instead from changes in autophagy and stem cell maintenance within the gut, resulting in decreased midgut permeability (Hauck et al., 2013).

Overexpression of Plasmodium *Parasite Opsonins*

Overexpression of factors known to bind to the surface of *Plasmodium* spp. may boost parasite recognition, its opsonization, and subsequent killing. Overexpression of the immunolectin FBN9 (see the section Antimalarial Immunity Against Parasites Traversing the Midgut Epithelium and references therein) under the control of the *Vg* promoter displayed increased resistance to *P. berghei* as well as Gram-positive and Gram-negative bacteria. Transgenic mosquitoes resulted in decreased *Plasmodium* oocyst counts and enhanced survival after *Escherichia coli* and *Staphylococcus aureus* immune challenge as compared to wild type. However, neither infection prevalence nor intensity was affected when these transgenics were infected with *P. falciparum* (Simões et al., 2016).

Dscam, a *Plasmodium*-binding molecule was found previously to affect *Plasmodium* development through knockdown experiments (see the section Antimalarial Immunity Against Parasites Traversing the Midgut Epithelium and references therein). Three transgenic lines were generated placing distinct splice isoforms, the *P. falciparum* infection-responsive *Dscam1.17.21* (Pf-L) and *Dscam1.17* (Pf-S), and the *P. berghei* infection-responsive *Dscam14.2* (Pb-S), under the control of the *Cp* promoter (Dong et al., 2012a). Pf-L and Pf-S lines were indeed more resistant to *P. falciparum* infection than *P. berghei*. The Pb-S had significantly lower *P. berghei* oocyst loads, however did not impact *P. falciparum* infection. The overall strongest impact was observed in the Pf-L line, reducing *P. falciparum* parasite load by 50%, rendering a significant proportion of mosquitoes refractory to infection.

THE CHALLENGES AND OPPORTUNITIES FOR BOOSTING MOSQUITO IMMUNITY IN THE FIELD

Generating refractory mosquitoes by boosting basal immunity in the laboratory is, in principle, achievable and thus warrants further research. One line of investigation should focus on the identification and refinement of novel targets and their deployment through genetic engineering. The majority of current lines do not completely eliminate the parasite. However, complete elimination is likely not necessary to achieve significant reduction in the basic reproduction number of the parasite (Blagborough et al., 2013). Therefore, systematic evaluation of existing and additional targets for their impact on parasite demography, extending to RNAi constructs to knockdown/knockout immune factors should be considered. In addition, better control may be achieved by stacking multiple resistance transgenes in one line. The efficacy of existing resistance factors may also be improved by expanding the repertoire of promoters driving transgene expression, including the generation of synthetic promoters that are induced by the presence of the parasite prior to reaching the midgut epithelium.

The outcome of the interactions of parasites and vectors in the field is influenced by many ecological factors. Immunology is influenced by abiotic factors such as temperature, as well as biological factors, including the mosquito's microbiota, trade-off systems, and the coevolutionary history between parasite and vector (Lambrechts et al., 2006; Mitri and Vernick, 2012; Tripet, 2009). The evaluation of resistance factors therefore needs to take these factors into account to be successfully employed. Some studies have indeed included fitness and/or microbiome analyses of their transgenic lines (Corby-Harris et al., 2010; Dong et al., 2011). Going forward, it will be helpful to formalize an evaluation scheme for putative resistance factors that incorporates the important ecological drivers of immunity.

References

Abraham, E.G., Pinto, S.B., Ghosh, A., Vanlandingham, D.L., Budd, A., Higgs, S., Kafatos, F.C., Jacobs-Lorena, M., Michel, K., 2005. An immune-responsive serpin, SRPN6, mediates mosquito defense against malaria parasites. Proc. Natl. Acad. Sci. U.S.A. 102, 16327–16632.

Alavi, Y., Arai, M., Mendoza, J., Tufet-Bayona, M., Sinha, R., Fowler, K., Billker, O., Franke-Fayard, B., Janse, C.J., Waters, A., Sinden, R.E., 2003. The dynamics of interactions between *Plasmodium* and the mosquito: a study of the infectivity of *Plasmodium berghei* and *Plasmodium gallinaceum*, and their transmission by *Anopheles stephensi*, *Anopheles gambiae*, and *Aedes aegypti*. Int. J. Parasitol. 33, 933–943.

An, C., Budd, A., Kanost, M.R., Michel, K., 2011. Characterization of a regulatory unit that controls melanization and affects longevity of mosquitoes. Cell. Mol. Life Sci. 68, 1929–1939.

Anderson, K.V., Nüsslein-Volhard, C., 1984. Information for the dorsal-ventral pattern of the *Drosophila* embryo is stored as maternal mRNA. Nature 311, 223–227.

Antonova, Y., Alvarez, K.S., Kim, Y.J., Kokoza, V., Raikhel, A.S., 2009. The role of NF-κB factor REL2 in the *Aedes aegypti* immune response. Insect Biochem. Mol. Biol. 39, 303–314.

Arca, B., Lombardo, F., Valenzuela, J.G., Francischetti, I.M., Marinotti, O., Coluzzi, M., Ribeiro, J.M., 2005. An updated catalogue of salivary gland transcripts in the adult female mosquito, *Anopheles gambiae*. J. Exp. Biol. 208, 3971–3986.

Awono-Ambene, H., Robert, V., 1998. Estimation of *Plasmodium falciparum* oocyst survival in *Anopheles arabiensis*. Ann. Trop. Med. Parasitol. 92, 889–890.

Baeg, G., Zhou, R., Perrimon, N., 2005. Genome-wide RNAi analysis of JAK/STAT signaling components in *Drosophila*. Genes Dev. 19, 1861–1870.

Bahia, A.C., Kubota, M.S., Tempone, A.J., Araújo, H.R.C., Guedes, B.A.M., Orfanó, A.S., Tadei, W.P., Ríos-Velásquez, C.M., Han, Y.S., Secundino, N.F.C., Barillas-Mury, C., Pimenta, P.F.P., Traub-Csekö, Y.M., 2011. The JAK-STAT pathway controls *Plasmodium vivax* load in early stages of *Anopheles aquasalis* infection. PLoS Negl. Trop. Dis. 5, e1317.

Barillas-Mury, C., Kumar, S., 2005. *Plasmodium*-mosquito interactions: a tale of dangerous liaisons. Cell. Microbiol. 7, 1539–1545.

Barillas-Mury, C., Han, Y.S., Seeley, D., Kafatos, F.C., 1999. *Anopheles gambiae* Ag-STAT, a new insect member of the STAT family, is activated in response to bacterial infection. EMBO J. 18, 959–967.

Barreau, C., Touray, M., Pimenta, P.F., Miller, L.H., Vernick, K.D., 1995. *Plasmodium gallinaceum*: sporozoite invasion of *Aedes aegypti* salivary glands is inhibited by anti-gland antibodies and by lectins. Exp. Parasitol. 81, 332–343.

Bartholomay, L.C., Waterhouse, R.M., Mayhew, G.F., Campbell, C.L., Michel, K., Zou, Z., Ramirez, J.L., Das, S., Alvarez, K.S., Arensburger, P., Bryant, B., Chapman, S.B., Dong, Y., Erickson, S.M., Karunaratne, S.H., Kokoza, V., Kodira, C.D., Pignatelli, P., Shin, S.W., Vanlandingham, D.L., Atkinson, P.W., Birren, B., Christophides, G.K., Clem, R.J., Hemingway, J., Higgs, S., Megy, K., Ranson, H., Zdobnov, E.M., Raikhel, A.S., Christensen, B.M., Dimopoulos, G., Muskavitch, M.A., 2010. Pathogenomics of *Culex quinquefasciatus* and meta-analysis of infection responses to diverse pathogens. Science 330, 88–90.

Bastianelli, G., Bignami, A., 1900. Malaria and mosquitoes. Lancet 155, 79–80.

Baton, L.A., Ranford-Cartwright, L.C., 2005. Spreading the seeds of million-murdering death: metamorphoses of malaria in the mosquito. Trends Parasitol. 21, 573–580.

Baton, L.A., Robertson, A., Warr, E., Strand, M.R., Dimopoulos, G., 2009. Genome-wide transcriptomic profiling of *Anopheles gambiae* hemocytes reveals pathogen-specific signatures upon bacterial challenge and *Plasmodium berghei* infection. BMC Genomics 10, 257.

Baxter, R.H.G., Chang, C.I., Chelliah, Y., Blandin, S., Levashina, E.A., Deisenhofer, J., 2007. Structural basis for conserved complement factor-like function in the antimalarial protein TEP1. Proc. Natl. Acad. Sci. 104, 11615–11620.

Beier, J.C., 1998. Malaria parasite development in mosquitoes. Annu. Rev. Entomol. 43, 519–543.

Bennink, S., Kiesow, M.J., Pradel, G., 2016. The development of malaria parasites in the mosquito midgut. Cell. Microbiol. 18, 905–918.

Bian, G., Shin, S.W., Cheon, H.-M., Kokoza, V., Raikhel, A.S., 2005. Transgenic alteration of Toll immune pathway in the female mosquito *Aedes aegypti*. Proc. Natl. Acad. Sci. U.S.A. 102, 13568–13573.

Bischoff, V., Vignal, C., Boneca, I.G., Michel, T., Hoffmann, J.A., Royet, J., 2004. Function of the *Drosophila* pattern-recognition receptor PGRP-SD in the detection of Gram-positive bacteria. Nat. Immunol. 5, 1175–1180.

Blagborough, A.M., Churcher, T.S., Upton, L.M., Ghani, A.C., Gething, P.W., Sinden, R.E., 2013. Transmission-blocking interventions eliminate malaria from laboratory populations. Nat. Commun. 4, 1812.

Blandin, S., Shiao, S.H., Moita, L.F., Janse, C.J., Waters, A.P., Kafatos, F.C., Levashina, E.A., 2004. Complement-like protein TEP1 is a determinant of vectorial capacity in the malaria vector *Anopheles gambiae*. Cell 116, 661–670.

Blandin, S.A., Marois, E., Levashina, E.A., 2008. Antimalarial responses in *Anopheles gambiae*: from a complement-like protein to a complement-like pathway. Cell Host Microbe 3, 364–374.

Bryant, W.B., Michel, K., 2014. Blood feeding induces hemocyte proliferation and activation in the African malaria mosquito, *Anopheles gambiae* Giles. J. Exp. Biol. 217, 1238–1245.

Bryant, W.B., Michel, K., 2016. *Anopheles gambiae* hemocytes exhibit transient states of activation. Dev. Comp. Immunol. 55, 119–129.

Buchon, N., Broderick, N.A., Lemaitre, B., 2013. Gut homeostasis in a microbial world: insights from *Drosophila melanogaster*. Nat. Rev. Microbiol. 11, 615–626.

Buchon, N., Silverman, N., Cherry, S., 2014. Immunity in *Drosophila melanogaster*—from microbial recognition to whole-organism physiology. Nat. Rev. Immunol. 14, 796–810.

Calvo, E., Sanchez-Vargas, I., Favreau, A.J., Barbian, K.D., Pham, V.M., Olson, K.E., Ribeiro, J.M., 2010. An insight into the sialotranscriptome of the West Nile mosquito vector, *Culex tarsalis*. BMC Genomics 11.

Carissimo, G., Pondeville, E., McFarlane, M., Dietrich, I., Mitri, C., Bischoff, E., Antoniewski, C., Bourgouin, C., Failloux, A.-B., Kohl, A., Vernick, K.D., 2015. Antiviral immunity of *Anopheles gambiae* is highly compartmentalized, with distinct roles for RNA interference and gut microbiota. Proc. Natl. Acad. Sci. 112, E176–E185.

Casci, T., Vinós, J., Freeman, M., 1999. Sprouty, an intracellular inhibitor of Ras signaling. Cell 96, 655–665.

Castellano, E., Downward, J., 2011. RAS interaction with PI3K: more than just another effector pathway. Genes Cancer 2, 261–274.

Chen, Y., Dong, Y., Sandiford, S., Dimopoulos, G., 2012. Transcriptional mediators Kto and Skd are involved in the regulation of the IMD pathway and anti-*Plasmodium* defense in *Anopheles gambiae*. PLoS One 7, e45580.

Choe, K.-M.M., Werner, T., Stöven, S., Hultmark, D., Anderson, K.V., 2002. Requirement for a peptidoglycan recognition protein (PGRP) in Relish activation and antibacterial immune responses in *Drosophila*. Science 296, 359–362.

Christensen, B.M., Li, J., Chen, C.C., Nappi, A.J., 2005. Melanization immune responses in mosquito vectors. Trends Parasitol. 21, 192–199.

Christophides, G.K., Zdobnov, E., Barillas-Mury, C., Birney, E., Blandin, S., Blass, C., Brey, P.T., Collins, F.H., Danielli, A., Dimopoulos, G., Hetru, C., Hoa, N.T., Hoffmann, J.A., Kanzok, S.M., Letunic, I., Levashina, E.A., Loukeris, T.G., Lycett, G., Meister, S., Michel, K., Moita, L.F., Müller, H.-M., Osta, M.A., Paskewitz, S.M., Reichhart, J.-M., Rzhetsky, A., Troxler, L., Vernick, K.D., Vlachou, D., Volz, J., von Mering, C., Xu, J., Zheng, L., Bork, P., Kafatos, F.C., 2002. Immunity-related genes and gene families in *Anopheles gambiae*. Science 298, 159–165.

Clayton, A.M., Cirimotich, C.M., Dong, Y., Dimopoulos, G., 2013. Caudal is a negative regulator of the *Anopheles* IMD pathway that controls resistance to *Plasmodium falciparum* infection. Dev. Comp. Immunol. 39, 323–332.

Cohuet, A., Osta, M.A., Morlais, I., Awono-Ambene, P.H., Michel, K., Simard, F., Christophides, G.K., Fontenille, D., Kafatos, F.C., 2006. *Anopheles* and *Plasmodium*: from laboratory models to natural systems in the field. EMBO Rep. 7, 1285–1289.

Collins, F.H., Sakai, R.K., Vernick, K.D., Paskewitz, S., Seeley, D.C., Miller, L.H., Collins, W.E., Campbell, C.C., Gwadz, R.W., 1986. Genetic selection of a *Plasmodium*-refractory strain of the malaria vector *Anopheles gambiae*. Science 234, 607–610.

Corby-Harris, V., Drexler, A., de Jong, L.W., Antonova, Y., Pakpour, N., Ziegler, R., Ramberg, F., Lewis, E.E., Brown, J.M., Luckhart, S., Riehle, M.A., 2010. Activation of Akt signaling reduces the prevalence and intensity of malaria parasite infection and lifespan in *Anopheles stephensi* mosquitoes. PLoS Pathog. 6, e1001003.

Crompton, P.D., Moebius, J., Portugal, S., Waisberg, M., Hart, G., Garver, L.S., Miller, L.H., Barillas-Mury, C., Pierce, S.K., 2014. Malaria immunity in man and mosquito: insights into unsolved mysteries of a deadly infectious disease. Annu. Rev. Immunol. 32, 157–187.

Daniels, C.W., 1898. On transmission of Proteosoma to birds by the mosquito. Proc. R. Soc. London 64, 443–454.

Dimopoulos, G., Seeley, D., Wolf, A., Kafatos, F.C., 1998. Malaria infection of the mosquito *Anopheles gambiae* activates immune-responsive genes during critical transition stages of the parasite life cycle. EMBO J. 17, 6115–6123.

Dong, Y., Dimopoulos, G., 2009. *Anopheles* fibrinogen-related proteins provide expanded pattern recognition capacity against bacteria and malaria parasites. J. Biol. Chem. 284, 9835–9844.

Dong, Y., Taylor, H.E., Dimopoulos, G., 2006. AgDscam, a hypervariable immunoglobulin domain-containing receptor of the *Anopheles gambiae* innate immune system. PLoS Biol. 4, e229.

Dong, Y., Das, S., Cirimotich, C., Souza-Neto, J.A., McLean, K.J., Dimopoulos, G., 2011. Engineered *Anopheles* immunity to *Plasmodium* infection. PLoS Pathog. 7, e1002458.

Dong, Y., Cirimotich, C., Pike, A., Chandra, R., Dimopoulos, G., 2012a. *Anopheles* NF-κB-regulated splicing factors direct pathogen-specific repertoires of the hypervariable pattern recognition receptor AgDscam. Cell Host Microbe 12, 521–530.

Dong, Y., Morton, J.C.J., Ramirez, J.L., Souza-Neto, J.A., Dimopoulos, G., 2012b. The entomopathogenic fungus *Beauveria bassiana* activate toll and JAK-STAT pathway-controlled effector genes and anti-dengue activity in *Aedes aegypti*. Insect Biochem. Mol. Biol. 42, 126–132.

Drexler, A., Nuss, A., Hauck, E., Glennon, E., Cheung, K., Brown, M., Luckhart, S., 2013. Human IGF1 extends lifespan and enhances resistance to *Plasmodium falciparum* infection in the malaria vector *Anopheles stephensi*. J. Exp. Biol. 216, 208–217.

Drexler, A.L., Pietri, J.E., Pakpour, N., Hauck, E., Wang, B., Glennon, E.K.K., Georgis, M., Riehle, M.A., Luckhart, S., 2014. Human IGF1 regulates midgut oxidative stress and epithelial homeostasis to balance lifespan and *Plasmodium falciparum* resistance in *Anopheles stephensi*. PLoS Pathog. 10, e1004231.

Edwards, M.J., Jacobs-Lorena, M., 2000. Permeability and disruption of the peritrophic matrix and caecal membrane from *Aedes aegypti* and *Anopheles gambiae* mosquito larvae. J. Insect Physiol. 46, 1313–1320.

El Chamy, L., Leclerc, V., Caldelari, I., Reichhart, J.-M., 2008. Sensing of "danger signals" and pathogen-associated molecular patterns defines binary signaling pathways "upstream" of Toll. Nat. Immunol. 9, 1165–1170.

Ertürk-Hasdemir, D., Broemer, M., Leulier, F., Lane, W.S., Paquette, N., Hwang, D., Kim, C.-H., Stöven, S., Meier, P., Silverman, N., 2009. Two roles for the *Drosophila* IKK complex in the activation of Relish and the induction of antimicrobial peptide genes. Proc. Natl. Acad. Sci. U.S.A. 106, 9779–9784.

Fraiture, M., Baxter, R.H.G., Steinert, S., Chelliah, Y., Frolet, C., Quispe-Tintaya, W., Hoffmann, J.A., Blandin, S.A., Levashina, E.A., 2009. Two mosquito LRR proteins function as complement control factors in the TEP1-mediated killing of *Plasmodium*. Cell Host Microbe 5, 273–284.

Fraser, M.J., Ciszczon, T., Elick, T., Bauser, C., 1996. Precise excision of TTAA-specific lepidopteran transposons *piggyBac* (IFP2) and *tagalong* (TFP3) from the baculovirus genome in cell lines from two species of *Lepidoptera*. Insect Mol. Biol. 5, 141–151.

Frolet, C., Thoma, M., Blandin, S., Hoffmann, J.A., Levashina, E.A., 2006. Boosting NF-κB-dependent basal immunity of *Anopheles gambiae* aborts development of *Plasmodium berghei*. Immunity 25, 677–685.

Fullaondo, A., García-Sánchez, S., Sanz-Parra, A., Recio, E., Lee, S.Y., Gubb, D., 2011. Spn1 regulates the GNBP3-dependent Toll signaling pathway in *Drosophila melanogaster*. Mol. Cell. Biol. 31, 2960–2972.

Garsin, D.A., Villanueva, J.M., Begun, J., Kim, D.H., Sifri, C.D., Calderwood, S.B., Ruvkun, G., Ausubel, F.M., 2003. Long-lived *C. elegans daf-2* mutants are resistant to bacterial pathogens. Science 300, 1921.

Garver, L.S., Xi, Z., Dimopoulos, G., 2008. Immunoglobulin superfamily members play an important role in the mosquito immune system. Dev. Comp. Immunol. 32, 519–531.

Garver, L.S., Dong, Y., Dimopoulos, G., 2009. Caspar controls resistance to *Plasmodium falciparum* in diverse anopheline species. PLoS Pathog. 5, e1000335.

Garver, L.S., Bahia, A.C., Das, S., Souza-Neto, J.A., Shiao, J., Dong, Y., Dimopoulos, G., 2012. *Anopheles* Imd pathway factors and effectors in infection intensity-dependent anti-*Plasmodium* action. PLoS Pathog. 8, e1002737.

Garver, L.S., Oliveira, G.A., Barillas-Mury, C., 2013. The JNK pathway is a key mediator of *Anopheles gambiae* antiplasmodial immunity. PLoS Pathog. 9, e1003622.

Gay, N.J., Keith, F.J., 1991. *Drosophila* Toll and IL-1 receptor. Nature 351, 355–356.

Gobert, V., Gottar, M., Matskevich, A.A., Rutschmann, S., Royet, J., Belvin, M.P., Hoffmann, J.A., Ferrandon, D., 2003. Dual activation of the *Drosophila* Toll pathway by two pattern recognition receptors. Science 302, 2126–2130.

Gorman, M.J., Severson, D.W., Cornel, A.J., Collins, F.H., Paskewitz, S.M., 1997. Mapping a quantitative trait locus involved in melanotic encapsulation of foreign bodies in the malaria vector, *Anopheles gambiae*. Genetics 146, 965–971.

Gorman, M.J., An, C., Kanost, M.R., 2007. Characterization of tyrosine hydroxylase from *Manduca sexta*. Insect Biochem. Mol. Biol. 37, 1327–1337.

Gottar, M., Gobert, V., Michel, T., Belvin, M., Duyk, G., Hoffmann, J.A., Ferrandon, D., Royet, J., 2002. The *Drosophila* immune response against Gram-negative bacteria is mediated by a peptidoglycan recognition protein. Nature 416, 640–644.

Gottar, M., Gobert, V., Matskevich, A.A., Reichhart, J.M., Wang, C., Butt, T.M., Belvin, M., Hoffmann, J.A., Ferrandon, D., 2006. Dual detection of fungal infections in *Drosophila* via recognition of glucans and sensing of virulence factors. Cell 127, 1425–1437.

Gouagna, L.C., Bonnet, S., Gounoue, R., Verhave, J.P., Eling, W., Sauerwein, R., Boudin, C., 2004. Stage-specific effects of host plasma factors on the early sporogony of autologous *Plasmodium falciparum* isolates within *Anopheles gambiae*. Trop. Med. Int. Heal 9, 937–948.

Goulielmaki, E., Sidén-Kiamos, I., Loukeris, T.G., 2014. Functional characterization of *Anopheles* matrix metalloprotease 1 reveals its agonistic role during sporogonic development of malaria parasites. Infect. Immun. 82, 4865–4877.

Groth, A.C., Fish, M., Nusse, R., Calos, M.P., 2004. Construction of transgenic *Drosophila* by using the site-specific integrase from phage phiC31. Genetics 166, 1775–1782.

Gupta, L., Kumar, S., Han, Y.S., Pimenta, P.F.P., Barillas-Mury, C., 2005. Midgut epithelial responses of different mosquito-*Plasmodium* combinations: the actin cone zipper repair mechanism in *Aedes aegypti*. Proc. Natl. Acad. Sci. U.S.A. 102, 4010–4015.

Gupta, L., Molina-Cruz, A., Kumar, S., Rodrigues, J., Dixit, R., Zamora, R.E., Barillas-Mury, C., 2009. The STAT pathway mediates late-phase immunity against *Plasmodium* in the mosquito *Anopheles gambiae*. Cell Host Microbe 5, 498–507.

Gwadz, R.W., Kaslow, D., Lee, J.Y., Maloy, W.L., Zasloff, M., Miller, L.H., 1989. Effects of magainins and cecropins on the sporogonic development of malaria parasites in mosquitoes. Infect. Immun. 57, 2628–2633.

Habtewold, T., Povelones, M., Blagborough, A.M., Christophides, G.K., 2008. Transmission blocking immunity in the malaria non-vector mosquito *Anopheles quadriannulatus* species A. PLoS Pathog. 4, e1000070.

Han, Y.S., Thompson, J., Kafatos, F.C., Barillas-Mury, C., 2000. Molecular interactions between *Anopheles stephensi* midgut cells and *Plasmodium berghei*: the time bomb theory of ookinete invasion. Mem. Inst. Oswaldo Cruz 19, 6030–6040.

Hashimoto, C., Hudson, K.L., Anderson, K.V., 1988. The *Toll* gene of *Drosophila*, required for dorsal-ventral embryonic polarity, appears to encode a transmembrane protein. Cell 52, 269–279.

Hauck, E.S., Antonova-Koch, Y., Drexler, A., Pietri, J., Pakpour, N., Liu, D., Blacutt, J., Riehle, M.A., Luckhart, S., 2013. Overexpression of phosphatase and tensin homolog improves fitness and decreases *Plasmodium falciparum* development in *Anopheles stephensi*. Microbe. Infect. 15, 775–787.

Hay, N., 2011. Interplay between FOXO, TOR, and Akt. Biochim. Biophys. Acta 1813, 1965–1970.

Hegedus, D., Erlandson, M., Gillott, C., Toprak, U., 2009. New insights into peritrophic matrix synthesis, architecture, and function. Annu. Rev. Entomol. 54, 285–302.

Hernández-Martínez, S., Lanz, H., Rodríguez, M.H., González-Ceron, L., Tsutsumi, V., 2002. Cellular-mediated reactions to foreign organisms inoculated into the hemocoel of *Anopheles albimanus* (Diptera: Culicidae). J. Med. Entomol. 39, 61–69.

Hillyer, J.F., Schmidt, S.L., Christensen, B.M., 2003. Rapid phagocytosis and melanization of bacteria and *Plasmodium* sporozoites by hemocytes of the mosquito *Aedes aegypti*. J. Parasitol. 89, 62–69.

Hillyer, J.F., Barreau, C., Vernick, K.D., 2007. Efficiency of salivary gland invasion by malaria sporozoites is controlled by rapid sporozoite destruction in the mosquito haemocoel. Int. J. Parasitol. 37, 673–681.

Horton, A.A., Wang, B., Camp, L., Price, M.S., Arshi, A., Nagy, M., Nadler, S.A., Faeder, J.R., Luckhart, S., 2011. The mitogen-activated protein kinome from *Anopheles gambiae*: identification, phylogeny and functional characterization of the ERK, JNK and p38 MAP kinases. BMC Genomics 12, 1.

Hotamisligil, G.S., Davis, R.J., 2016. Cell signaling and stress responses. Cold Spring Harb. Perspect. Biol. 8, a006072.

Huber, M., Cabib, E., Miller, L.H., 1991. Malaria parasite chitinase and penetration of the mosquito peritrophic membrane. Proc. Natl. Acad. Sci. U.S.A. 88, 2807–2810.

Huff, C.G., 1927. Studies on the infectivity of plasmodia of birds for mosquitoes, with special reference to the problem of immunity in the mosquito. Am. J. Epidemiol. 7, 706–734.

Huff, C.G., 1929. The effects of selection upon susceptibility to bird malaria in *Culex Pipiens* Linn. Ann. Trop. Med. Parasitol. 23, 427–442.

Huff, C.G., 1930. Individual immunity and susceptibility to *Culex pipiens* to various species of bird malaria as studied by means of double infectious feedings. Am. J. Epidemiol. 12, 424–441.

Huff, C.G., 1931. The inheritance of natural immunity to *Plasmodium cathemerium* in two species of *Culex*. J. Prev. Med. 5, 249–259.

Huff, C.G., 1935. Natural immunity and susceptibility of culicine mosquitoes to avian malaria. Am. J. Trop. Med. 15, 427–434.

Hurd, H., Taylor, P.J., Adams, D., Underhill, A., Eggleston, P., 2005. Evaluating the costs of mosquito resistance to malaria parasites. Evolution 59, 2560–2572.

Jaramillo-Gutierrez, G., Molina-Cruz, A., Kumar, S., Barillas-Mury, C., 2010. The *Anopheles gambiae* oxidation resistance 1 (OXR1) gene regulates expression of enzymes that detoxify reactive oxygen species. PLoS One 5, e11168.

Ji, S., Sun, M., Zheng, X., Li, L., Sun, L., Chen, D., Sun, Q., 2014. Cell-surface localization of Pellino antagonizes Toll-mediated innate immune signalling by controlling MyD88 turnover in *Drosophila*. Nat. Commun. 5.

Kanehisa, M., Goto, S., Furumichi, M., Tanabe, M., Hirakawa, M., 2010. KEGG for representation and analysis of molecular networks involving diseases and drugs. Nucleic Acids Res. 38, D355–D360.

Kaneko, T., Yano, T., Aggarwal, K., Lim, J.-H., Ueda, K., Oshima, Y., Peach, C., Erturk-Hasdemir, D., Goldman, W.E., Oh, B.-H., Kurata, S., Silverman, N., 2006. PGRP-LC and PGRP-LE have essential yet distinct functions in the *Drosophila* immune response to monomeric DAP-type peptidoglycan. Nat. Immunol. 7, 715–723.

Kim, W., Koo, H., Richman, A.M., Seeley, D., Vizioli, J., Klocko, A.D., O'Brochta, D.A., 2004. Ectopic expression of a cecropin transgene in the human malaria vector mosquito *Anopheles gambiae* (Diptera: Culicidae): effects on susceptibility to *Plasmodium*. J. Med. Entomol. 41, 447–455.

Kim, M., Lee, J.H., Lee, S.Y., Kim, E., Chung, J., 2006. Caspar, a suppressor of antibacterial immunity in *Drosophila*. Proc. Natl. Acad. Sci. U.S.A. 103, 16358–16363.

Kingsolver, M.B., Huang, Z., Hardy, R.W., 2013. Insect antiviral innate immunity: pathways, effectors, and connections. J. Mol. Biol. 425, 4921–4936.

Kleino, A., Silverman, N., 2014. The *Drosophila* IMD pathway in the activation of the humoral immune response. Dev. Comp. Immunol. 42, 25–35.

Kleino, A., Valanne, S., Ulvila, J., Kallio, J., Myllymäki, H., Enwald, H., Stöven, S., Poidevin, M., Ueda, R., Hultmark, D., Lemaitre, B., Rämet, M., 2005. Inhibitor of apoptosis 2 and TAK1-binding protein are components of the *Drosophila* Imd pathway. EMBO J. 24, 3423–3434.

Knowles, R., Basu, B.C., 1933. The nature of the so-called "black spores" of Ross in malaria-transmitting mosquitoes. Indian J. Med. Res. 20, 757–776.

Kokoza, V., Ahmed, A., Cho, W.L., Jasinskiene, N., James, A.A., Raikhel, A., 2000. Engineering blood meal-activated systemic immunity in the yellow fever mosquito, *Aedes aegypti*. Proc. Natl. Acad. Sci. U.S.A. 97, 9144–9149.

Kokoza, V., Ahmed, A., Woon Shin, S., Okafor, N., Zou, Z., Raikhel, A.S., 2010. Blocking of *Plasmodium* transmission by cooperative action of Cecropin A and Defensin A in transgenic *Aedes aegypti* mosquitoes. Proc. Natl. Acad. Sci. U.S.A. 107, 8111–8116.

Kumar, S., Gupta, L., Han, Y.S., Barillas-Mury, C., 2004. Inducible peroxidases mediate nitration of *Anopheles* midgut cells undergoing apoptosis in response to *Plasmodium* invasion. J. Biol. Chem. 279, 53475–53482.

Kumar, S., Molina-Cruz, A., Gupta, L., Rodrigues, J., Barillas-Mury, C., 2010. A peroxidase/dual oxidase system modulates midgut epithelial immunity in *Anopheles gambiae*. Science 327, 1644–1648.

Lai, S.-C., Chen, C.-C., Hou, R.F., 2002. Immunolocalization of prophenoloxidase in the process of wound healing in the mosquito *Armigeres subalbatus* (Diptera: Culicidae). J. Med. Entomol. 39, 266–274.

Lambrechts, L., Halbert, J., Durand, P., Gouagna, L.C., Koella, J.C., 2005. Host genotype by parasite genotype interactions underlying the resistance of anopheline mosquitoes to *Plasmodium falciparum*. Malar. J. 4, 1.

Lambrechts, L., Chavatte, J.-M., Snounou, G., Koella, J.C., 2006. Environmental influence on the genetic basis of mosquito resistance to malaria parasites. Proc. Biol. Sci. 273, 1501–1506.

Lee, K.-Z., Ferrandon, D., 2011. Negative regulation of immune responses on the fly. EMBO J. 30, 988–990.

Lemaitre, B., Kromer-Metzger, E., Michaut, L., Nicolas, E., Meister, M., Georgel, P., Reichhart, J.M., Hoffmann, J.A., 1995. A recessive mutation, immune deficiency (*imd*), defines two distinct control pathways in the *Drosophila* host defense. Proc. Natl. Acad. Sci. U.S.A. 92, 9465–9469.

Lemaitre, B., Nicolas, E., Michaut, L., Reichhart, J.M., Hoffmann, J.A., 1996. The dorsoventral regulatory gene cassette *spätzle/Toll/cactus* controls the potent antifungal response in *Drosophila* adults. Cell 86, 973–983.

Lensen, A.H., Bolmer-Van de Vegte, M., van Gemert, G.J., Eling, W.M., Sauerwein, R.W., 1997. Leukocytes in a *Plasmodium falciparum*-infected blood meal reduce transmission of malaria to *Anopheles* mosquitoes. Infect. Immun. 65, 3834–3837.

Leulier, F., Rodriguez, A., Khush, R.S., Abrams, J.M., Lemaitre, B., 2000. The *Drosophila* caspase Dredd is required to resist Gram-negative bacterial infection. EMBO Rep. 1, 353–358.

Leulier, F., Vidal, S., Saigo, K., Ueda, R., Lemaitre, B., 2002. Inducible expression of double-stranded RNA reveals a role for dFADD in the regulation of the antibacterial response in *Drosophila* adults. Curr. Biol. 12, 996–1000.

Leulier, F., Parquet, C., Pili-Floury, S., Ryu, J.H., Caroff, M., Lee, W.J., Mengin-Lecreulx, D., Lemaitre, B., 2003. The *Drosophila* immune system detects bacteria through specific peptidoglycan recognition. Nat. Immunol. 4, 478–484.

Levashina, E.A., Langley, E., Green, C., Gubb, D., Ashburner, M., Hoffmann, J.A., Reichhart, J.M., 1999. Constitutive activation of Toll-mediated antifungal defense in serpin-deficient *Drosophila*. Science 285, 1917–1919.

Levashina, E.A., Moita, L.F., Blandin, S., Vriend, G., Lagueux, M., Kafatos, F.C., 2001. Conserved role of a complement-like protein in phagocytosis revealed by dsRNA knockout in cultured cells of the mosquito, *Anopheles gambiae*. Cell 104, 709–718.

Lhocine, N., Ribeiro, P.S., Buchon, N., Wepf, A., Wilson, R., Tenev, T., Lemaitre, B., Gstaiger, M., Meier, P., Leulier, F., 2008. PIMS modulates immune tolerance by negatively regulating *Drosophila* innate immune signaling. Cell Host Microbe 4, 147–158.

Lindberg, B.G., Merritt, E.A., Rayl, M., Liu, C., Parmryd, I., Olofsson, B., Faye, I., 2013. Immunogenic and antioxidant effects of a pathogen-associated prenyl pyrophosphate in *Anopheles gambiae*. PLoS One 8, e73868.

Liongue, C., Ward, A.C., 2013. Evolution of the JAK-STAT pathway. JAK-STAT 2, e22756.

Liu, B., Zheng, Y., Yin, F., Yu, J., Silverman, N., Pan, D., 2016. Toll receptor-mediated Hippo signaling controls innate immunity in *Drosophila*. Cell 164, 406–419.

Lu, Y., Wu, L.P., Anderson, K.V., 2001. The antibacterial arm of the *Drosophila* innate immune response requires an IκB kinase. Genes Dev. 15, 104–110.

Luckhart, S., Vodovotz, Y., Cui, L., Rosenberg, R., 1998. The mosquito *Anopheles stephensi* limits malaria parasite development with inducible synthesis of nitric oxide. Proc. Natl. Acad. Sci. U.S.A. 95, 5700–5705.

Luckhart, S., Crampton, A.L., Zamora, R., Lieber, M.J., Dos Santos, P.C., Peterson, T., Emmith, N., Lim, J., Wink, D.A., Vodovotz, Y., 2003. Mammalian transforming growth factor β1 activated after ingestion by *Anopheles stephensi* modulates mosquito immunity. Infect. Immun. 71, 3000–3009.

Luckhart, S., Giulivi, C., Drexler, A.L., Antonova-Koch, Y., Sakaguchi, D., Napoli, E., Wong, S., Price, M.S., Eigenheer, R., Phinney, B.S., Pakpour, N., Pietri, J.E., Cheung, K., Georgis, M., Riehle, M., 2013. Sustained activation of Akt elicits mitochondrial dysfunction to block *Plasmodium falciparum* infection in the mosquito host. PLoS Pathog. 9, e1003180.

Lyon, I.P., 1900. The inoculation of malaria by the mosquito: a review of the literature. Med. Rec. 57.

Margos, G., Navarette, S., Butcher, G., Davies, A., Willers, C., Sinden, R.E., Lachmann, P.J., 2001. Interaction between host complement and mosquito-midgut-stage *Plasmodium berghei*. Infect. Immun. 69, 5064–5071.

Marinotti, O., Calvo, E., Nguyen, Q.K., Dissanayake, S., Ribeiro, J.M.C., James, A.A., 2006. Genome-wide analysis of gene expression in adult *Anopheles gambiae*. Insect Mol. Biol. 15, 1–12.

Martín-Blanco, E., Gampel, A., Ring, J., Virdee, K., Kirov, N., Tolkovsky, A.M., Martinez-Arias, A., 1998. *Puckered* encodes a phosphatase that mediates a feedback loop regulating JNK activity during dorsal closure in *Drosophila*. Genes Dev. 12, 557–570.

Martín-Blanco, E., 2000. p38 MAPK signalling cascades: ancient roles and new functions. BioEssays 22, 637–645.

Mason, J.M., Morrison, D.J., Basson, A.M., Licht, J.D., 2006. Sprouty proteins: multifaceted negative-feedback regulators of receptor tyrosine kinase signaling. Trends Cell Biol. 16, 45–54.

McEwen, D.G., Peifer, M., 2005. Puckered, a *Drosophila* MAPK phosphatase, ensures cell viability by antagonizing JNK-induced apoptosis. Development 132, 3935–3946.

Meinander, A., Runchel, C., Tenev, T., Chen, L., Kim, C.-H., Ribeiro, P.S., Broemer, M., Leulier, F., Zvelebil, M., Silverman, N., Meier, P., 2012. Ubiquitylation of the initiator caspase DREDD is required for innate immune signalling. EMBO J. 31, 2770–2783.

Meister, S., Koutsos, A.C., Christophides, G.K., 2004. The *Plasmodium* parasite - a "new" challenge for insect innate immunity. Int. J. Parasitol. 34, 1473–1482.

Meister, S., Kanzok, S., Zheng, X., Luna, C., Li, T., Hoa, N., Clayton, J., White, K., Kafatos, F., Christophides, G., Zheng, L., 2005. Immune signaling pathways regulating bacterial and malaria parasite infection of the mosquito *Anopheles gambiae*. Proc. Natl. Acad. Sci. 102, 11420–11425.

Meister, S., Agianian, B., Turlure, F., Relógio, A., Morlais, I., Kafatos, F.C., Christophides, G.K., 2009. *Anopheles gambiae* PGRPLC-mediated defense against bacteria modulates infections with malaria parasites. PLoS Pathog. 5, e1000542.

Mellroth, P., Karlsson, J., Steiner, H., 2003. A scavenger function for a *Drosophila* peptidoglycan recognition protein. J. Biol. Chem. 278, 7059–7064.

Michel, T., Reichhart, J.M., Hoffmann, J.A., Royet, J., 2001. *Drosophila* Toll is activated by Gram-positive bacteria through a circulating peptidoglycan recognition protein. Nature 414, 756–759.

Michel, K., Budd, A., Pinto, S., Gibson, T.J., Kafatos, F.C., 2005. *Anopheles gambiae* SRPN2 facilitates midgut invasion by the malaria parasite *Plasmodium berghei*. EMBO Rep. 6, 891–897.

Michel, K., Suwanchaichinda, C., Morlais, I., Lambrechts, L., Cohuet, A., Awono-Ambene, P.H., Simard, F., Fontenille, D., Kanost, M.R., Kafatos, F.C., 2006. Increased melanizing activity in *Anopheles gambiae* does not affect development of *Plasmodium falciparum*. Proc. Natl. Acad. Sci. U.S.A. 103, 16858–16863.

Micks, D.W., 1949. Investigations on the mosquito transmission of *Plasmodium elongatum*. J. Natl. Malar. Soc. 8, 206–218.

Ming, M., Obata, F., Kuranaga, E., Miura, M., 2014. Persephone/Spätzle pathogen sensors mediate the activation of toll receptor signaling in response to endogenous danger signals in apoptosis-deficient *Drosophila*. J. Biol. Chem. 289, 7558–7568.

Mitri, C., Vernick, K.D., 2012. *Anopheles gambiae* pathogen susceptibility: the intersection of genetics, immunity and ecology. Curr. Opin. Microbiol. 15, 285–291.

Mitri, C., Jacques, J.C., Thiery, I., Riehle, M.M., Xu, J., Bischoff, E., Morlais, I., Nsango, S.E., Vernick, K.D., Bourgouin, C., 2009. Fine pathogen discrimination within the APL1 gene family protects *Anopheles gambiae* against human and rodent malaria species. PLoS Pathog. 5, e1000576.

Morales, J., Falanga, Y., Depcrynski, A., Fernando, J., Ryan, J., 2010. Mast cell homeostasis and the JAK–STAT pathway. Genes Immun. 11, 599–608.

Mueller, A.-K., Kohlhepp, F., Hammerschmidt, C., Michel, K., 2010. Invasion of mosquito salivary glands by malaria parasites: prerequisites and defense strategies. Int. J. Parasitol. 40, 1229–1235.

Myllymäki, H., Rämet, M., 2014. JAK/STAT pathway in *Drosophila* immunity. Scand. J. Immunol. 79, 377–385.

Myllymäki, H., Valanne, S., Rämet, M., 2014. The *Drosophila* Imd signaling pathway. J. Immunol. 192, 3455–3462.

Neafsey, D.E., Waterhouse, R.M., Abai, M.R., Aganezov, S.S., Alekseyev, M.A., Allen, J.E., Amon, J., Arcà, B., Arensburger, P., Artemov, G., Assour, L.A., Basseri, H., Berlin, A., Birren, B.W., Blandin, S.A., Brockman, A.I., Burkot, T.R., Burt, A., Chan, C.S., Chauve, C., Chiu, J.C., Christensen, M., Costantini, C., Davidson, V.L.M., Deligianni, E., Dottorini, T., Dritsou, V., Gabriel, S.B., Guelbeogo, W.M., Hall, A.B., Han, M.V., Hlaing, T., Hughes, D.S.T., Jenkins, A.M., Jiang, X., Jungreis, I., Kakani, E.G., Kamali, M., Kemppainen, P., Kennedy, R.C., Kirmitzoglou, I.K., Koekemoer, L.L., Laban, N., Langridge, N., Lawniczak, M.K.N., Lirakis, M., Lobo, N.F., Lowy, E., MacCallum, R.M., Mao, C., Maslen, G., Mbogo, C., McCarthy, J., Michel, K., Mitchell, S.N., Moore, W., Murphy, K.A., Naumenko, A.N., Nolan, T., Novoa, E.M., O'Loughlin, S., Oringanje, C., Oshaghi, M.A., Pakpour, N., Papathanos, P.A., Peery, A.N., Povelones, M., Prakash, A., Price, D.P., Rajaraman, A., Reimer, L.J., Rinker, D.C., Rokas, A., Russell, T.L., Sagnon, N., Sharakhova, M.V., Shea, T., Simão, F.A., Simard, F., Slotman, M.A., Somboon, P., Stegniy, V., Struchiner, C.J., Thomas, G.W.C., Tojo, M., Topalis, P., Tubio, J.M.C., Unger, M.F., Vontas, J., Walton, C., Wilding, C.S., Willis, J.H., Wu, Y.-C., Yan, G., Zdobnov, E.M., Zhou, X., Catteruccia, F., Christophides, G.K., Collins, F.H., Cornman, R.S., Crisanti, A., Donnelly, M.J., Emrich, S.J., Fontaine, M.C., Gelbart, W., Hahn, M.W., Hansen, I.A., Howell, P.I., Kafatos, F.C., Kellis, M., Lawson, D., Louis, C., Luckhart, S., Muskavitch, M.A.T., Ribeiro, J.M., Riehle, M.A., Sharakhov, I.V., Tu, Z., Zwiebel, L.J., Besansky, N.J., 2015. Highly evolvable malaria vectors: the genomes of 16 *Anopheles* mosquitoes. Science 347, 1258522.

Niaré, O., Markianos, K., Volz, J., Oduol, F., Touré, A., Bagayoko, M., Sangaré, D., Traoré, S.F., Wang, R., Blass, C., Dolo, G., Bouaré, M., Kafatos, F.C.C., Kruglyak, L., Touré, Y.T., Vernick, K.D.D., 2002. Genetic loci affecting resistance to human malaria parasites in a West African mosquito vector population. Science 298, 213–216.

Oliveira, G.A., Lieberman, J., Barillas-Mury, C., 2012. Epithelial nitration by a peroxidase/NOX5 system mediates mosquito antiplasmodial immunity. Science 335, 856–859.

Osta, M.A., Christophides, G.K., Kafatos, F.C., 2004a. Effects of mosquito genes on *Plasmodium* development. Science 303, 2030–2032.

Osta, M.A., Christophides, G.K., Vlachou, D., Kafatos, F.C., 2004b. Innate immunity in the malaria vector *Anopheles gambiae*: comparative and functional genomics. J. Exp. Biol. 207, 2551–2563.

Paquette, N., Broemer, M., Aggarwal, K., Chen, L., Husson, M., Ertürk-Hasdemir, D., Reichhart, J.M., Meier, P., Silverman, N., 2010. Caspase-mediated cleavage, IAP binding, and ubiquitination: linking three mechanisms crucial for *Drosophila* NF-κB signaling. Mol. Cell 37, 172–182.

Paradkar, P.N., Trinidad, L., Voysey, R., Duchemin, J.B., Walker, P.J., 2012. Secreted Vago restricts West Nile virus infection in *Culex* mosquito cells by activating the Jak-STAT pathway. Proc. Natl. Acad. Sci. U.S.A. 109, 18915–18920.

Paskewitz, S.M., Andreev, O., Shi, L., 2006. Gene silencing of serine proteases affects melanization of Sephadex beads in *Anopheles gambiae*. Insect Biochem. Mol. Biol. 36, 701–711.

Pietri, J.E., Pietri, E.J., Potts, R., Riehle, M.A., Luckhart, S., 2015. *Plasmodium falciparum* suppresses the host immune response by inducing the synthesis of insulin-like peptides (ILPs) in the mosquito *Anopheles stephensi*. Dev. Comp. Immunol. 53, 134–144.

Pike, A., Vadlamani, A., Sandiford, S.L., Gacita, A., Dimopoulos, G., 2014. Characterization of the Rel2-regulated transcriptome and proteome of *Anopheles stephensi* identifies new anti-*Plasmodium* factors. Insect Biochem. Mol. Biol. 52, 82–93.

Pinto, S.B., Kafatos, F.C., Michel, K., 2008. The parasite invasion marker SRPN6 reduces sporozoite numbers in salivary glands of *Anopheles gambiae*. Cell. Microbiol. 10, 891–898.

Povelones, M., Waterhouse, R.M., Kafatos, F.C., Christophides, G.K., 2009. Leucine-rich repeat protein complex activates mosquito complement in defense against *Plasmodium* parasites. Science 324, 258–261.

Povelones, M., Bhagavatula, L., Yassine, H., Tan, L.A., Upton, L.M., Osta, M.A., Christophides, G.K., 2013. The CLIP-domain serine protease homolog SPCLIP1 regulates complement recruitment to microbial surfaces in the malaria mosquito *Anopheles gambiae*. PLoS Pathog. 9, e1003623.

Ramirez, J.L., Garver, L.S., Brayner, F.A., Alves, L.C., Rodrigues, J., Molina-Cruz, A., Barillas-Mury, C., 2014. The role of hemocytes in *Anopheles gambiae* antiplasmodial immunity. J. Innate Immun. 6, 119–128.

Ramphul, U.N., Garver, L.S., Molina-Cruz, A., Canepa, G.E., Barillas-Mury, C., 2015. *Plasmodium falciparum* evades mosquito immunity by disrupting JNK-mediated apoptosis of invaded midgut cells. Proc. Natl. Acad. Sci. 112, 1273–1280.

Richards, A.G., Richards, P.A., 1977. The peritrophic membranes of insects. Annu. Rev. Entomol. 22, 219–240.

Riehle, M.M., Markianos, K., Niaré, O., Xu, J., Li, J., Touré, A.M., Podiougou, B., Oduol, F., Diawara, S., Diallo, M., Coulibaly, B., Ouatara, A., Kruglyak, L., Traoré, S.F., Vernick, K.D., 2006. Natural malaria infection in *Anopheles gambiae* is regulated by a single genomic control region. Science 312, 577–579.

Riehle, M.M., Xu, J., Lazzaro, B.P., Rottschaefer, S.M., Coulibaly, B., Sacko, M., Niare, O., Morlais, I., Traore, S.F., Vernick, K.D., 2008. *Anopheles gambiae* APL1 is a family of variable LRR proteins required for Rel1-mediated protection from themalaria parasite, *Plasmodium berghei*. PLoS One 3, e3672.

Rodrigues, J., Oliveira, G.A., Kotsyfakis, M., Dixit, R., Molina-Cruz, A., Jochim, R., Barillas-Mury, C., 2012. An epithelial serine protease, AgESP, is required for *Plasmodium* invasion in the mosquito *Anopheles gambiae*. PLoS One 7, e35210.

Rosenberg, R., Rungsiwongse, J., 1991. The number of sporozoites produced by individual malaria oocysts. Am. J. Trop. Med. Hyg. 45, 574–577.

Rosinski-Chupin, I., Briolay, J., Brouilly, P., Perrot, S., Gomez, S.M., Chertemps, T., Roth, C.W., Keime, C., Gandrillon, O., Couble, P., Brey, P.T., 2007. SAGE analysis of mosquito salivary gland transcriptomes during *Plasmodium* invasion. Cell. Microbiol. 9, 708–724.

Ross, R., Smyth, J., 1897. On some peculiar pigmented cells found in two mosquitoes fed on malarial blood. Indian J. Malariol. 34, 47.

Ross, R., 1898. Report on the cultivation of proteosoma, Labbe, in Grey mosquitos. Ind. Med. Gaz. 401–408.

Ross, R., 1899. Inaugural lecture on the possibility of extirpating malaria from certain localities by a new method. Br. Med. J. 2, 1–4.

Ryu, J.-H., Kim, S.-H., Lee, H.-Y., Bai, J.Y., Nam, Y.-D., Bae, J.-W., Lee, D.G., Shin, S.C., Ha, E.-M., Lee, W.-J., 2008. Innate immune homeostasis by the homeobox gene *Caudal* and commensal-gut mutualism in *Drosophila*. Science 319, 777–782.

Saraiva, R.G., Kang, S., Simões, M.L., Angleró-Rodríguez, Y.I., Dimopoulos, G., 2016. Mosquito gut antiparasitic and antiviral immunity. Dev. Comp. Immunol. 64, 53–64.

Schnitger, A.K.D., Kafatos, F.C., Osta, M.A., 2007. The melanization reaction is not required for survival of *Anopheles gambiae* mosquitoes after bacterial infections. J. Biol. Chem. 282, 21884–21888.

Schwartz, A., Koella, J.C., 2002. Melanization of *Plasmodium falciparum* and C-25 Sephadex beads by field-caught *Anopheles gambiae* (Diptera: Culicidae) from Southern Tanzania. J. Med. Entomol. 39, 84–88.

Shahabuddin, M., Cociancich, S., Zieler, H., 1998. The search for novel malaria transmission-blocking targets in the mosquito midgut. Parasitol. Today 14, 493–497.

Shin, S.W., Kokoza, V., Ahmed, A., Raikhel, A.S., 2002. Characterization of three alternatively spliced isoforms of the Rel/NF-κB transcription factor Relish from the mosquito *Aedes aegypti*. Proc. Natl. Acad. Sci. U.S.A. 99, 9978–9983.

Shin, S.W., Bian, G., Raikhel, A.S., 2006. A toll receptor and a cytokine, Toll5A and Spz1C, are involved in toll antifungal immune signaling in the mosquito *Aedes aegypti*. J. Biol. Chem. 281, 39388–39395.

Silverman, N., Zhou, R., Stöven, S., Pandey, N., Hultmark, D., Maniatis, T., 2000. A *Drosophila* IκB kinase complex required for relish cleavage and antibacterial immunity. Genes Dev. 14, 2461–2471.

Simões, M.L., Dong, Y., Hammond, A., Hall, A., Crisanti, A., Nolan, T., Dimopoulos, G., 2016. The *Anopheles* FBN9 immune factor mediates *Plasmodium* species-specific defense through transgenic fat body expression. Dev. Comp. Immunol. 67, 257–265.

Simon, N., Lasonder, E., Scheuermayer, M., Kuehn, A., Tews, S., Fischer, R., Zipfel, P.F., Skerka, C., Pradel, G., 2013. Malaria parasites co-opt human factor H to prevent complement-mediated lysis in the mosquito midgut. Cell Host Microbe 13, 29–41.

Sinden, R.E., Garnham, P.C.C., 1973. A comparative study on the ultrastructure of *Plasmodium* sporozoites within the oocyst and salivary glands, with particular reference to the incidence of the micropore. Trans. R. Soc. Trop. Med. Hyg. 67, 631–637.

Smidler, A.L., Terenzi, O., Soichot, J., Levashina, E.A., Marois, E., 2013. Targeted mutagenesis in the malaria mosquito using TALE nucleases. PLoS One 8, e74511.

Smith, R.C., Barillas-Mury, C., 2016. *Plasmodium* oocysts: overlooked targets of mosquito immunity. Trends Parasitol. 32, 979–990.

Smith, R.C., Vega-Rodríguez, J., Jacobs-Lorena, M., 2014. The *Plasmodium* bottleneck: malaria parasite losses in the mosquito vector. Mem. Inst. Oswaldo Cruz 109, 644–661.

Smith, R.C., Barillas-mury, C., Jacobs-Lorena, M., 2015. Hemocyte differentiation mediates the mosquito late-phase immune response against *Plasmodium* in *Anopheles gambiae*. Proc. Natl. Acad. Sci. U.S.A. 112, E3412–E3420.

Souza-Neto, J.A., Sim, S., Dimopoulos, G., 2009. An evolutionary conserved function of the JAK-STAT pathway in anti-dengue defense. Proc. Natl. Acad. Sci. U.S.A. 106, 17841–17846.

Steelman, L.S., Chappell, W.H., Abrams, S.L., Kempf, R.C., Long, J., Laidler, P., Mijatovic, S., Maksimovic-Ivanic, D., Stivala, F., Mazzarino, M.C., Donia, M., Fagone, P., Malaponte, G., Nicoletti, F., Libra, M., Milella, M., Tafuri, A., Bonati, A., Bäsecke, J., Cocco, L., Evangelisti, C., Martelli, A.M., Montalto, G., Cervello, M., McCubrey, J.A., 2011. Roles of the Raf/MEK/ERK and PI3K/PTEN/Akt/mTOR pathways in controlling growth and sensitivity to therapy-implications for cancer and aging. Aging (Albany, NY) 3, 192–222.

Stoven, S., Silverman, N., Junell, A., Hedengren-Olcott, M., Erturk, D., Engstrom, Y., Maniatis, T., Hultmark, D., 2003. Caspase-mediated processing of the *Drosophila* NF-κB factor Relish. Proc. Natl. Acad. Sci. U.S.A. 100, 5991–5996.

Trager, W., 1942. A strain of the mosquito *Aedes aegypti* selected for susceptibility to the avian malaria parasite *Plasmodium lophurae*. J. Parasitol. 28, 457–465.

Tripet, F., 2009. Ecological immunology of mosquito-malaria interactions: of non-natural versus natural model systems and their inferences. Parasitology 136, 1935–1942.

Tsai, Y.L., Hayward, R.E., Langer, R.C., Fidock, D.A., Vinetz, J.M., 2001. Disruption of *Plasmodium falciparum* chitinase markedly impairs parasite invasion of mosquito midgut. Infect. Immun. 69, 4048–4054.

Valanne, S., Wang, J.-H., Rämet, M., 2011. The *Drosophila* Toll signaling pathway. J. Immunol. 186, 649–656.

Valentino, L., Pierre, J., 2006. JAK/STAT signal transduction: regulators and implication in hematological malignancies. Biochem. Pharmacol. 71, 713–721.

Vaughan, J.A., Hensley, L., Beier, J.C., 1994. Sporogonic development of *Plasmodium yoelii* in five anopheline species. J. Parasitol. 80, 674–681.

Vernick, K.D., Fujioka, H., Seeley, D.C., Tandler, B., Aikawa, M., Miller, L.H., 1995. *Plasmodium gallinaceum*: a refractory mechanism of ookinete killing in the mosquito, *Anopheles gambiae*. Exp. Parasitol. 80, 583–595.

Viljakainen, L., 2015. Evolutionary genetics of insect innate immunity. Brief. Funct. Genomics 14, 407–412.

Volz, J., Osta, M.A., Kafatos, F.C., Müller, H.-M., 2005. The roles of two clip domain serine proteases in innate immune responses of the malaria vector *Anopheles gambiae*. J. Biol. Chem. 280, 40161–40168.

Volz, J., Müller, H.M., Zdanowicz, A., Kafatos, F.C., Osta, M.A., 2006. A genetic module regulates the melanization response of *Anopheles* to *Plasmodium*. Cell. Microbiol. 8, 1392–1405.

Wang, B., Pakpour, N., Napoli, E., Drexler, A., Glennon, E.K.K., Surachetpong, W., Cheung, K., Aguirre, A., Klyver, J.M., Lewis, E.E., Eigenheer, R., Phinny, B.S., Giulivi, C., Luckhart, S., 2015. *Anopheles stephensi* p38 MAPK signaling regulates innate immunity and bioenergetics during *Plasmodium falciparum* infection. Parasit. Vectors 8, 424.

Warr, E., Lambrechts, L., Koella, J.C., Bourgouin, C., Dimopoulos, G., 2006. *Anopheles gambiae* immune responses to Sephadex beads: involvement of anti-*Plasmodium* factors in regulating melanization. Insect Biochem. Mol. Biol. 36, 769–778.

Warren, W.D., Atkinson, P.W., O'Brochta, D.A., 1994. The hermes transposable element from the house fly, *Musca domestica*, is a short inverted repeat-type element of the hobo, Ac, and Tam3 (hAT) element family. Genet. Res. 64, 87–97.

Waterhouse, R.M., Kriventseva, E.V., Meister, S., Xi, Z., Alvarez, K.S., Bartholomay, L.C., Barillas-Mury, C., Bian, G., Blandin, S., Christensen, B.M., Dong, Y., Jiang, H., Kanost, M.R., Koutsos, A.C., Levashina, E.A., Li, J., Ligoxygakis, P., Maccallum, R.M., Mayhew, G.F., Mendes, A., Michel, K., Osta, M.a, Paskewitz, S., Shin, S.W., Vlachou, D., Wang, L., Wei, W., Zheng, L., Zou, Z., Severson, D.W., Raikhel, A.S., Kafatos, F.C., Dimopoulos, G., Zdobnov, E.M., Christophides, G.K., 2007. Evolutionary dynamics of immune-related genes and pathways in disease-vector mosquitoes. Science 316, 1738–1743.

Whitten, M.M.A., Shiao, S.H., Levashina, E.A., 2006. Mosquito midguts and malaria: cell biology, compartmentalization and immunology. Parasite Immunol. 28, 121–130.

Wojtowicz, W.M., Flanagan, J.J., Millard, S.S., Zipursky, S.L., Clemens, J.C., 2004. Alternative splicing of *Drosophila* Dscam generates axon guidance teceptors that exhibit isoform-specific homophilic binding. Cell 118, 619–633.

Yassine, H., Kamareddine, L., Osta, M.A., 2012. The mosquito melanization response is implicated in defense against the entomopathogenic fungus *Beauveria bassiana*. PLoS Pathog. 8, e1003029.

Yassine, H., Kamareddine, L., Chamat, S., Christophides, G.K., Osta, M.A., 2014. A serine protease homolog negatively regulates TEP1 consumption in systemic infections of the malaria vector *Anopheles gambiae*. J. Innate Immun. 6, 806–818.

Zarubin, T., Han, J., 2005. Activation and signaling of the p38 MAP kinase pathway. Cell Res. 15, 11–18.

Zeidler, M.P., Bausek, N., 2013. The *Drosophila* JAK-STAT pathway. JAK-STAT 2, e25353.

Zeke, A., Misheva, M., Reményi, A., Bogoyevitch, M.A., 2016. JNK signaling: regulation and functions based on complex protein-protein partnerships. Microbiol. Mol. Biol. Rev. 80, 793–835.

Zhang, G., Niu, G., Franca, C.M., Dong, Y., Wang, X., Butler, N.S., Dimopoulos, G., Li, J., 2015. *Anopheles* midgut FREP1 mediates *Plasmodium* invasion. J. Biol. Chem. 290, 16490–16501.

Zhao, H.-F., Wang, J., Tony To, S.-S., 2015. The phosphatidylinositol 3-kinase/Akt and c-Jun N-terminal kinase signaling in cancer: alliance or contradiction? (Review). Int. J. Oncol. 47, 429–436.

Zheng, L., Cornel, A.J., Wang, R., Erfle, H., Voss, H., Ansorge, W., Kafatos, F.C., Collins, F.H., 1997. Quantitative trait loci for refractoriness of *Anopheles gambiae* to *Plasmodium cynomolgi* B. Science 276, 425–428.

Zhou, R., Silverman, N., Hong, M., Liao, D.S., Chung, Y., Chen, Z.J., Maniatis, T., 2005. The role of ubiquitination in *Drosophila* innate immunity. J. Biol. Chem. 280, 34048–34055.

Zieler, H., Dvorak, J.A., 2000. Invasion in vitro of mosquito midgut cells by the malaria parasite proceeds by a conserved mechanism and results in death of the invaded midgut cells. Proc. Natl. Acad. Sci. U.S.A. 97, 11516–11521.

Zou, Z., Souza-Neto, J., Xi, Z., Kokoza, V., Shin, S.W., Dimopoulos, G., Raikhel, A., 2011. Transcriptome analysis of *Aedes aegypti* transgenic mosquitoes with altered immunity. PLoS Pathog. 7, e1002394.

CHAPTER

5

Molecular Mechanisms Mediating Immune Priming in *Anopheles gambiae* Mosquitoes

Jose L. Ramirez[1,2], *Ana Beatriz F. Barletta*[1], *Carolina V. Barillas-Mury*[1]

[1]**National Institutes of Health, Rockville, MD, United States;** [2]**U.S. Department of Agriculture, Peoria, IL, United States**

INTRODUCTION

Vertebrates have the ability to "learn" from a previous encounter with a pathogen and this allows them to mount highly specific and effective immune responses to subsequent infections (Hirano et al., 2011; Rivera et al., 2016). The dogma for many years had been that these hallmarks of adaptive immunity were not present in invertebrates. However, recent studies have shown that insects are capable of mounting effective memory-like responses following an immune challenge with eukaryotic parasites, bacteria, or viruses (Moret and Siva-Jothy, 2003; Tidbury et al., 2011; Mikonranta et al., 2014; Rodrigues et al., 2010; Moret and Schmid-Hempel, 2001; Pham et al., 2007).

Some of these protective responses are nonspecific. For instance, challenge of the mealworm beetle, *Tenebrio molitor*, with lipopolysaccharides (LPS) provided a long-lasting antimicrobial response against the entomopathogenic fungus *Metarhizium anisopliae* (Moret and Siva-Jothy, 2003). However, there is also evidence of highly specific immune priming. For example, exposure of the red flour beetle to different bacterial species provided strain-specific protection to subsequent infections (Roth et al., 2009), and challenge of the bumble bee (*Bombus terrestris*) with several bacterial strains showed highly specific protection several weeks following the initial challenge (Sadd and Schmid-Hempel, 2006). Within the Diptera, exposure of the fruit fly (*Drosophila melanogaster*) to sublethal doses of *Streptococcus pneumoniae* protected them from subsequent lethal doses with the same pathogen, an effect that was found to be specific and long-lasting (Pham et al., 2007).

Recent advances have been made in understanding the molecular mechanisms behind memory-like response in the malaria vector *Anopheles gambiae* following infection with

Arthropod Vector: Controller of Disease Transmission, Volume 1
http://dx.doi.org/10.1016/B978-0-12-805350-8.00005-2

Plasmodium parasites. In *A. gambiae*, immune priming is the result of complex interaction between the parasite, the gut microbiota, and the mosquito vector, involving a range of bioactive molecules and immune signaling pathways that allow the mosquito to respond more effectively to subsequent infections (Rodrigues et al., 2010; Ramirez et al., 2014, 2015). This chapter details our current understanding of the molecular mechanisms that mediate immune priming in *A. gambiae* mosquitoes in response to *Plasmodium* infection.

ESSENTIAL COMPONENTS IN THE ESTABLISHMENT OF IMMUNE MEMORY

The *Plasmodium* parasite undergoes a complex life cycle in the human and mosquito hosts. In the mosquito, *Plasmodium* gametocytes fuse soon after ingestion of the blood meal, forming a motile ookinete. The ookinete invades the midgut epithelium and develops into an oocyst. The parasite will multiply continuously and successful oocysts will eventually rupture, releasing thousands of sporozoites that invade the mosquito salivary glands, allowing for transmission when the mosquito takes the next blood meal (Vlachou et al., 2006; Angrisano et al., 2012).

In an elegant study, Rodrigues et al. (2010) investigated whether preexposure of mosquitoes to *Plasmodium* enhanced their immune response to subsequent infections with the same parasite. To address this, two groups of mosquitoes were provided with the same infectious blood meal. In one group, parasites were allowed to develop and invade the mosquito midgut while the other group was placed at a nonpermissive temperature that prevented infection. Seven days postexposure, both groups of mosquitoes were allowed to feed on another infectious blood meal. This time, both groups were maintained at conditions that were permissive to parasite development in the mosquito midgut. They observed a significant and long-lasting reduction in the intensity of infection in the mosquitoes that had been preexposed to the malaria parasite, indicative of heightened immunity. Several essential components in the establishment of immune priming were identified, including the mosquito gut microbiota and circulating hemocytes.

Gut Bacteria and the Establishment of Immune Priming

Bacteria and other microorganisms are prevalent members of the mosquito gut microbiome, and their effect on digestion, immunity, and pathogen protection continues to be documented. The mosquito gut bacterial microbiota increases 1000-fold following a blood meal (Oliveira et al., 2011). The blood meal is surrounded by a chitinous membrane known as the peritrophic matrix. This thin membrane is formed soon after blood is ingested, and is thought to provide protection from abrasion and to contain any microorganisms that might have been ingested. Hence, the peritrophic matrix prevents the gut microbiota from coming in direct contact with midgut epithelial cells and constitutes one of the first barriers that the *Plasmodium* parasite must traverse.

Ookinetes disrupt the peritrophic matrix as they invade the midgut, allowing bacteria to come in contact with the midgut, and this is a critical step for the establishment of immune priming. Elimination of the mosquito gut flora, via oral administration of antibiotics before the first infection, prevents the establishment of immune priming. The microbiota must also be present during the second infection to elicit enhanced immunity in mosquitoes that had been challenged with a previous infection. Further assays solidified this assertion, as the increase in granulocytes and the transcriptional activation of antimicrobial effector genes were lost when mosquitoes were devoid of bacteria prior to *Plasmodium* challenge. Furthermore, Rodrigues et al. (2010) showed that immune priming is a systemic response that also affects the mosquito gut microbiota, as challenged mosquitoes harbored significantly lower bacterial levels 24h after the second *Plasmodium* infection than naïve mosquitoes.

If the normal barriers are intact, proliferation of bacteria in the midgut lumen following a blood meal does not elicit priming, nor does disruption of the barriers following *Plasmodium* invasion in the absence of bacteria. Taken together, these findings indicate that the direct contact of gut bacteria with midgut cells is the signal that triggers the establishment and elicitation of the priming response and that the effect of priming on *Plasmodium* infection is due to an indirect bystander effect.

Hemocytes as Key Mediators in Early- and Late-phase Antiplasmodial Immunity

Hemocytes are immune effector cells that participate in cellular defenses such as phagocytosis, encapsulation, and clotting. In addition, they mediate systemic immune responses (Lavine and Strand, 2002; Strand, 2008) and are actively involved in synthesizing antimicrobial peptides following an immune challenge. These immune cells circulate in the mosquito hemocoel and are critical mediators of innate immune memory.

Distinct types of hemocytes have been described in several insect species based on their morphology and function. In mosquitoes, three major types of hemocytes have been identified: prohemocytes, granulocytes, and oenocytoids (Strand, 2008; Castillo et al., 2006) (Fig. 5.1A). It is believed that the prohemocytes are progenitor cells that give rise to granulocytes and oenocytoids (Strand, 2008); granulocytes are strongly adhesive professional phagocytes with abundant granules in their cytoplasm (Strand, 2008; Castillo et al., 2006); and oenocytoids are round cells that are less phagocytic and are characterized by their constitutive expression of components of the phenoloxidase cascade (Castillo et al., 2006).

Important hemocyte-mediated antimalarial immune responses have been recognized in the mosquito *A. gambiae*. The thioester-containing protein 1 (Tep1) is one of the most critical anti-*Plasmodium* effectors. Tep1 is a complement-like molecule that binds to the surface of ookinetes as they emerge from the midgut and come in contact with the mosquito hemolymph and marks them for subsequent lysis or melanization. It has been shown that Tep1 is synthesized by hemocytes and is secreted into the hemolymph, where it circulates as a stable complex with two leucine-rich proteins, the leucine-rich repeat protein 1 and the *Anopheles Plasmodium*–responsive leucine-rich repeat protein 1 (Blandin et al., 2004; Povelones et al., 2011).

Plasmodium infection induces the synthesis and secretion of effector molecules by hemocytes (Frolet et al., 2006; Müller et al., 1999) and results in functional differences, such as changes in lectin binding properties on the surface of granulocytes and enhanced association of hemocytes with the midgut in response to subsequent infections, that appear to be essential to reduce parasite burden (Rodrigues et al., 2010; Ramirez et al., 2014; Baton et al., 2009).

One of the hallmarks of immune priming in *A. gambiae* is the drastic increase in the proportion of circulating granulocytes. This is a long-lasting response, as high granulocytes levels are maintained for up to 14days postinfection (Rodrigues et al., 2010; Ramirez et al., 2014). This is concomitant with a decrease in the proportion of prohemocytes, suggesting that priming leads to hemocyte differentiation of prohemocytes to granulocytes (Rodrigues et al., 2010). This modulation of the hemocyte population, with an increase in granulocyte, is essential for enhanced antiplasmodial immunity, as disruption or depletion of granulocytes by Sephadex

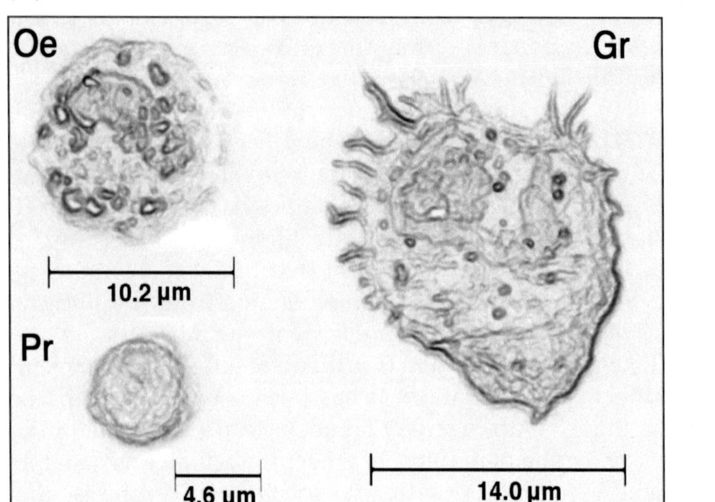

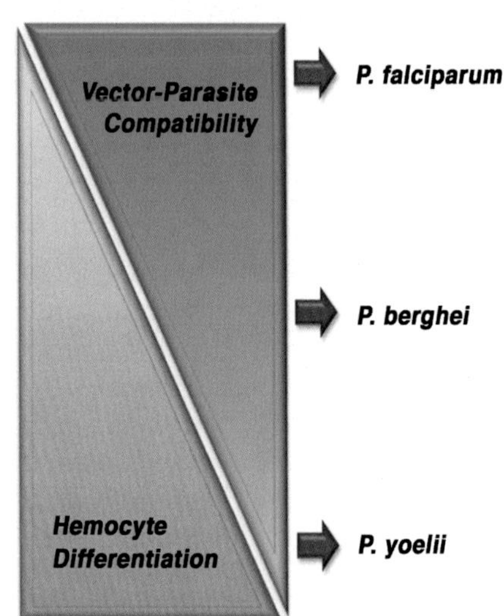

FIGURE 5.1 Hemocytes are critical components of the mosquito immune priming response. (A) Three distinct hemocyte populations are observed in the mosquito *Anopheles gambiae*: prohemocytes (Pr), granulocytes (Gr), and oenocytoids (Oe). (B) The magnitude of the hemocyte differentiation response is also dependent on the degree of vector–parasite compatibility. In the most compatible system, *A. gambiae–Plasmodium falciparum*, a mild hemocyte differentiation response is observed while a stronger response is exhibited in the least compatible system, *A. gambiae–Plasmodium yoelii*.

beads injection abrogates the priming phenotype (Rodrigues et al., 2010).

While complement activation and deposition targets early stages of infection, hemocytes also play important functions at later stages and limit oocyst survival. For instance, an LPS-induced TNFα transcription factor–like 3 (LL3) is essential for hemocyte differentiation mediating late-phase immune responses to *Plasmodium* infection (Smith et al., 2012, 2015). Although LL3 is constitutively expressed in hemocytes, it is also highly induced in the midgut in response to ookinete invasion, in a mechanism that is independent of the gut microbiota. Interestingly, LL3 antiplasmodial effects are not exerted at the ookinete stage, but limit oocyst survival and maturation (Smith et al., 2015). These studies underscore that hemocyte differentiation following a priming response also affects later stages of parasites from the initial infection.

MOSQUITO–PARASITE COMPATIBILITY AND THE STRENGTH OF THE PRIMING RESPONSE

Although bacteria trigger immune priming as they come in contact with injured midgut cells, it appears that vector–parasite compatibility also modulates the strength of the priming response (Ramirez et al., 2014). In assays in which *A. gambiae* mosquitoes were infected with three *Plasmodium* species, the strongest response was observed with the least compatible combination (*A. gambiae–Plasmodium yoelli*), followed by the intermediate phenotype (*A. gambiae–Plasmodium berghei*) and the mildest response with the most compatible combination (*A. gambiae–Plasmodium falciparum*) (Ramirez et al., 2014) (Fig. 5.1B). This variation in the magnitude of the priming response in diverse vector–parasite compatibilities could result from differences in the level of activation of the complement-like system (Povelones et al., 2011; Molina-Cruz et al., 2012, 2013). For example, some *P. falciparum* parasites evade the mosquito complement-like system by preventing the induction of two enzymes, NOX5 and HPX2 (Molina-Cruz et al., 2013; Ramphul et al., 2015), that potentiate midgut epithelial nitration in response to ookinete invasion, a reaction that promotes complement activation (Oliveira et al., 2012). This may explain the reduced immune priming response observed in the *A. gambiae–P. falciparum* combination.

MOLECULAR FACTORS MEDIATING THE ESTABLISHMENT AND MAINTENANCE OF INNATE IMMUNE PRIMING

Immune priming triggers the production of a soluble factor in the hemolymph, and transfer of cell-free hemolymph from challenged females also triggers the priming response in naïve recipient mosquitoes (Rodrigues et al., 2010; Ramirez et al., 2014). This soluble factor, coined hemocyte differentiation factor (HDF), is synthesized soon after parasite infection and persists for the rest of the life of challenged mosquitoes (Rodrigues et al., 2010; Ramirez et al., 2014). A series of assays to explore the biochemical nature of HDF revealed that it was a very stable compound with ambivalent properties in terms of hydrophobicity (Ramirez et al., 2015). Subsequent biochemical purification from mosquito hemolymph using hydrophilic interaction liquid chromatography (HILIC) confirmed that HDF is a protein/lipid complex.

Evokin, a Bioactive Lipid Carrier, Is Critical in Immune Priming

Hemolymph extracts from *Plasmodium*-infected mosquitoes were fractionated using HILIC and assayed for hemocyte differentiation activity by injection into naïve mosquitoes. Mass spectrometry analysis identified a protein with homology to vertebrate lipocalins in the fraction with the highest biological activity. Expression of this ApoD-like lipocalin protein, named Evokin, was shown to be induced in the fat body of infected mosquitoes at 24h postinfection and to remain high for up to 7days postinfection (Ramirez et al., 2015). Evokin is an essential component of HDF, as RNAi-mediated silencing of this lipocalin abolished HDF activity in hemolymph extract.

Lipocalins are known to be protein carriers of small hydrophobic molecules such as steroids, retinoids, and other bioactive lipids (Ruiz et al., 2012, 2013; Morais Cabral et al., 1995; Flower, 1996). It has been suggested that lipocalins aid in providing target specificity to their ligand, protect them from oxidative degradation, and regulate their release from the site of synthesis (Flower, 1996). It was thus quite plausible that Evokin was serving similar function as a lipid carrier, given the high stability and the long-lasting effect of HDF in the mosquito hemolymph. A novel strategy was developed to identify the lipid cargo of Evokin.

Eicosanoids in Hemocyte Differentiation and Activation

Lipid extraction followed by lipid mediator lipidomics (LC/MS/MS analysis) of HILIC-purified Evokin uncovered the lipid ligand of Evokin as an eicosanoid, with similar mass spectrum signatures as vertebrate lipoxins (Ramirez et al., 2015). Comparative lipid mediator metabololipidomics between hemolymph from challenged and naïve mosquitoes that had been injected with deuterium (D8) labeled arachidonic acid unequivocally established that challenged mosquitoes synthesize significantly higher levels of Lipoxin A4 than naïve controls (Ramirez et al., 2015). Arachidonic acid is a well-established substrate for synthesis of all eicosanoids, but lipoxygenases are not present in the insect genomes, and the biochemical pathway(s) mediating Lipoxin A4 synthesis remain to be defined (Stanley-Samuelson et al., 1991; Stanley-Samuelson and Dadd, 1983; Stanley et al., 2012; Dean et al., 2002). The metabololipidomic results demonstrated that mosquitoes can actively use arachidonic acid as a substrate to synthesize eicosanoids, more precisely lipoxins, which are bioactive lipids not previously known to be synthesized by insects (Ramirez et al., 2015).

The biological activity of chemically synthesized lipoxins as final mediators of immune priming was also evaluated. These assays further corroborated the lipidomics findings, as mosquitoes that received lipoxin exhibited elevated numbers of granulocytes and enhanced antiplasmodial immune response. The interaction between Evokin and lipoxins was also confirmed, as the biological activity of synthetic lipoxins was abrogated in the absence of Evokin in recipient mosquitoes, suggesting a protective role for this lipid carrier (Ramirez et al., 2015). In essence, synthetic lipoxins recapitulate HDF activity and require Evokin as a carrier.

The Role of Immune Signaling Pathways in the Establishment of Immune Priming

The *Toll*, *IMD*, *JNK*, and *Jak-STAT* pathways are critical in the regulation of antiplasmodial immune responses (Cirimotich et al., 2010; Dong et al., 2006; Garver et al., 2009, 2012; Gupta et al., 2009; Jaramillo-Gutierrez et al., 2009) and have been previously implicated in hemocyte differentiation and proliferation in *Drosophila* (Agaisse et al., 2003; Owusu-Ansah and Banerjee, 2009). Functional assays to elucidate the participation of these signaling cascades in HDF synthesis and hemocyte responses to HDF highlighted the complexity of the system. While none of these four immune signaling pathways were essential for HDF synthesis, the *Toll*, *JNK*, and *Jak-STAT* pathways are critical for hemocyte differentiation in response to HDF (Ramirez et al., 2014).

The Effect of Priming on Vectorial Capacity

The effect of immune priming on malaria transmission depends, to some extent, on the frequency and propensity for mosquitoes to ingest an infectious second blood meal. Field data have shown that *Anopheles* females can blood feed at least once every 2–4 days (Gillies, 1953; Mala et al., 2014). In fact, it has been shown that about 22% of the *A. gambiae* field-caught mosquitoes have gone through at least two gonotrophic cycles and 6.3% had survived for four or more gonotrophic cycles (Mala et al., 2014; Charlwood et al., 2000). These multiple feeding events could potentially expose the mosquitoes to multiple infectious blood meals in their lifetime, especially in highly endemic areas where malaria transmission is intense. Hence, a priming mechanism could effectively suppress subsequent infections, preventing a superinfection process in the mosquito that could translate into higher malaria transmission potential. Both the establishment and elicitation of immune priming require the presence of the gut microbiota and disruption of the barriers that prevent direct contact between bacteria in the gut lumen and epithelial cells; thus, it is likely that this mechanism is present as a safeguard against invasive pathogens that the mosquitoes might be exposed to during their life cycle. The mosquito digestive system is in direct contact with many fungi and bacteria during larval states, and exposure to invasive microbes might also trigger a priming response during this developmental stage. This is an area that needs further investigation. One can envision that exposure to invasive pathogens during larval stages could prime a long-lasting response that may persist in the adult stage and reduce the mosquito's vectorial capacity.

CONCLUSIONS AND FUTURE PERSPECTIVES

Establishment of immune priming in *A. gambiae* mosquitoes involves a complex network of interaction between the *Plasmodium* parasite, the mosquito immune system, and the gut microbiota. A previous encounter with the malaria parasite activates the constitutive release of a soluble HDF into the hemolymph that changes the functional state of the mosquito immune system. HDF consists of a lipocalin/lipoxin (Evokin/Lipoxin A4) complex that drives hemocyte differentiation and enhances antiplasmodial immunity (Fig. 5.2). Although strides have been made in understanding immune memory in mosquitoes, many unanswered questions remain: What is the specific role of each immune pathway in hemocyte differentiation? Which immune signaling pathways are driving the differences in immune priming when mosquitoes are challenged with different parasites? Are different molecules driving the observed differences? Is HDF activating these pathways directly in hemocytes and/or in other immune-competent tissues? Are there cytokines that drive differentiation of prohemocytes into granulocytes? Are all bacteria equally capable of priming the immune system? What sustains the constitutive increase in Lipoxin A4 synthesis? The possibility of other mechanisms of immune priming in response

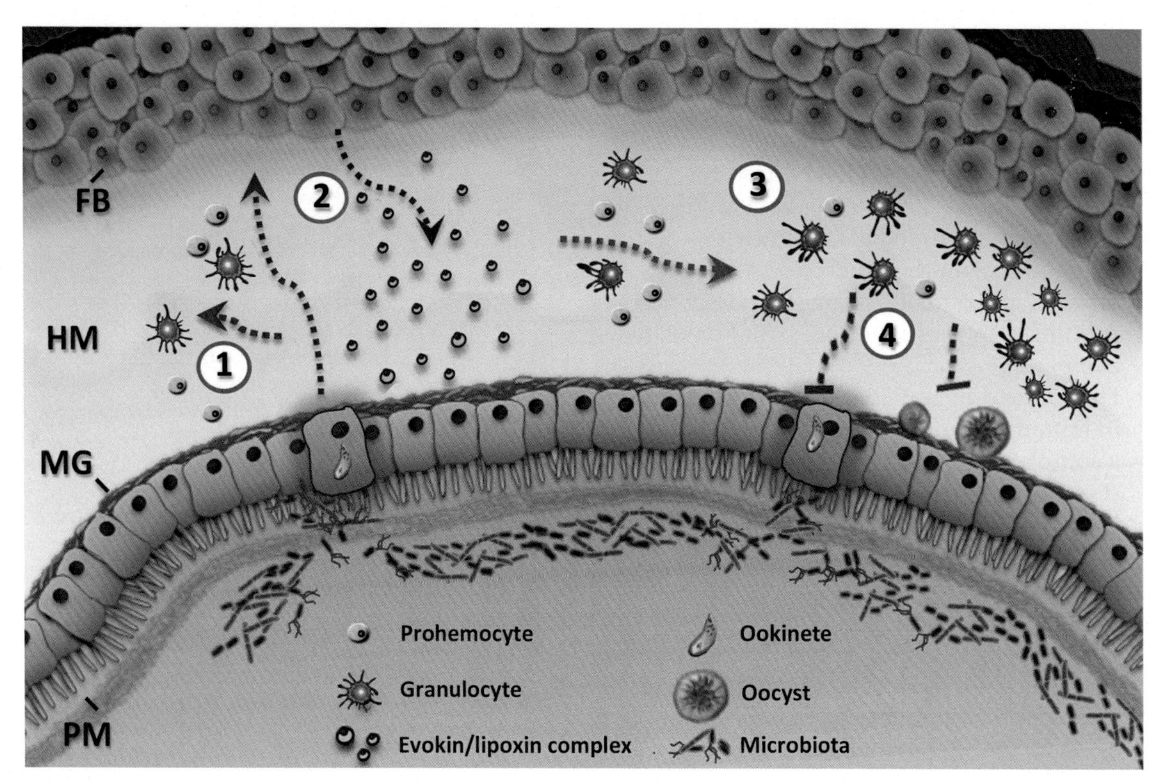

FIGURE 5.2 Immune priming in *Anopheles gambiae*. (1) The *Plasmodium* ookinete disrupts the peritrophic matrix allowing bacteria from the midgut lumen to enter in contact with the injured midgut cells. This stimulates the immune system of the mosquito. (2) The mosquito produces a soluble Evokin/Lipoxin complex with an active participation of the fat body. (3) This soluble bioactive complex induces hemocyte differentiation leading to an increase in granulocyte proportions. (4) Once the priming response is established the mosquito presents an enhanced antiplasmodial immune response that limits subsequent infections. *FB*, fat body; *HM*, hemocoele; *MG*, midgut; *PM*, peritrophic matrix.

to arboviruses, fungi, or other pathogenic bacteria is also open in areas that remain to be explored.

TAKE-HOME MESSAGES

- Insects are capable of mounting effective memory-like responses following an immune challenge. These responses can be nonspecific or highly specific depending on the system and challenge and can last for a long time after the initial exposure to a microorganism.
- In mosquitoes, the gut microbiota and hemocytes are essential components of immune memory, where gut bacteria establish and recall immune priming as they come in direct contact with the mosquito midgut. Hemocytes undergo a differentiation response that increases the proportion of granulocytes, is long-lasting, and is critical to enhance antiplasmodial immunity.
- The initial antiplasmodial response appears to modulate the magnitude of the priming response in mosquito and parasite combinations.
- Hemocyte differentiation is a hallmark event in immune priming that is triggered by HDF, a complex composed by a bioactive lipid (Lipoxin) and its carrier, a lipocalin (Evokin). Long-lasting expression of these two components of HDF is critical to maintain immune memory.
- Besides the classical roles of the immune pathways in mosquito immunity, they are also critical for hemocyte differentiation in response to HDF, but none of them were essential for HDF synthesis.

Acknowledgments

This work was supported by the Intramural Research Program of the Division of Intramural Research, National Institute of Allergy and Infectious Diseases, and National Institutes of Health. J.L Ramirez received funding from the Intramural National Institutes of Allergy and Infectious Diseases Research Opportunities Program, and A.B. Barletta Ferreira from the Science Without Borders/CNPq-National Council of Scientific and Technological Development from Brazil.

References

Agaisse, H., Petersen, U., Boutros, M., Mathey-Prevot, B., Perrimon, N., 2003. Signaling role of hemocytes in *Drosophila* JAK/STAT-dependent response to septic injury. Dev. Cell. 5 (3), 441–450. http://dx.doi.org/10.1016/S1534-5807(03)00244-2. PubMed PMID: 12967563.

Angrisano, F., Tan, Y.-H., Sturm, A., McFadden, G.I., Baum, J., 2012. Malaria parasite colonisation of the mosquito midgut – placing the *Plasmodium ookinete* centre stage. Int. J. Parasitol. 42 (6), 519–527.

Baton, L., Robertson, A., Warr, E., Strand, M., Dimopoulos, G., 2009. Genome-wide transcriptomic profiling of *Anopheles gambiae* hemocytes reveals pathogen-specific signatures upon bacterial challenge and *Plasmodium berghei* infection. BMC Genomics 10 (1), 257. http://dx.doi.org/10.1186/1471-2164-10-257. PubMed PMID: 19500340.

Blandin, S., Shiao, S., Moita, L., Janse, C., Waters, A., Kafatos, F., et al., 2004. Complement-like protein TEP1 is a determinant of vectorial capacity in the malaria vector *Anopheles gambiae*. Cell. 116 (5), 661–670. http://dx.doi.org/10.1016/S0092-8674(04)00173-4. PubMed PMID: 15006349.

Castillo, J.C., Robertson, A.E., Strand, M.R., 2006. Characterization of hemocytes from the mosquitoes *Anopheles gambiae* and *Aedes aegypti*. Insect Biochem. Mol. Biol. 36 (12), 891–903.

Charlwood, J.D., Vij, R., Billingsley, P.F., 2000. Dry season refugia of malaria-transmitting mosquitoes in a dry savannah zone of east Africa. Am. J. Trop. Med. Hyg. 62 (6), 726–732.

Cirimotich, C.M., Dong, Y., Garver, L.S., Sim, S., Dimopoulos, G., 2010. Mosquito immune defenses against *Plasmodium* infection. Dev. Comp. Immunol. 34 (4), 387–395. http://dx.doi.org/10.1016/j.dci.2009.12.005.

Dean, P., Gadsden, J.C., Richards, E.H., Edwards, J.P., Keith Charnley, A., Reynolds, S.E., 2002. Modulation by eicosanoid biosynthesis inhibitors of immune responses by the insect *Manduca sexta* to the pathogenic fungus *Metarhizium anisopliae*. J. Invertebr. Pathol. 79 (2), 93–101. http://dx.doi.org/10.1016/S0022-2011(02)00014-9.

Dong, Y., Aguilar, R., Xi, Z., Warr, E., Mongin, E., Dimopoulos, G., 2006. *Anopheles gambiae* immune responses to human and rodent *Plasmodium* parasite species. PLoS Pathog. 2 (6), e52. http://dx.doi.org/10.1371/journal.ppat.0020052. PubMed PMID: 16789837.

Flower, D.R., 1996. The lipocalin protein family: structure and function. Biochem. J. 318 (1), 1–14.

Frolet, C., Thoma, M., Blandin, S., Hoffmann, J., Levashina, E., 2006. Boosting NF-kappa B-dependent basal immunity of *Anopheles gambiae* aborts development of *Plasmodium berghei*. Immunity 25 (4), 677–685. http://dx.doi.org/10.1016/j.immuni.2006.08.019. PubMed PMID: 17045818.

Garver, L., Dong, Y., Dimopoulos, G., 2009. Caspar controls resistance to *Plasmodium falciparum* in diverse anopheline species. PLoS Pathog. 5 (3), e1000335. http://dx.doi.org/10.1371/journal.ppat.1000335. PubMed PMID: 19282971.

Garver, L.S., Bahia, A.C., Das, S., Souza-Neto, J.A., Shiao, J., Dong, Y., et al., 2012. Anopheles Imd pathway factors and effectors in infection intensity-dependent anti-*Plasmodium* action. PLoS Pathog. 8 (6), e1002737. http://dx.doi.org/10.1371/journal.ppat.1002737.

Gillies, M.T., 1953. The duration of the gonotrophic cycle in *Anopheles gambiae* and *Anopheles funestus*, with a note on the efficiency of hand catching. East Afr. Med. J. 30 (4), 129–135.

Gupta, L., Molina-Cruz, A., Kumar, S., Rodrigues, J., Dixit, R., Zamora, R.E., et al., 2009. The STAT pathway mediates late-phase immunity against *Plasmodium* in the mosquito *Anopheles gambiae*. Cell Host Microbe 5 (5), 498–507.

Hirano, M., Das, S., Guo, P., Cooper, M.D., Frederick, W.A., 2011. The Evolution of Adaptive Immunity in Vertebrates. Advances in Immunology, vol. 109. Academic Press, pp. 125–157 (Chapter 4).

Jaramillo-Gutierrez, G., Rodrigues, J., Ndikuyeze, G., Povelones, M., Molina-Cruz, A., Barillas-Mury, C., 2009. Mosquito immune responses and compatibility between *Plasmodium* parasites and anopheline mosquitoes. BMC Microbiol. 9 (1), 154. http://dx.doi.org/10.1186/1471-2180-9-154. PubMed PMID: 19643026.

Lavine, M.D., Strand, M.R., 2002. Insect hemocytes and their role in immunity. Insect Biochem. Mol. Biol. 32 (10), 1295–1309.

Mala, A.O., Irungu, L.W., Mitaki, E.K., Shililu, J.I., Mbogo, C.M., Njagi, J.K., et al., 2014. Gonotrophic cycle duration, fecundity and parity of *Anopheles gambiae* complex mosquitoes during an extended period of dry weather in a semi arid area in Baringo County, Kenya. Int. J. Mosq. Res. 1 (2), 28–34.

Mikonranta, L., Mappes, J., Kaukoniitty, M., Freitak, D., 2014. Insect immunity: oral exposure to a bacterial pathogen elicits free radical response and protects from a recurring infection. Front. Zool. 11 (1), 1–7. http://dx.doi.org/10.1186/1742-9994-11-23.

Molina-Cruz, A., DeJong, R.J., Ortega, C., Haile, A., Abban, E., Rodrigues, J., et al., 2012. Some strains of *Plasmodium falciparum*, a human malaria parasite, evade the complement-like system of *Anopheles gambiae* mosquitoes. Proc. Natl. Acad. Sci. 109 (28), E1957–E1962. http://dx.doi.org/10.1073/pnas.1121183109.

Molina-Cruz, A., Garver, L.S., Alabaster, A., Bangiolo, L., Haile, A., Winikor, J., et al., 2013. The human malaria parasite Pfs47 gene mediates evasion of the mosquito immune system. Science 340 (6135), 984–987. http://dx.doi.org/10.1126/science.1235264.

Morais Cabral, J.H., Atkins, G.L., Sánchez, L.M., López-Boado, Y.S., López-Otin, C., Sawyer, L., 1995. Arachidonic acid binds to apolipoprotein D: implications for the protein's function. FEBS Lett. 366 (1), 53–56. http://dx.doi.org/10.1016/0014-5793(95)00484-Q.

Moret, Y., Schmid-Hempel, P., 2001. Immune defence in bumble-bee offspring. Nature 414. http://dx.doi.org/10.1038/35107138.

Moret, Y., Siva-Jothy, M.T., 2003a. Adaptive innate immunity? Responsive-mode prophylaxis in the mealworm beetle, *Tenebrio molitor*. Proc. R. Soc. B 270. http://dx.doi.org/10.1098/rspb.2003.2511.

Moret, Y., Siva-Jothy, M.T., 2003b. Adaptive innate Immunity? Responsive-mode prophylaxis in the mealworm beetle, *Tenebrio molitor*. Proc. Biol. Sci. B 270 (1532), 2475–2480. http://dx.doi.org/10.2307/3592217.

Müller, H.-M., Dimopoulos, G., Blass, C., Kafatos, F.C., 1999. A hemocyte-like cell line established from the malaria vector *Anopheles gambiae* expresses six prophenoloxidase genes. J. Biol. Chem. 274 (17), 11727–11735. http://dx.doi.org/10.1074/jbc.274.17.11727.

Oliveira, J.H.M., Gonçalves, R.L.S., Lara, F.A., Dias, F.A., Gandara, A.C.P., Menna-Barreto, R.F.S., et al., 2011. Blood meal-derived heme decreases ROS levels in the midgut of *Aedes aegypti* and allows proliferation of intestinal microbiota. PLoS Pathog. 7 (3), e1001320. http://dx.doi.org/10.1371/journal.ppat.1001320.

Oliveira, GdA., Lieberman, J., Barillas-Mury, C., 2012. Epithelial nitration by a peroxidase/NOX5 system mediates mosquito antiplasmodial immunity. Science 335 (6070), 856–859. http://dx.doi.org/10.1126/science.1209678.

Owusu-Ansah, E., Banerjee, U., September 04, 2009. Reactive oxygen species prime *Drosophila* haematopoietic progenitors for differentiation. Nature 461 (7263), 537–541. http://dx.doi.org/10.1038/nature08313. PubMed PMID: 19727075.

Pham, L.N., Dionne, M.S., Shirasu-Hiza, M., Schneider, D.S., 2007. A specific primed immune response in *Drosophila* is dependent on phagocytes. PLoS Pathog. 3 (3), e26. http://dx.doi.org/10.1371/journal.ppat.0030026.

Povelones, M., Upton, L.M., Sala, K.A., Christophides, G.K., 2011. Structure-function analysis of the *Anopheles gambiae* LRIM1/APL1C complex and its interaction with complement C3-Like protein TEP1. PLoS Pathog. 7 (4), e1002023. http://dx.doi.org/10.1371/journal.ppat.1002023.

Ramirez, J.L., Garver, L.S., Brayner, F.A., Alves, L.C., Rodrigues, J., Molina-Cruz, A., et al., 2014. The role of hemocytes in *Anopheles gambiae* antiplasmodial immunity. J. Innate Immun. 6 (2), 119–128.

Ramirez, J.L., de Almeida Oliveira, G., Calvo, E., Dalli, J., Colas, R.A., Serhan, C.N., et al., 2015. A mosquito lipoxin/lipocalin complex mediates innate immune priming in *Anopheles gambiae*. Nat. Commun. 6. http://dx.doi.org/10.1038/ncomms8403.

Ramphul, U.N., Garver, L.S., Molina-Cruz, A., Canepa, G.E., Barillas-Mury, C., 2015. *Plasmodium falciparum* evades mosquito immunity by disrupting JNK-mediated apoptosis of invaded midgut cells. Proc. Natl. Acad. Sci. 112 (5), 1273–1280. http://dx.doi.org/10.1073/pnas.1423586112.

Rivera, A., Siracusa, M.C., Yap, G.S., Gause, W.C., 2016. Innate cell communication kick-starts pathogen-specific immunity. Nat. Immunol. 17 (4), 356–363.

Rodrigues, J., Brayner, F.A., Alves, L.C., Dixit, R., Barillas-Mury, C., 2010. Hemocyte differentiation mediates innate immune memory in *Anopheles gambiae* mosquitoes. Science 329 (5997), 1353–1355. http://dx.doi.org/10.1126/science.1190689.

Roth, O., Sadd, B.M., Schmid-Hempel, P., Kurtz, J., 2009. Strain-specific priming of resistance in the red flour beetle, *Tribolium castaneum*. Proc. R. Soc. London B: Biol. Sci. 276 (1654), 145–151. http://dx.doi.org/10.1098/rspb.2008.1157.

Ruiz, M., Wicker-Thomas, C., Sanchez, D., Ganfornina, M.D., 2012. Grasshopper Lazarillo, a GPI-anchored lipocalin, increases *Drosophila* longevity and stress resistance, and functionally replaces its secreted homolog NLaz. Insect Biochem. Mol. Biol. 42 (10), 776–789. http://dx.doi.org/10.1016/j.ibmb.2012.07.005.

Ruiz, M., Sanchez, D., Correnti, C., Strong, R.K., Ganfornina, M.D., 2013. Lipid-binding properties of human ApoD and Lazarillo-related lipocalins: functional implications for cell differentiation. FEBS J. 280 (16), 3928–3943. http://dx.doi.org/10.1111/febs.12394.

Sadd, B.M., Schmid-Hempel, P., 2006. Insect immunity shows specificity in protection upon secondary pathogen exposure. Curr. Biol. 16 (12), 1206–1210. http://dx.doi.org/10.1016/j.cub.2006.04.047. PubMed PMID: 16782011.

Smith, R.C., Eappen, A.G., Radtke, A.J., Jacobs-Lorena, M., 2012. Regulation of anti-*Plasmodium* immunity by a LITAF-like transcription factor in the malaria vector *Anopheles gambiae*. PLoS Pathog. 8 (10), e1002965. http://dx.doi.org/10.1371/journal.ppat.1002965.

Smith, R.C., Barillas-Mury, C., Jacobs-Lorena, M., 2015. Hemocyte differentiation mediates the mosquito late-phase immune response against *Plasmodium* in *Anopheles gambiae*. Proc. Natl. Acad. Sci. 112 (26), E3412–E3420. http://dx.doi.org/10.1073/pnas.1420078112.

Stanley, D., Haas, E., Miller, J., 2012. Eicosanoids: exploiting insect immunity to improve biological control programs. Insects 3 (2), 492–510. http://dx.doi.org/10.3390/insects3020492. PubMed PMID: 26466540.

Stanley-Samuelson, D.W., Dadd, R.H., 1983. Long-chain polyunsaturated fatty acids: patterns of occurrence in insects. Insect Biochem. 13 (5), 549–558. http://dx.doi.org/10.1016/0020-1790(83)90014-8.

Stanley-Samuelson, D.W., Jensen, E., Nickerson, K.W., Tiebel, K., Ogg, C.L., Howard, R.W., 1991. Insect immune response to bacterial infection is mediated by eicosanoids. Proc. Natl. Acad. Sci. 88 (3), 1064–1068.

Strand, M.R., 2008. 2-Insect Hemocytes and Their Role in Immunity. A2-Beckage, Nancy E. Insect Immunology. Academic Press, San Diego, pp. 25–47.

Tidbury, H.J., Pedersen, A.B., Boots, M., 2011. Within and transgenerational immune priming in an insect to a DNA virus. Proc. R. Soc. London B: Biol. Sci. 278 (1707), 871–876. http://dx.doi.org/10.1098/rspb.2010.1517.

Vlachou, D., Schlegelmilch, T., Runn, E., Mendes, A., Kafatos, F., 2006. The developmental migration of *Plasmodium* in mosquitoes. Curr. Opin. Genet. Dev. 16 (4), 384–391. http://dx.doi.org/10.1016/j.gde.2006.06.012. PubMed PMID: 16793259.

CHAPTER

6

The Mosquito Immune System and Its Interactions With the Microbiota: Implications for Disease Transmission

Faye H. Rodgers, Mathilde Gendrin, George K. Christophides

Imperial College London, London, United Kingdom

INTRODUCTION

As obligate blood feeders, female mosquitoes are responsible for the transmission of protozoan and viral pathogens that have a high impact on human mortality and morbidity throughout the world. Malaria, transmitted by mosquitoes of the *Anopheles* genus, was responsible for approximately 438,000 deaths in 2015, mainly children under the age of five in sub-Saharan Africa (WHO, World Malaria Report, 2015). Similarly, *Aedes* mosquitoes transmit viral diseases such as dengue, yellow fever, and chikungunya, while the *Culex* genus is responsible for the transmission of lymphatic filariasis, Japanese encephalitis, and West Nile fever. All of these diseases impose considerable burdens on human populations.

It was understood that these infections require a blood sucking arthropod for their transmission in the late 19th century, after which time vector control became a primary means of tackling these diseases. The case of malaria in early 20th century Italy exemplifies the early success of this approach. Today, control of malaria vector populations through insecticides remains an essential means of disease control, though its efficacy is receding at an alarming rate due to the evolution of insecticide resistance. The same is true for the efficacy of drugs that target the malaria parasite. As such, the development of novel interventions, derived from both new and old technologies, is of paramount importance if global malaria eradication is to be conceivable in the coming years. Such new approaches under investigation include the development of transgenic mosquitoes that involve engineering population crashes or producing malaria refractory strains, and the use of natural and/or engineered resistance mediated by the mosquito microbiota.

The introduction of transgenes via the genome of a microbial symbiont is referred to as paratransgenesis (reviewed in Wilke and Marrelli, 2015). In addition to using symbionts as a delivery tool, several microorganisms have inherent characteristics that make their introduction to mosquito populations desirable. For instance,

Arthropod Vector: Controller of Disease Transmission, Volume 1
http://dx.doi.org/10.1016/B978-0-12-805350-8.00006-4

Wolbachia is a genus of intracellular, maternally inherited bacteria that naturally infects about 40% of all terrestrial arthropod species (Zug and Hammerstein, 2012). It has great potential as a biological control agent for two reasons: it can be reproductively manipulative, inducing cytoplasmic incompatibility and therefore ensuring its persistence in a population, and it can affect the host's ability to transmit pathogens via effects on longevity and resistance (reviewed in Bourtzis et al., 2014). *Wolbachia* infections do not occur naturally in many disease vector mosquitoes; however, they have been stably introduced in *Aedes aegypti* (McMeniman et al., 2009), and shown to be able to invade a wild Australian population (Hoffmann et al., 2011). *Wolbachia* has also been stably introduced to an *Anopheles stephensi* colony, where it confers increased resistance to *Plasmodium falciparum* infection (Bian et al., 2013). Interestingly, it was long thought that anophelines do not harbor *Wolbachia* infections; however, recently, West African populations of *Anopheles gambiae* and *Anopheles coluzzii* were found to have maternally transmissible *Wolbachia* infections in their reproductive tissues (Baldini et al., 2014). This finding further supports the prospect of exploiting *Wolbachia* for population replacement or suppression strategies in African anophelines.

Finally, several genera that are common components of the anopheline gut microbiota have been identified as having the potential to reduce vector competence via effects on mosquito longevity, immune induction, and direct parasite killing (Angleró-Rodríguez et al., 2016; Bahia et al., 2014; Bando et al., 2013). In particular, an *Enterobacter* sp. isolated from the guts of a Zambian *A. gambiae* population confers resistance to *P. falciparum* infection by producing reactive oxygen species (ROS) in the gut lumen (Cirimotich et al., 2011). Similarly, a *Chromobacterium* species is able to dominantly colonize both *A. aegypti* and *A. gambiae* guts when introduced in a sugar meal, where it reduces mosquito life span and increases resistance to dengue and *Plasmodium* infection, respectively (Ramirez et al., 2014a). Conversely, a study has identified an *Anopheles*-associated ascomycete fungus, *Penicillin chrysogenum*, which increases susceptibility to *Plasmodium* infection by suppressing the mosquito antiparasite response (Angleró-Rodríguez et al., 2016).

The manipulation of gut bacteria as a disease intervention is exceptionally appealing; such methods can be tangible as bacteria can be introduced via sugar baits or the larval aquatic habitats, are potentially safe as species in question may already be naturally present in the environment, and are regulatory attainable as they may avoid the necessity of ethical and regulatory approval associated with transgenic mosquito releases. With these in mind, a renewed focus on the mosquito immune system is now paramount to understand how the native gut bacteria population is shaped and maintained, and, reciprocally, how the gut bacteria affect immune induction and host physiology in a manner that influences parasite load and vector competence.

THE MOSQUITO INNATE IMMUNE SYSTEM

Overview

For the purposes of this review, we examine the two main compartments in the mosquito body; the nominally sterile hemocoel that houses the fat body, the main metabolic and immune organ of the insect; and the gut that is open to the environment, receives and digests blood and sugar meals, and houses the microbiota. These two compartments are separated by the gut barrier that is crucial in maintaining their interaction and homeostasis. The insect blood, called hemolymph, contains circulating immunosurveillant cells that are equivalent of the white blood cells and are known as hemocytes. Both compartments are vulnerable to infection; the former either via cuticular damage or pathogen penetration through the gut epithelium and the latter via microbe ingestion. As all metazoans, the mosquito innate immune system can distinguish non-self from self through the binding of pathogen- or microbe-associated molecular patterns (PAMPs or MAMPs; here we use the latter) to pattern recognition receptors (PRRs). This recognition can trigger transcriptional responses via immune signaling pathways including the Toll, Imd (Immune deficiency), and JAK/STAT (Janus kinase/signal transducers and activators of transcription), which either direct pathogen killing through antimicrobial peptide (AMP) or ROS production or modulate the host tolerance to infection through epithelial stem cell renewal and other mechanisms. Alternatively, the detection of MAMPs can elicit extracellular signaling cascades that orchestrate localized attack of the pathogen that triggered the pathway. In all cases, the outcome for the pathogen is one of the three methods of killing: lysis, melanization, or phagocytosis.

Pattern Recognition and Immune Signaling

The primary MAMP in insect immunity is peptidoglycan, a major cell wall component of bacteria. Briefly, peptidoglycan is composed of sugar polymers [alternating residues of β-(1,4) linked N-acetylglucosamine (NAG) and N-acetylmuramic acid (NAM)], cross-linked by short peptides. A peptidoglycan monomer, or muropeptide, is composed of NAG-NAM pair bound to a peptide chain of three to five amino acids (Dworkin, 2014). Two types of peptidoglycans are distinguishable by the molecular composition of their peptide chain, with Gram-negative bacteria and Gram-positive bacilli containing a *meso*-diaminopimelic acid (DAP) residue in the third amino acid position and Gram-positive cocci containing a lysine residue. The insect Toll and Imd pathways are both stimulated by peptidoglycan.

The intracellular signaling pathways Toll and Imd were identified upon the sequencing of the *A. gambiae* genome (Holt et al., 2002) by comparison to their homologous pathways in the model organism *Drosophila melanogaster* (Fig. 6.1) (Christophides et al., 2002). As experimental characterization of these two signaling pathways is considerably more advanced in *Drosophila* and the intracellular components of the pathway are highly conserved between the two organisms, the *Drosophila* model for each pathway is presented before considering what is known in *A. gambiae*.

The *Drosophila* Toll pathway recognizes Gram-positive bacteria and fungi (Lemaitre et al., 1996). The peptidoglycan recognition protein (PGRP) PGRP-SA and the glucan binding protein (formerly known as Gram-negative binding protein) GNBP1 are PRRs of the Toll pathway secreted in the hemocoel. Together they bind a muropeptide dimer of Lys-type peptidoglycan (Filipe et al., 2005). Genetic evidence additionally points to a role for GNBP3 in binding fungal glucans (Gottar et al., 2006). Ligand binding triggers an extracellular proteolytic cascade that culminates in the cleavage of the procytokine pro-Spaetzle into Spaetzle, which binds the Toll receptor, inducing Toll dimerization (Weber et al., 2003). Intracellularly, Toll recruits a protein complex consisting of Tube, Myd88, and Pelle, the latter being a kinase that then phosphorylates the negative regulator Cactus, causing its degradation by the proteasome (reviewed in Belvin and Anderson, 1996). Cactus degradation releases the transcription factor Dif that translocates to the nucleus initiating the transcription of effector genes such as AMPs. The Toll pathway is also involved in defining the dorsal-ventral axis during *Drosophila* development, where the transcription factor Dorsal is activated instead of Dif.

With the exception of Dif, the *A. gambiae* genome encodes 1:1 orthologs for all of the intracellular components of the Toll pathway, likely reflecting the conserved role of this pathway

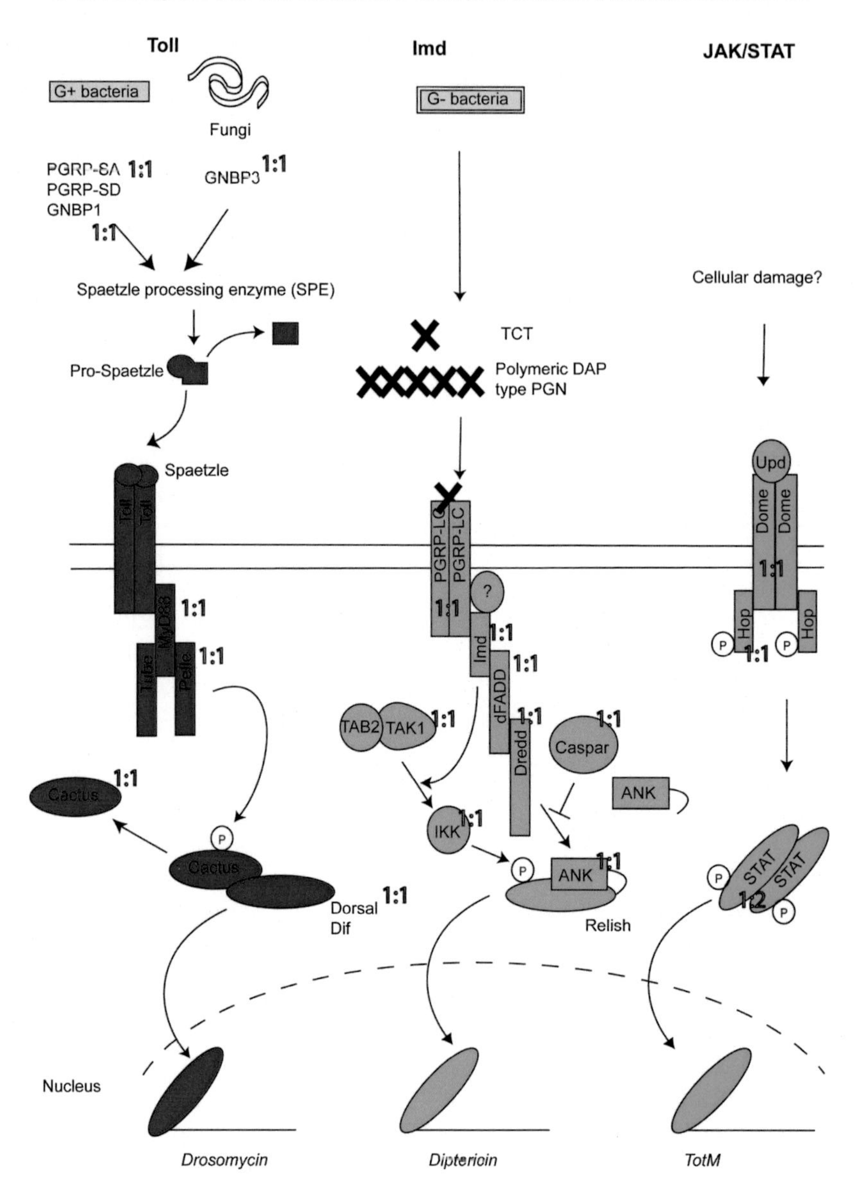

FIGURE 6.1 The Toll, Imd, and JAK/STAT immune pathways. In *Drosophila*, the Toll pathway recognizes Gram-positive bacteria and fungi, the Imd pathway recognizes Gram-negative bacteria, and the JAK/STAT pathway is likely triggered by cellular damage. All result in the transcription of immune effectors. The *Drosophila–Anopheles* ortholog numbers are indicated in light green, except if there is no identified orthologs or if there are multiple possible orthologs in each species.

in dorsalventral patterning (Christophides et al., 2002). Extracellularly, *A. gambiae* encodes a 1:1 ortholog of *D. melanogaster* PGRP-SA, PGRPS1, though the GNBPs have undergone a mosquito-specific expansion, with seven GNBPs in *A. gambiae* and three in *D. melanogaster* (Waterhouse et al., 2007). Similarly, the *D. melanogaster* Toll that has been recruited for immune function does not have a 1:1 ortholog but resembles four *A. gambiae* TOLLs (1A, 1B, 5A, and 5B). No clear equivalents to Spaetzle and the proteases that cleave it have been identified so far (Waterhouse et al., 2007). Functionally, the situation is better resolved in *A. aegypti*, where TOLL5A and SPZ1C have been characterized as functional equivalents to *Drosophila* Toll-1 and Spaetzle (Shin et al., 2006), while the serine protease CLIPB15 likely mediates the proteolytic processing of the Spaetzle cytokine (Zou et al., 2010). The divergence of mosquito and *Drosophila* recognition modules may reflect the different microbial challenges that these genera have evolved with.

Functionally, the *Anopheles* Toll pathway has been implicated in limiting *Plasmodium berghei* but not *P. falciparum* infection (Frolet et al., 2006; Garver et al., 2009) via a mechanism that may involve inducing basal levels of complement-like factors in the hemolymph (Frolet et al., 2006) and shaping hemocyte populations (Ramirez et al., 2014b). Although there is no functional evidence for a protective role in *P. falciparum* defense, one study did identify two single nucleotide polymorphisms in the *TOLL5B* gene that are associated with *P. falciparum* susceptibility (Horton et al., 2010). Considering antibacterial defense, silencing PGRPS1 has no significant effect on survival to *Escherichia coli* and *Staphylococcus aureus* challenge (Meister et al., 2009), while silencing *A. gambiae* GNBPs negatively affects mosquito survival to an *E. coli–S. aureus* co-challenge and increases the bacterial load in the hemolymph (Warr et al., 2008). In *A. aegypti*, the Toll pathway has been implicated in immunity to fungal infection (Shin et al., 2005) and dengue virus infection (Pan et al., 2012; Ramirez and Dimopoulos, 2010; Xi et al., 2008).

The *Drosophila* Imd pathway is responsible for defense against Gram-negative bacteria (reviewed in Lemaitre and Hoffmann, 2007). The stimulating ligand is a DAP-type peptidoglycan, which can be recognized in its monomeric form (notably tracheal cytotoxin, TCT) or as a polymer. Two PRRs are responsible for Imd activation: PGRP-LC and PGRP-LE. PGRP-LC is a transmembrane receptor that detects extracellular TCT and polymeric peptidoglycan, while PGRP-LE is a TCT receptor that is able to act both extracellularly, where it circulates and recruits TCT to membrane-bound PGRP-LC, and intracellularly, where it acts independently as a receptor for intracellular TCT (Kaneko et al., 2006; Lim et al., 2006). As with the Toll receptor, ligand binding induces receptor dimerization and activates a downstream signaling cascade consisting of the Imd protein, the dFADD adaptor and the caspase Dredd. In parallel, the IKK complex is activated by TAB2/TAK1 in an Imd-dependent manner to phosphorylate the transcription factor Relish. Phosphorylated Relish is a target for cleavage by Dredd, which cleaves off the Ankyrin domain, allowing the Rel domain to translocate to the nucleus and activate transcription. The negative regulator Caspar inhibits Dredd-dependent Relish cleavage to suppress the pathway (Kim et al., 2006). Similar to the Toll pathway, the *A. gambiae* genome encodes 1:1 orthologs of all intracellular components of the Imd pathway except TAB2, though it only encodes one receptor, PGRP-LC, with no PGRP-LE ortholog (Christophides et al., 2002).

Functionally, the *A. gambiae* Imd pathway appears to be the main pathway involved in antibacterial defense against both Gram-positive and Gram-negative bacteria in *A. gambiae* and *A. aegypti* (Cooper et al., 2009; Dong et al., 2011; Magalhaes et al., 2010; Meister et al., 2005, 2009). The *A. gambiae* genome encodes two isoforms of REL2, the homolog of *Drosophila* Relish; the short isoform, REL2-S, lacks the Ankyrin domain-encoding region and appears to be responsible for the response against Gram-negative bacteria, while the full length isoform, REL2-F, includes the Ankyrin domain and is implicated in resistance to Gram-positive bacteria (Meister et al., 2005). Furthermore, the Imd pathway is now well established as playing a major role in the defense against *P. falciparum*

(Clayton et al., 2013; Dong et al., 2011; Garver et al., 2012, 2009; Meister et al., 2009). REL2-dependent parasite killing occurs in both the gut lumen and the hemocoel (Dong et al., 2011; Garver et al., 2012) and is at least partly dependent on the presence of the gut microbiota for its activation (Meister et al., 2009). The downstream mechanism responsible for parasite killing is unclear, though REL2 is known to induce the transcription of numerous effectors that have been shown to have anti-*Plasmodium* activity, including the thioester-containing protein 1 (TEP1) and the *Anopheles–Plasmodium* resistance locus 1 (APL1C) (Dong et al., 2011; Meister et al., 2005).

The JAK/STAT pathway was first identified for its role in vertebrate immune signaling (reviewed in Villarino et al., 2015) and was later characterized in *Drosophila* as a key signaling pathway in development, immunity, and stem cell maintenance (Arbouzova and Zeidler, 2006; Myllymäki and Rämet, 2014). Briefly, binding of the Unpaired (Upd) ligand to the transmembrane receptor Dome leads to receptor dimerization, causing the receptor-associated JAKs to phosphorylate both themselves and Dome. Cytoplasmic STAT transcription factors are then recruited to the membrane-associated complex; their phosphorylation induces STAT dimerization and subsequent nuclear translocation. Rather than direct recognition of a microbial elicitor, the trigger of the pathway is notably cell damage causing cells to produce the Upd ligand, which then acts as a cytokine to induce the pathway in neighboring cells (Buchon et al., 2009).

The *A. gambiae* genome encodes two homologs of the STAT transcription factor, STAT1 and STAT2 (Waterhouse et al., 2007). In *Drosophila*, the JAK/STAT pathway appears to contribute to antiviral immunity, where it induces a set of effector genes distinct from those induced by the Toll and Imd pathways (Dostert et al., 2005). The same seems to be true in *A. aegypti* mosquitoes, where JAK/STAT has been shown to play a role in anti-dengue defense (Souza-Neto et al., 2009). In *A. gambiae*, the JAK/STAT pathway does not appear to affect systemic O'nyong nyong virus infection of the hemolymph (Waldock et al., 2012). There has also been early evidence that the JAK/STAT pathway responds to bacterial infection in *A. gambiae* (Barillas-Mury et al., 1999), but this response does not appear to be necessary for survival to systemic bacterial infection (Gupta et al., 2009). Interestingly, the pathway plays a dual role in *Plasmodium* infection, with STAT signaling supporting the survival of early stage *P. berghei* oocysts but mediating lysis of late stage oocysts (Gupta et al., 2009).

Pathogen Killing: Lysis, Phagocytosis, and Melanization

One of the main outcomes of activation of the above pathways is the killing of the infectious agent by lysis, phagocytosis, or melanization (Fig. 6.2). AMPs are small, stable peptides believed to kill pathogens by acting at the membrane, causing lysis. They are synthesized by the fat body, hemocytes, and epithelia and generally show specificity in their microbicidal activity (reviewed in Lemaitre and Hoffmann, 2007). The *A. gambiae* genome encodes at least four families of AMPs: cecropins, defensins, attacin, and gambicin (Waterhouse et al., 2007). The cecropins have broad spectrum activity against both Gram types of bacteria, filamentous fungi, yeast, viruses, *Leishmania*, and *P. berghei* (Kim et al., 2004; Luplertlop et al., 2011; Vizioli et al., 2000), and gambicin, which is a mosquito-specific AMP, is similarly active against both Gram types of bacteria, fungi, and *P. berghei* ookinetes (Vizioli et al., 2001a). Defensin seems to be more specific in its activity, killing Gram-positive bacteria and filamentous fungi but with no effect on yeast, Gram-negatives, or *Plasmodium* (Blandin et al., 2002; Vizioli et al., 2001b). The *A. gambiae* genome also encodes several C-type lysozymes that act by hydrolyzing peptidoglycan. Interestingly, in addition to

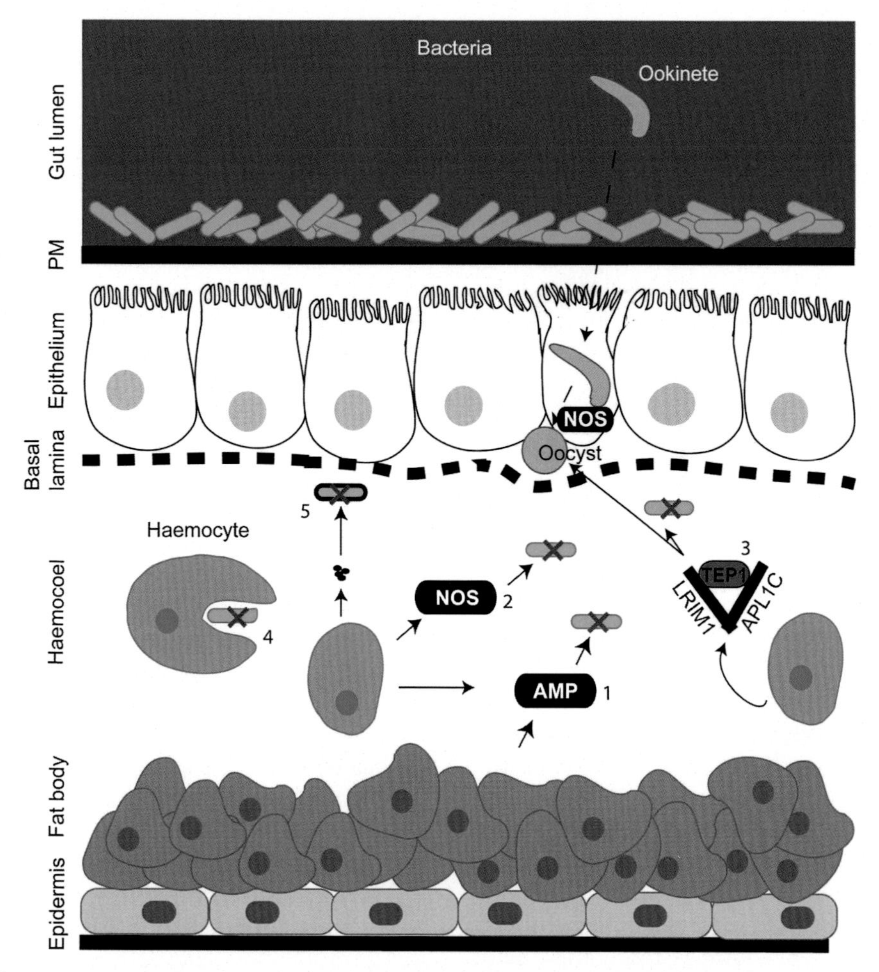

FIGURE 6.2 The major mechanisms of pathogen elimination in *Anopheles gambiae*. (1) The fat body and hemocytes produce antimicrobial peptides (AMPs) that cause pathogen lysis. (2) Nitric oxide synthase (NOS) transcription is induced in epithelial cells by invading ookinetes and in hemocytes by systemic bacterial infection. (3) Hemocytes secrete a complement-like complex that is deposited on the surface of bacterial and *Plasmodium* parasites, resulting in their lysis or melanization. (4) Pathogens are phagocytosed by a subpopulation of hemocytes, the granulocytes. (5) A prophenoloxidase cascade results in the deposition of melanin on the pathogen surface.

its antibacterial role, LYSC1 appears to facilitate *Plasmodium* development and has been shown to bind both *P. falciparum* and *P. berghei* oocysts in vivo (Kajla et al., 2011).

In addition to AMPs, immune signaling induces the production of reactive nitrogen species and ROS. Nitric oxide synthase (NOS) catalyzes the conversion of L-arginine to L-citrulline, producing nitric oxide in the process. NOS is transcriptionally induced in hemocytes in response to systemic bacterial infection (Hillyer and Estévez-Lao, 2010) and in midgut epithelial cells in response to *Plasmodium* infection (Luckhart et al., 1998). Midgut NOS is induced through sensing of the parasite glycosylphosphatidylinositol (GPI) anchor and hemozoin by the insulin pathway (Akman-Anderson et al., 2007; Lim et al., 2005) and through JAK/STAT signaling (Gupta et al., 2009). Nitric oxide is widely cytotoxic, limiting in particular bacterial proliferation, *Plasmodium* development, and dengue infection (Hillyer and Estévez-Lao, 2010; Luckhart et al., 1998; Ramos-Castañeda et al., 2008). Similarly, resistance to both systemic bacterial infection and *Plasmodium* infection correlates with systemic levels of ROS, and the presence of ROS is essential for effective antibacterial and anti-*Plasmodium* responses (Kumar et al., 2003; Molina-Cruz et al., 2008). Mitochondrial and membrane-bound NADPH oxidases produce the highly toxic superoxide anion; levels of superoxide are kept under close control by detoxifying enzymes such as superoxide dismutase, which converts superoxide to the less toxic hydrogen peroxide, and catalase, which further breaks down hydrogen peroxide. Interestingly, *P. falciparum* infection of *A. stephensi* has been shown to activate p38 mitogen-activated protein kinase (MAPK) signaling, which has the effect of reducing ROS generation and, broadly, immune gene induction, resulting in longer mosquito life span and greater permissiveness to parasite development (Wang et al., 2015).

One of the best-characterized pathogen killing pathways in the mosquito is the complement-like response. TEP1 is a soluble hemolymph protein that is similar in structure to vertebrate C3 (Baxter et al., 2007). In the vertebrate system, proteolytic cleavage leads to deposition of C3 on the pathogen surface, opsonizing it for clearance by lysis or phagocytosis. TEP1 appears to play an analogous role in mosquitoes: it is cleaved in response to septic injury and binds bacterial surfaces via a thioester bond, promoting phagocytosis (Levashina et al., 2001). It similarly binds *P. berghei* ookinetes in the midgut basal lamina, resulting in parasite death via lysis and/or melanization (Blandin et al., 2004). TEP1 deposition is controlled by two leucine-rich repeat containing proteins (LRRs), LRIM1 and APL1C (Osta et al., 2004; Riehle et al., 2006, 2008), which circulate in the hemolymph as a disulphide-bonded dimer and are necessary for TEP1 binding to *P. berghei* ookinetes (Povelones et al., 2009). The mechanics of this pathway are under continuing investigation, with both positive and negative regulators being identified now (Povelones et al., 2013; Yassine et al., 2014). Interestingly, the role of the complement response against *P. falciparum* infection has so far been unclear, with one field study reporting no effect of LRIM1 on *P. falciparum* infection (Cohuet et al., 2006) and another laboratory study suggesting that APL1C may be replaced by a paralog, APL1A, in the context of *P. falciparum* infection (Mitri et al., 2009). However, TEP1 itself does appear to be active against human malaria parasites (Dong et al., 2006; Garver et al., 2009). A study suggests that *P. falciparum* may evade the complement response by repressing midgut nitration, which "tags" the parasite for targeting by the complement pathway (Molina-Cruz et al., 2013; Ramphul et al., 2015). This evasion capacity has been subject to strong selection pressure during the globalization of malaria, as parasites encountered new vector species when reaching remote countries (Molina-Cruz et al., 2015).

Phagocytosis occurs solely in the hemolymph by a highly abundant population of hemocytes, the granulocytes. The pathogen is either opsonized by circulating humoral factors or recognized directly by transmembrane hemocyte receptors, then internalized by the hemocyte into a phagosome, which fuses with the lysosome before pathogen digestion (reviewed in Hillyer and Strand, 2014). This defense mechanism is known to be effective against bacteria (Hillyer et al., 2004), yeast (Da Silva et al., 2000), and, with a species-dependent efficiency, *Plasmodium* sporozoites (Hillyer et al., 2003, 2007). Several mosquito proteins have been implicated in initiating phagocytosis, including the transmembrane Imd pathway receptor PGRP-LC, the complement-like factors TEP1, 3, and 4, and their associated LRRs (Moita et al., 2005).

Oenocytoids are a separate subpopulation of hemocytes and produce the enzymes involved in melanization. Insect melanization is essential for several physiological processes besides immunity, including cuticle hardening and egg darkening, and involves the deposition of melanin and cross-linked proteins on the surface of a pathogen, causing its death. The rate-limiting step in melanization is the conversion of tyrosine to melanin precursors, which is catalyzed by phenoloxidase. This enzyme is produced by cleavage of its zymogen, prophenoloxidase, downstream a proteolytic cascade that is activated upon pathogen recognition and regulated by CLIP domain serine proteases and serine protease inhibitors (Yassine and Osta, 2010). There was considerable interest in melanization with the observation that a *Plasmodium*-refractory strain of *A. gambiae*, L3-5, kills ookinetes by melanization (Collins et al., 1986). However, it has since transpired that nonselected laboratory strains and wild-caught mosquitoes rarely melanize ookinetes, and indeed that melanization may only be a relevant response in allopatric mosquito–parasite species combinations (Michel et al., 2006; Schwartz and Koella, 2002). Similarly, although mosquitoes are observed to melanize bacteria in the hemolymph, genetic disruption of the melanization response does not compromise their ability to resist the infection (Schnitger et al., 2007). More evidence, however, suggests that melanization is an important response in conferring resistance to fungal infection (Yassine et al., 2012).

Many studies have focused on pathogen killing in the hemocoel; more interest has arisen in better understanding how microbes in the gut lumen interact with and are controlled by the immune system. A genome-wide association study using an oral bacterial infection model identified several novel immune modulators that control the outcome of infection, including a gustatory receptor that links immunity to feeding behavior for the first time in the mosquito (Stathopoulos et al., 2014). This study highlighted that there is still much to understand about the gut immune response; especially, how it is modulated to tolerate constant exposure to microbial elicitors in the gut, in the form of the microbiota.

THE MOSQUITO MICROBIOTA

Now there is a rapidly growing interest in microbiota, defined as the complement of microorganisms that live in symbiosis with a host organism (Sommer and Bäckhed, 2013). Studies have focused both on describing the species of microorganisms living in animal hosts (Ellis et al., 2013; Mathieu et al., 2013) and their impact on host physiology (Devaraj et al., 2013; Pope et al., 2012), resulting in increasingly complex models of the microbiota not as an immunological threat but with an important, if not essential, role in host health. The microbiome (the collective genomes of the microbiota) is a dynamic entity that changes in size and composition in response to changes in host lifestyle (diet, threat of infection, use of antibiotics), unlike the host genome, which must remain static (Rosenberg and Zilber-Rosenberg, 2011). This arrangement is mutually beneficial to host and bacteria, with the host providing a stable, nutrient-rich environment and bacteria aiding in digestion, protection against opportunistic pathogens, and maturation of the immune system. Indeed, the impact of the microbiota on host physiology cannot be understated: in humans, microbiota perturbation has been associated with not only gastrointestinal and metabolic disorders (such as obesity, irritable bowel disease, colorectal cancer, and diabetes), but also several neuronal disorders such as attention deficit hypersensitivity disorder, multiple sclerosis, and depression (Marchesi et al., 2015; Petra et al., 2015). Similarly, in the mosquito the microbiota are increasingly understood to influence host susceptibility to protozoan and viral infection, as well as several other aspects of mosquito physiology with relevance to disease transmission.

Mosquito microbiota composition has been studied primarily by sequence analysis of the *16S* rRNA gene, using both culture-dependent and culture-independent methods. In *Anopheles*, over 100 bacterial genera have now been identified in different developmental stages of both field-caught mosquitoes and laboratory colonies (reviewed in Villegas and Pimenta, 2014), though no genera have been reported in all studies, suggesting that *Anopheles* mosquitoes do not harbor obligate symbionts. The most commonly reported genera are *Pseudomonas, Aeromonas, Asaia, Elizabethkingia, Enterobacter, Klebsiella, Pantoea,* and *Serratia* (Gendrin and Christophides, 2013), all of which are Gram-negative bacteria, and all of which are *Proteobacteria* with the exception of *Elizabethkingia*, a *Bacteroidetes*. The frequency with which these genera are reported may suggest they constitute something of a "core *Anopheles* microbiota." Among them, *Pseudomonas, Aeromonas,* and *Asaia* have also been found in *Aedes, Culex,* and *Mansonia* mosquitoes (Minard et al., 2013). Although a majority of bacteria found in *Anopheles* are *Proteobacteria*, the microbiota in mosquito populations is highly diverse, including bacteria from other phyla namely *Bacteroidetes, Actinobacteria, Firmicutes, Tenericutes, Cyanobacteria, Fusobacteria, Deinococcus-Thermus* (Gendrin and Christophides, 2013; Minard et al., 2013), illustrating the diversity of bacterial phyla identified in *A. gambiae* and *A. stephensi* larvae and adults.

The majority of mosquito microbiota research to date has been bacteria-centric, while knowledge on fungal and viral components of the gut ecosystem remains limited. Early evidence does suggest, however, that mosquitoes harbor complex viral and fungal communities (Chandler et al., 2015). In all studies to date looking at field or field-derived African anopheline mosquitoes, *Proteobacteria* have dominated the adult gut (Boissière et al., 2012; Gimonneau et al., 2014; Osei-Poku et al., 2012; Tchioffo et al., 2013; Wang et al., 2011). It has also been reported that gut community structure changes drastically upon maturation, with a decrease in microbial diversity from larvae to sugar-fed adults, and the lowest diversity in blood-fed adult females (Wang et al., 2011). Several studies have noted the large variation in composition between individuals within a mosquito species and the fact that each individual hosts a small number of dominant taxa (Boissière et al., 2012; Osei-Poku et al., 2012). For example, one study in Kenyan populations estimated individuals to harbor a median of 42 operational taxonomic units, though on average the four most abundant of these constitute 90% of the total population (Osei-Poku et al., 2012). Furthermore, there are indications that the microbiota composition is influenced by the larval breeding site (Boissière et al., 2012; Gimonneau et al., 2014), although in another study that used adult mosquitoes collection site could not predict microbiota composition, suggesting that breeding site–acquired compositions may be masked by bacteria acquired during adulthood (Osei-Poku et al., 2012). A study of the microbiota composition in adult mosquitoes collected in three neighboring villages in Burkina Faso suggests that the life history of an individual mosquito, including its migrations from one village to the other, may leave some marks in the composition of its microbiota (Buck et al., 2016).

Importantly, these studies have additionally given some insight into the extent by which the microbiota of laboratory mosquito colonies reflects that of its field counterparts. Two reports have addressed this directly for *A. gambiae*, with one concluding that although laboratory mosquitoes have lower overall diversity than field mosquitoes, the majority of genera in laboratory mosquitoes are also found in the field (Wang et al., 2011). The other found the Ngousso colony to be dominated by *Elizabethkingia* that is also frequently found in field mosquitoes (Boissière et al., 2012). In agreement with both of these conclusions, other microbiota-based analyses from laboratory colonies have reported similar taxa to those identified in field studies, though in some cases with an over dominance of *Bacteroidetes* (in particular *Flavobacteriaceae*) instead of *Proteobacteria* (Coon et al., 2014; Gendrin et al., 2015; Ngwa et al., 2013; Stathopoulos et al., 2014). The evidence for maintenance of the native microbiota upon mosquito colonization reinforces the notion of a "core microbiota." Whether this is due to host selectivity, bacterial adaptation, or bacterial heritability, or some combination of the three, remains unclear. Nevertheless, it does also appear to be the case that the bacteria found in the mosquito gut are widespread in the environment, so opportunistic colonization by any species that is able to persist in the gut environment (considering its pH, redox potential, and substrate availability notably) is likely to shape the community to a great extent. The variation seen between laboratory colonies in the *Bacteroidetes:Proteobacteria* ratio could stem from differences in husbandry, such as larval diet, water purity, and container cleaning procedures, between laboratories.

Indeed, one question that remains poorly addressed is how adult mosquitoes acquire their microbiota. It is known that some bacterial genera such as *Asaia* are transmissible both horizontally (between individuals of the same generation) and vertically (from parents to offspring); in the latter case this may be via smearing of the bacteria onto eggs in the reproductive tract (Damiani et al., 2010). Larvae feed on microorganisms in their aquatic environment, providing a clear mechanism for gut colonization; however, upon metamorphosis the gut is thought to undergo two processes of drastic reduction in the microbiota load, leaving the newly emerged adult gut almost free of bacteria. First, the fourth instar larvae egest the contents of the alimentary canal and the peritrophic matrix before pupation and, second, emerging adults ingest exuvial fluid that has bactericidal properties (Moll et al., 2001). Nevertheless, this process is incomplete and adult guts are able to be colonized to some extent transstadially (Coon et al., 2014). Adult mosquitoes are also thought to recolonize their guts by feeding on larval water (Lindh et al., 2008), and they likely acquire new bacteria from the environment whilst taking sugar meals from flowers (Manda et al., 2007).

The most striking characteristic of the mosquito microbiota is its behavior following a blood meal, when bacterial load increases from 10 to 1000 fold within the first 24h following the blood meal (Kumar et al., 2010; Pumpuni et al., 1996). The cause of this bacterial growth is not clear, though two main, nonmutually exclusive, explanations have been proposed. First, the ingestion of blood may provide gut bacteria with the nutritional sustenance for an increased rate of proliferation. Second, following the blood meal the gut may reduce its level of immune control, essentially permitting bacteria that are under tight control in the non–blood-fed gut to proliferate (Kumar et al., 2010; Oliveira et al., 2011). The potential mechanisms behind this are discussed further below. By 48–72h post blood meal, bacterial load has receded back to preblood meal-levels; again, the mechanism that brings bacterial load back to homeostatic levels has not been elucidated, but it is thought that excretion of the blood bolus that is encased by the peritrophic matrix plays a key role.

The contribution of the microbiota to mosquito physiology is also an open question. In other insects, the microbiota have been shown to play diverse roles in host nutrition including digestion of plant polymers (Engel et al., 2012; Kaufman and Klug, 1991), provisioning nutrients such as essential amino acids and B vitamins (Douglas, 1998; Nikoh et al., 2011; Schaub and Eichler, 1998) and neutralizing dietary toxins (Dowd and Shen, 1990). The role of the microbiota in providing nutritional sustenance to adult mosquitoes is unclear. In *A. aegypti*, a 90%–100% depletion of the microbiota was found to cause a reduction or delay of red blood cell lysis that was concomitant with a reduction in mosquito fecundity (Gaio et al., 2011). Similarly, in *Aedes albopictus* two isolated

Acinetobacter species were found to have specifically gained the capacity to metabolize amino acids and common plant components, suggesting that strains found within the mosquito host have adapted to digest blood and nectar respectively (Minard et al., 2013). What is more, it has been noted in one study that blood-feeding induces a shift in microbiota composition favoring *Enterobacteriaceae* (Wang et al., 2011). Genera within this family contain more genes known to be involved in combating oxidative and nitrosative stress than genera prevalent in the sugar-fed gut, appropriate for the environment rich in ROS and nitric oxide species that is associated with blood meal-catabolism.

More broadly speaking, some components of the microbiota appear to adversely affect mosquito fitness and reproduction. Treatment of adult female *A. gambiae* with penicillin and streptomycin is shown to increase their life span and fecundity (Gendrin et al., 2015), but treatment with azithromycin had an opposite effect (Gendrin et al., 2016). Similarly, bacterial clearance by antibiotics is shown to have a mildly detrimental effect on *A. stephensi* longevity and a more severe negative effect on fecundity (Sharma et al., 2013), altogether pointing to complex interactions that are highly dependent on the microbiota composition.

The *Anopheles* microbiota additionally have a fundamental impact on larval development. Several studies have shown retardation of larval development with antibiotic treatment (Chouaia et al., 2012; Mitraka et al., 2013), and another that fully axenic larvae cannot survive beyond the 1st instar (Coon et al., 2014). These effects appear not to be due to the loss of bacteria as a food source, but maybe rather on bacterial metabolites, as even when provided with heat-killed bacteria, the larvae did not develop. Despite this, the existence of an aseptic colony has been reported; however, the nutritional composition of the larval diet that supports this colony is not publically available (Lyke et al., 2010). A similar developmental delay is observed in *Drosophila* larvae and has been ascribed to two mechanisms: first, *Acetobacter pomorum* promotes insulin signaling by production and metabolism of acetic acid (Shin et al., 2011) and secondly *Lactobacillus plantarum* facilitates dietary amino acid accumulation, which signals via the TOR kinase, again promoting insulin signaling (Storelli et al., 2011).

MICROBIOTA–IMMUNE SYSTEM INTERACTIONS

Despite the microbiota playing a key role in the physiology of the mosquito, mainly at the larval stage, the proximity of bacteria to epithelial membranes and their shear numbers present a challenge to the host immune system, which must develop systems to balance the ability to recognize pathogenic organisms and manage microbiota growth while avoiding chronic activation. This conundrum has led to the concepts of immune resistance and tolerance; resistance responses combat infection by reducing microbial load, while tolerance responses aim to reduce the impact of the infection on the host. The latter may involve either reducing or avoiding the induction of an immune response that can itself be harmful to the host or repairing microbe or immunity-derived tissue damage.

An understanding of both resistance and tolerance mechanisms in the *Anopheles* gut could open up novel approaches for transmission blocking. Targeting resistance responses could introduce dysbiosis or uncontrolled microbiota growth in the gut, resulting in shortened mosquito life span. Equally, targeting a tolerance response could also affect host life span (excessive immune activation is known to be detrimental to health) while having the added effect of increasing resistance to *Plasmodium* infection. It is paramount to study these responses specifically in the gut for two reasons: first, this is the tissue that houses the microbiota, as well as being the site of a vulnerable stage of parasite development, the ookinete to oocyst transition. Second, any intervention must be able to access its target, rendering molecules that are either secreted into the gut lumen or situated in the luminal membrane of the epithelium as ideal targets for orally administered drugs.

Understanding of microbiota–gut interactions remains in its infancy in the mosquito. It is known that midgut bacterial load is limited by the Imd pathway in both sugar-fed and blood-fed guts (Clayton et al., 2013; Dong et al., 2009, 2011; Meister et al., 2009). Indeed, microbiota-dependent Imd pathway induction is thought to be one of the reasons for the consistent observation that antibiotic-treated mosquitoes have increased susceptibility to *Plasmodium* infection (Dong et al., 2009; Meister et al., 2009). A transcriptional analysis of the effect of microbiota depletion on whole mosquitoes revealed that considerably fewer genes are regulated by the presence of the microbiota than by oral bacterial infection, and identified a set of immune genes that are responsive to the gut microbiota including AMPs, PRRs, and signal transducing serine proteases (Dong et al., 2009). Nevertheless, a more complete picture of the gut immune system has emerged in *Drosophila*: this review is therefore based around the *Drosophila* system, with reference to *Anopheles* where information is available.

There are two main resistance mechanisms in gut epithelial immunity in *Drosophila*: AMP production, exclusively driven by the Imd pathway, and ROS production, driven in the gut by the dual oxidase (DUOX) system. Both of these are now understood to be negatively regulated at multiple levels, preventing their overstimulation by gut commensals.

First, PGRPs with amidase activity act at the level of the peptidoglycan ligand to reduce Imd pathway stimulation. *Drosophila* PGRP-LB is able to hydrolyze both polymeric DAP-type peptidoglycan and its monomer, TCT, to nonimmunostimulatory compounds (Zaidman-Rémy et al., 2006). It is induced by the Imd pathway and has a major role in the gut in preventing excessive microbiota-dependent induction of AMPs and intestinal stem cell activity (Paredes et al., 2011; Zaidman-Rémy et al., 2006). PGRP-SC1 and 2 appear to play a similar role, though in the response to systemic infection (Paredes et al., 2011) and have also been implicated in the epithelial response in larvae (Bischoff et al., 2006). PGRP-SB1 and 2 have been shown to have enzymatic activity against DAP-type peptidoglycan, but so far no significant effect in vivo has been demonstrated in terms of resistance to infection or immune regulation (Paredes et al., 2011; Zaidman-Rémy et al., 2011). A role as an immune effector is possible, whose significance may so far be masked due to redundancy with other AMPs. *A. gambiae* encodes a 1:1 ortholog of PGRP-LB, along with two other catalytic PGRPs, PGRPS1 and S2. Our yet unpublished study of *Anopheles* PGRPs suggests that, as in *Drosophila*, PGRP-LB negatively regulates the Imd pathway, and that this tolerance response participates in the mosquito permissiveness to *P. falciparum* infections (Gendrin et al., unpublished).

Two negative regulators of the Imd pathway in *Drosophila*, PGRP-LF and Pirk, are directly involved in regulating PGRP-LC activation. PGRP-LF is a transmembrane protein thought to heterodimerize with PGRP-LC in the absence of elicitors (Basbous et al., 2011; Maillet et al., 2008; Persson et al., 2007). Pirk, transcriptionally induced in the gut by the microbiota, relocalizes PGRP-LC into intracellular vesicles, thereby downregulating signaling (Aggarwal et al., 2008; Kleino et al., 2008; Lhocine et al., 2008). Interestingly, *A. gambiae* does not encode orthologs of either PGRP-LF or Pirk, raising the question of which mechanisms, if any, are employed to downregulate PGRP-LC activation. Finally, Caudal is a homeobox transcription factor that is expressed exclusively in the gut where it specifically represses the NF-kB-dependent expression of AMPs but not of the negative regulators of the Imd pathway (Ryu et al., 2008). Similarly, in *Anopheles*, Caudal has been shown to repress

AMP expression, support *P. falciparum* development, and maintain microbiota load (Clayton et al., 2013).

The other main resistance mechanism in *Drosophila* is ROS production, driven by the DUOX system. DUOX is a membrane-bound NADPH oxidase that is indispensable for the defense against oral bacterial infection (Ha et al., 2005) and plays an essential role in limiting microbiota (primarily yeast) proliferation in noninfected flies (Ha et al., 2009a). DUOX is regulated at several levels; the DUOX activation pathway is dependent on phospholipase C-β (PLCβ) activity (Ha et al., 2009a), while *DUOX* expression is regulated via PGRP-LC and p38 MAPK. Furthermore, in noninfected conditions, low PLCβ activity inhibits *DUOX* expression via negative regulation of p38 MAPK phosphorylation, whilst in infected conditions high PLCβ activity activates the *DUOX* expression pathway upstream of p38 MAPK (Ha et al., 2009b). Both of these regulatory mechanisms are essential for survival to varying levels of microbial ligand: the former prevents commensal bacteria from inducing excessive oxidative activity, while the latter is essential for resistance to high bacterial loads that are indicative of infection. Interestingly, DUOX is regulated by two microbial ligands: peptidoglycan, which is recognized by PGRP-LC, and the nucleobase uracil, which activates the PLCβ pathway via hedgehog signaling (Lee et al., 2013, 2015). Interestingly, uracil was found to be secreted by pathogenic bacteria but not bacterial constituents of the *Drosophila* microbiota, revealing for the first time a mechanism by which the innate immune system is able to discriminate between pathogens and symbionts (Lee et al., 2013, 2015).

In *Anopheles*, DUOX has been characterized as playing a slightly different role in antimicrobial defense; it coordinates with a heme peroxidase, IMPer (also known as HPX15), to catalyze the formation of dityrosine bonds in a mucin protein layer between the peritrophic matrix and the gut lumen following blood-feeding (Kumar et al., 2010). This has the dual effect of reducing the elicitation of an immune response to both the gut microbiota and *Plasmodium* parasites, indicating one means by which ROS may partake in a tolerance response. The role of ROS as a microbiota resistance mechanism in the gut lumen is less clear. At least one bacteria strain (*Enterobacter* sp.) found in the gut of field-caught *Anopheles arabiensis* itself produces ROS, which are detrimental to *P. falciparum* development, as well as likely supporting *Enterobacter* sp. dominance within the gut ecosystem (Cirimotich et al., 2011). Interestingly, one study in *A. aegypti* has suggested that DUOX-produced ROS play a key role in preventing microbiota proliferation in the sugar-fed gut, but that the presence of heme in the blood meal triggers an immediate reduction in ROS generation, via protein kinase C signaling, with this being partially responsible for the microbiota growth observed upon blood-feeding (Oliveira et al., 2011). This is proposed to be an adaptation to the pro-oxidant effect of heme, which catalyzes the conversion of oxidized molecules to highly reactive intermediates in the lipid peroxidation chain.

PERSPECTIVE

Through the induction of immune responses and production of antimicrobial molecules such as ROS, it is now established that the mosquito microbiota overall decreases mosquito permissiveness to infections with human pathogens, while they also impact the mosquito fitness albeit adversely. Therefore, there is growing interest in the potential of manipulating the mosquito microbiota for disease transmission blocking. Several strategies have been proposed (Fig. 6.3). A first strategy is paratransgenesis, whereby a bacterial strain expressing effector molecules acting against parasites or viruses is introduced into the mosquito,

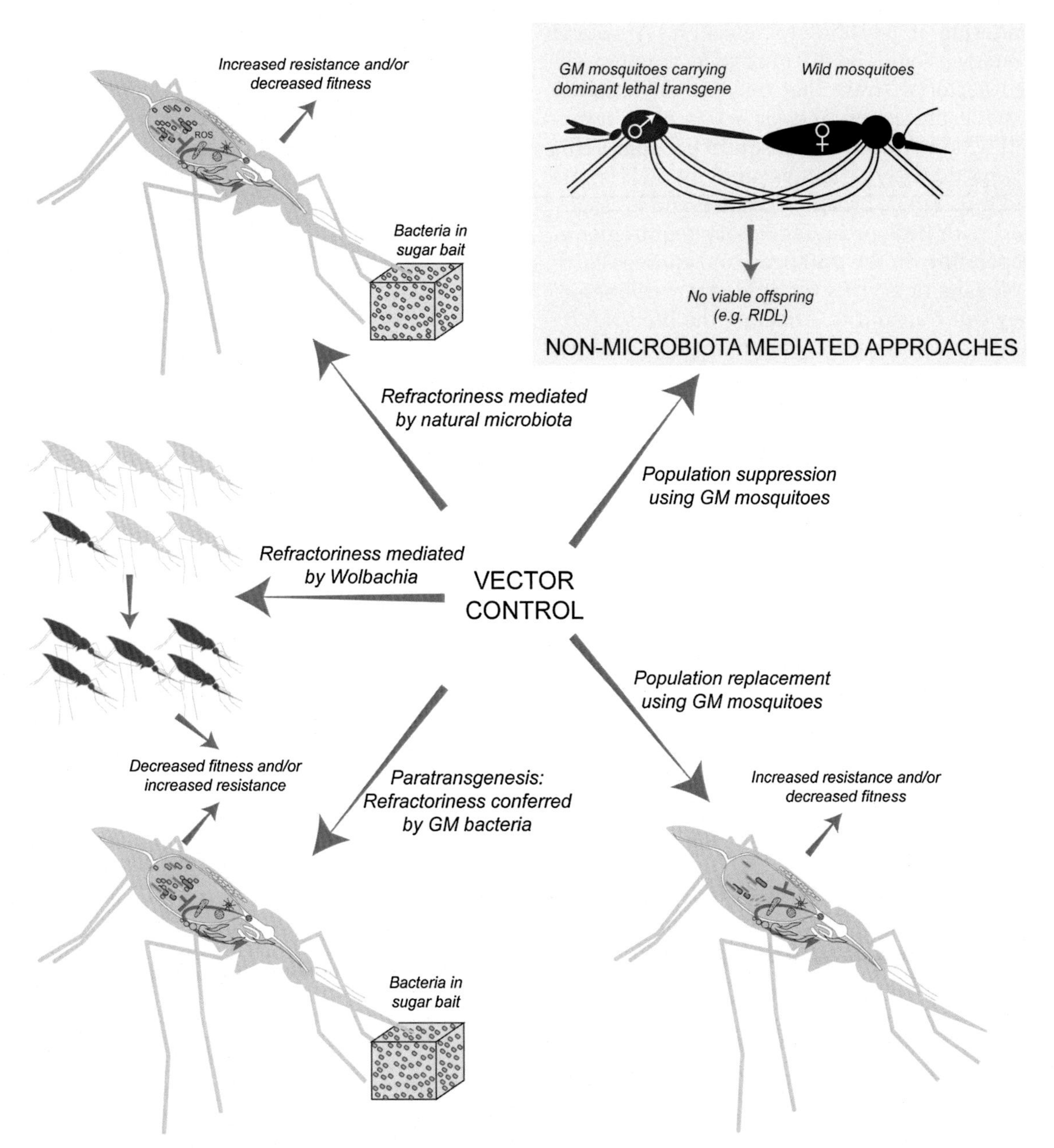

FIGURE 6.3 **Microbiota-mediated vector control interventions.** These strategies include paratransgenesis that is driving effectors in wild mosquito populations through genetically modified bacteria; *Wolbachia*-mediated control that involves spreading *Wolbachia* through mosquito populations to induce resistance to pathogens and/or decrease mosquito fitness; spreading of natural mosquito microbiota that can directly or indirectly (through mosquito immune reactions) affect pathogens; and mosquito population replacement with mosquitoes carrying transgenes that can disturb the natural mosquito microbiota causing resistance to pathogens and/or decrease mosquito fitness. Such strategies could be combined with efforts for mosquito population suppression using transgenic mosquitoes spreading dominant lethal transgenes.

rendering it resistant to infection. A second strategy would be to introduce a nonmodified bacterial strain that may either naturally provide protection and/or act as an entomopathogen to affect the mosquito life span. This is a particularly attractive strategy, as it is safe and thus can be broadly accepted and is associated with little or no regulatory requirements. Depending on the pathogen, mosquitoes carry a parasite or a virus for at least a week before they can transmit it. Affecting the life span by only a few days impacts a mosquito's number of blood meals and thus its chance to transmit the disease. It is suggested that targeting the late stages of the mosquito life would indeed not significantly affect the mosquito fitness and thus selection pressure to resist to the intervention is expected to be very small (Read et al., 2009). A similar result can be achieved by *Wolbachia*, which can rapidly spread within wild mosquito populations through cytoplasmic incompatibility and can confer increased resistance to the pathogen or decrease mosquito fitness including life span. Finally, a detailed understanding of the immune reactions against and control of the microbiota could identify additional approaches for that would involve driving transgenes in wild mosquito populations, which could cause microbiota dysbiosis or overgrowth again resulting in lowered mosquito fitness or increased resistance to infection. Such interventions in conjunction with current efforts for mosquito population suppression using transgenic male mosquitoes carrying dominant lethal transgenes could make a significant impact on the global effort to eliminate vectorborne diseases.

Acknowledgments

This chapter is modified from the doctoral thesis of FHR (Imperial College London, 2016), supported by a Biotechnology and Biological Sciences Research Council (BBSRC) DTP grant to the Department of Life Sciences. MG was supported by the BBSRC project grant BB/K009338/1 to GKC.

References

Aggarwal, K., Rus, F., Vriesema-Magnuson, C., Ertürk-Hasdemir, D., Paquette, N., Silverman, N., 2008. Rudra interrupts receptor signaling complexes to negatively regulate the IMD pathway. PLoS Pathog. 4, e1000120.

Akman-Anderson, L., Olivier, M., Luckhart, S., 2007. Induction of nitric oxide synthase and activation of signaling proteins in *Anopheles* mosquitoes by the malaria pigment, hemozoin. Infect. Immun. 75, 4012–4019.

Angleró-Rodríguez, Y.I., Blumberg, B.J., Dong, Y., Sandiford, S.L., Pike, A., Clayton, A.M., Dimopoulos, G., 2016. A natural *Anopheles*-associated *Penicillium chrysogenum* enhances mosquito susceptibility to *Plasmodium* infection. Sci. Rep. 6, 34084.

Arbouzova, N.I., Zeidler, M.P., 2006. JAK/STAT signalling in *Drosophila*: insights into conserved regulatory and cellular functions. Development 133, 2605–2616.

Bahia, A.C., Dong, Y., Blumberg, B.J., Mlambo, G., Tripathi, A., BenMarzouk-Hidalgo, O.J., Chandra, R., Dimopoulos, G., 2014. Exploring *Anopheles* gut bacteria for *Plasmodium* blocking activity. Environ. Microbiol. 16, 2980–2994.

Baldini, F., Segata, N., Pompon, J., Marcenac, P., Robert Shaw, W., Dabiré, R.K., Diabaté, A., Levashina, E.A., Catteruccia, F., 2014. Evidence of natural *Wolbachia* infections in field populations of *Anopheles gambiae*. Nat. Commun. 5, 3985.

Bando, H., Okado, K., Guelbeogo, W.M., Badolo, A., Aonuma, H., Nelson, B., Fukumoto, S., Xuan, X., Sagnon, N., Kanuka, H., 2013. Intra-specific diversity of *Serratia marcescens* in *Anopheles* mosquito midgut defines *Plasmodium* transmission capacity. Sci. Rep. 3, 1641.

Barillas-Mury, C., Han, Y.S., Seeley, D., Kafatos, F.C., 1999. *Anopheles gambiae* Ag-STAT, a new insect member of the STAT family, is activated in response to bacterial infection. EMBO J. 18, 959–967.

Basbous, N., Coste, F., Leone, P., Vincentelli, R., Royet, J., Kellenberger, C., Roussel, A., 2011. The *Drosophila* peptidoglycan-recognition protein LF interacts with peptidoglycan-recognition protein LC to downregulate the Imd pathway. EMBO Rep. 12, 327–333.

Baxter, R.H.G., Chang, C.-I., Chelliah, Y., Blandin, S., Levashina, E.A., Deisenhofer, J., 2007. Structural basis for conserved complement factor-like function in the antimalarial protein TEP1. Proc. Natl. Acad. Sci. 104, 11615–11620.

Belvin, M.P., Anderson, K.V., 1996. A conserved signaling pathway: the *Drosophila* toll-dorsal pathway. Annu. Rev. Cell Dev. Biol. 12, 393–416.

Bian, G., Joshi, D., Dong, Y., Lu, P., Zhou, G., Pan, X., Xu, Y., Dimopoulos, G., Xi, Z., 2013. Wolbachia invades *Anopheles stephensi* populations and induces refractoriness to *Plasmodium* infection. Science 340, 748–751.

Bischoff, V., Vignal, C., Duvic, B., Boneca, I.G., Hoffmann, J.A., Royet, J., 2006. Downregulation of the *Drosophila* immune response by peptidoglycan-recognition proteins SC1 and SC2. PLoS Pathog. 2, e14.

Blandin, S., Moita, L.F., Köcher, T., Wilm, M., Kafatos, F.C., Levashina, E.A., 2002. Reverse genetics in the mosquito *Anopheles gambiae*: targeted disruption of the Defensin gene. EMBO Rep. 3, 852–856.

Blandin, S., Shiao, S.-H., Moita, L.F., Janse, C.J., Waters, A.P., Kafatos, F.C., Levashina, E.A., 2004. Complement-like protein TEP1 is a determinant of vectorial capacity in the malaria vector *Anopheles gambiae*. Cell 116, 661–670.

Boissière, A., Tchioffo, M.T., Bachar, D., Abate, L., Marie, A., Nsango, S.E., Shahbazkia, H.R., Awono-Ambene, P.H., Levashina, E.A., Christen, R., et al., 2012. Midgut microbiota of the malaria mosquito vector *Anopheles gambiae* and interactions with plasmodium falciparum infection. PLoS Pathog. 8, e1002742.

Bourtzis, K., Dobson, S.L., Xi, Z., Rasgon, J.L., Calvitti, M., Moreira, L.A., Bossin, H.C., Moretti, R., Baton, L.A., Hughes, G.L., et al., 2014. Harnessing mosquito-Wolbachia symbiosis for vector and disease control. Acta Trop. 132 (Suppl.), S150–S163.

Buchon, N., Broderick, N.A., Chakrabarti, S., Lemaitre, B., 2009. Invasive and indigenous microbiota impact intestinal stem cell activity through multiple pathways in *Drosophila*. Genes Dev. 23, 2333–2344.

Buck, M., Nilsson, L.K.J., Brunius, C., Dabiré, R.K., Hopkins, R., Terenius, O., 2016. Bacterial associations reveal spatial population dynamics in *Anopheles gambiae* mosquitoes. Sci. Rep. 6, 22806.

Chandler, J.A., Liu, R.M., Bennett, S.N., 2015. RNA shotgun metagenomic sequencing of northern California (USA) mosquitoes uncovers viruses, bacteria, and fungi. Front. Microbiol. 6, 185.

Chouaia, B., Rossi, P., Epis, S., Mosca, M., Ricci, I., Damiani, C., Ulissi, U., Crotti, E., Daffonchio, D., Bandi, C., et al., 2012. Delayed larval development in *Anopheles* mosquitoes deprived of *Asaia* bacterial symbionts. BMC Microbiol. 12 (Suppl. 1), S2.

Christophides, G.K., Zdobnov, E., Barillas-Mury, C., Birney, E., Blandin, S., Blass, C., Brey, P.T., Collins, F.H., Danielli, A., Dimopoulos, G., et al., 2002. Immunity-related genes and gene families in *Anopheles gambiae*. Science 298, 159–165.

Cirimotich, C.M., Dong, Y., Clayton, A.M., Sandiford, S.L., Souza-Neto, J.A., Mulenga, M., Dimopoulos, G., 2011. Natural microbe-mediated refractoriness to *Plasmodium* infection in *Anopheles gambiae*. Science 332, 855–858.

Clayton, A.M., Cirimotich, C.M., Dong, Y., Dimopoulos, G., 2013. Caudal is a negative regulator of the *Anopheles* IMD pathway that controls resistance to *Plasmodium falciparum* infection. Dev. Comp. Immunol. 39, 323–332.

Cohuet, A., Osta, M.A., Morlais, I., Awono-Ambene, P.H., Michel, K., Simard, F., Christophides, G.K., Fontenille, D., Kafatos, F.C., 2006. *Anopheles* and *Plasmodium*: from laboratory models to natural systems in the field. EMBO Rep. 7, 1285–1289.

Collins, F.H., Sakai, R.K., Vernick, K.D., Paskewitz, S., Seeley, D.C., Miller, L.H., Collins, W.E., Campbell, C.C., Gwadz, R.W., 1986. Genetic selection of a *Plasmodium*-refractory strain of the malaria vector *Anopheles gambiae*. Science 234, 607–610.

Coon, K.L., Vogel, K.J., Brown, M.R., Strand, M.R., 2014. Mosquitoes rely on their gut microbiota for development. Mol. Ecol. 23, 2727–2739.

Cooper, D.M., Chamberlain, C.M., Lowenberger, C., 2009. Aedes FADD: a novel death domain-containing protein required for antibacterial immunity in the yellow fever mosquito, *Aedes aegypti*. Insect Biochem. Mol. Biol. 39, 47–54.

Da Silva, J.B., De Albuquerque, C.M., De Araújo, E.C., Peixoto, C.A., Hurd, H., 2000. Immune defense mechanisms of *Culex quinquefasciatus* (Diptera: Culicidae) against *Candida albicans* infection. J. Invertebr. Pathol. 76, 257–262.

Damiani, C., Ricci, I., Crotti, E., Rossi, P., Rizzi, A., Scuppa, P., Capone, A., Ulissi, U., Epis, S., Genchi, M., et al., 2010. Mosquito-bacteria symbiosis: the case of *Anopheles gambiae* and *Asaia*. Microb. Ecol. 60, 644–654.

Devaraj, S., Hemarajata, P., Versalovic, J., 2013. The human gut microbiome and body metabolism: implications for obesity and diabetes. Clin. Chem. 59, 617–628.

Dong, Y., Aguilar, R., Xi, Z., Warr, E., Mongin, E., Dimopoulos, G., 2006. *Anopheles gambiae* immune responses to human and rodent *Plasmodium* parasite species. PLoS Pathog. 2, e52.

Dong, Y., Manfredini, F., Dimopoulos, G., 2009. Implication of the mosquito midgut microbiota in the defense against malaria parasites. PLoS Pathog. 5, e1000423.

Dong, Y., Das, S., Cirimotich, C., Souza-Neto, J.A., McLean, K.J., Dimopoulos, G., 2011. Engineered *Anopheles* immunity to *Plasmodium* infection. PLoS Pathog. 7, e1002458.

Dostert, C., Jouanguy, E., Irving, P., Troxler, L., Galiana-Arnoux, D., Hetru, C., Hoffmann, J.A., Imler, J.-L., 2005. The Jak-STAT signaling pathway is required but not sufficient for the antiviral response of drosophila. Nat. Immunol. 6, 946–953.

Douglas, A.E., 1998. Nutritional interactions in insect-microbial symbioses: aphids and their symbiotic bacteria *Buchnera*. Annu. Rev. Entomol. 43, 17–37.

Dowd, P.F., Shen, S.K., 1990. The contribution of symbiotic yeast to toxin resistance of the cigarette beetle (*Lasioderma serricorne*). Entomol. Exp. Appl. 56, 241–248.

Dworkin, J., 2014. The medium is the message: interspecies and interkingdom signaling by peptidoglycan and related bacterial glycans. Annu. Rev. Microbiol. 68, 137–154.

Ellis, R.J., Bruce, K.D., Jenkins, C., Stothard, J.R., Ajarova, L., Mugisha, L., Viney, M.E., 2013. Comparison of the distal gut microbiota from people and animals in Africa. PLoS One 8, e54783.

Engel, P., Martinson, V.G., Moran, N.A., 2012. Functional diversity within the simple gut microbiota of the honey bee. Proc. Natl. Acad. Sci. 109, 11002–11007.

Filipe, S.R., Tomasz, A., Ligoxygakis, P., 2005. Requirements of peptidoglycan structure that allow detection by the *Drosophila* Toll pathway. EMBO Rep. 6, 327–333.

Frolet, C., Thoma, M., Blandin, S., Hoffmann, J.A., Levashina, E.A., 2006. Boosting NF-kappaB-dependent basal immunity of *Anopheles gambiae* aborts development of *Plasmodium berghei*. Immunity 25, 677–685.

Gaio, A.O., Gusmão, D.S., Santos, A.V., Berbert-Molina, M.A., Pimenta, P.F.P., Lemos, F.J.A., 2011. Contribution of midgut bacteria to blood digestion and egg production in *aedes aegypti* (diptera: culicidae) (L.). Parasit. Vectors 4, 105.

Garver, L.S., Dong, Y., Dimopoulos, G., 2009. Caspar controls resistance to *Plasmodium falciparum* in diverse anopheline species. PLoS Pathog. 5, e1000335.

Garver, L.S., Bahia, A.C., Das, S., Souza-Neto, J.A., Shiao, J., Dong, Y., Dimopoulos, G., 2012. *Anopheles* Imd pathway factors and effectors in infection intensity-dependent anti-Plasmodium action. PLoS Pathog. 8, e1002737.

Gendrin, M., Christophides, G.K., 2013. The *Anopheles* mosquito microbiota and their impact on pathogen transmission. In: *Anopheles* Mosquitoes - New Insights into Malaria Vectors. InTech.

Gendrin, M., Rodgers, F.H., Yerbanga, R.S., Ouédraogo, J.B., Basáñez, M.-G., Cohuet, A., Christophides, G.K., 2015. Antibiotics in ingested human blood affect the mosquito microbiota and capacity to transmit malaria. Nat. Commun. 6, 5921.

Gendrin, M., Yerbanga, R.S., Ouedraogo, J.B., Lefèvre, T., Cohuet, A., Christophides, G.K., 2016. Differential effects of azithromycin, doxycycline, and cotrimoxazole in ingested blood on the vectorial capacity of malaria mosquitoes. Open Forum Infect. Dis. 3, ofw074.

Gendrin, M., Turlure, F., Rodgers, F.H., Cohuet, A., Morlais, I., Christophides, G.K. The peptidoglycan recognition proteins PGRPLA and PGRPLB regulate Anopheles immunity to bacteria and affect infection by *Plasmodium*. J. Innate Immun., (Revis).

Gimonneau, G., Tchioffo, M.T., Abate, L., Boissière, A., Awono-Ambéné, P.H., Nsango, S.E., Christen, R., Morlais, I., 2014. Composition of *Anopheles coluzzii* and *Anopheles gambiae* microbiota from larval to adult stages. Infect. Genet. Evol. 28, 715–724.

Gottar, M., Gobert, V., Matskevich, A.A., Reichhart, J.-M., Wang, C., Butt, T.M., Belvin, M., Hoffmann, J.A., Ferrandon, D., 2006. Dual detection of fungal infections in *Drosophila* via recognition of glucans and sensing of virulence factors. Cell 127, 1425–1437.

Gupta, L., Molina-Cruz, A., Kumar, S., Rodrigues, J., Dixit, R., Zamora, R.E., Barillas-Mury, C., 2009. The STAT pathway mediates late-phase immunity against *Plasmodium* in the mosquito *Anopheles gambiae*. Cell Host Microbe 5, 498–507.

Ha, E.-M., Oh, C.-T., Bae, Y.S., Lee, W.-J., 2005. A direct role for dual oxidase in *Drosophila* gut immunity. Science 310, 847–850.

Ha, E.-M., Lee, K.-A., Park, S.H., Kim, S.-H., Nam, H.-J., Lee, H.-Y., Kang, D., Lee, W.-J., 2009a. Regulation of DUOX by the Galphaq-phospholipase Cbeta-Ca^{2+} pathway in *Drosophila* gut immunity. Dev. Cell 16, 386–397.

Ha, E.-M., Lee, K.-A., Seo, Y.Y., Kim, S.-H., Lim, J.-H., Oh, B.-H., Kim, J., Lee, W.-J., 2009b. Coordination of multiple dual oxidase-regulatory pathways in responses to commensal and infectious microbes in *Drosophila* gut. Nat. Immunol. 10, 949–957.

Hillyer, J.F., Estévez-Lao, T.Y., 2010. Nitric oxide is an essential component of the hemocyte-mediated mosquito immune response against bacteria. Dev. Comp. Immunol. 34, 141–149.

Hillyer, J.F., Strand, M.R., 2014. Mosquito hemocyte-mediated immune responses. Curr. Opin. Insect Sci. 3, 14–21.

Hillyer, J.F., Schmidt, S.L., Christensen, B.M., 2003. Rapid phagocytosis and melanization of bacteria and *Plasmodium* sporozoites by hemocytes of the mosquito *Aedes aegypti*. J. Parasitol. 89, 62–69.

Hillyer, J.F., Schmidt, S.L., Christensen, B.M., 2004. The antibacterial innate immune response by the mosquito *Aedes aegypti* is mediated by hemocytes and independent of Gram type and pathogenicity. Microbe. Infect. 6, 448–459.

Hillyer, J.F., Barreau, C., Vernick, K.D., 2007. Efficiency of salivary gland invasion by malaria sporozoites is controlled by rapid sporozoite destruction in the mosquito haemocoel. Int. J. Parasitol. 37, 673–681.

Hoffmann, A.A., Montgomery, B.L., Popovici, J., Iturbe-Ormaetxe, I., Johnson, P.H., Muzzi, F., Greenfield, M., Durkan, M., Leong, Y.S., Dong, Y., et al., 2011. Successful establishment of Wolbachia in *Aedes* populations to suppress dengue transmission. Nature 476, 454–457.

Holt, R.A., Subramanian, G.M., Halpern, A., Sutton, G.G., Charlab, R., Nusskern, D.R., Wincker, P., Clark, A.G., Ribeiro, J.M.C., Wides, R., et al., 2002. The genome sequence of the malaria mosquito *Anopheles gambiae*. Science 298, 129–149.

Horton, A.A., Lee, Y., Coulibaly, C.A., Rashbrook, V.K., Cornel, A.J., Lanzaro, G.C., Luckhart, S., 2010. Identification of three single nucleotide polymorphisms in *Anopheles gambiae* immune signaling genes that are associated with natural *Plasmodium falciparum* infection. Malar. J. 9, 160.

Kajla, M.K., Shi, L., Li, B., Luckhart, S., Li, J., Paskewitz, S.M., 2011. A new role for an old antimicrobial: lysozyme c-1 can function to protect malaria parasites in *Anopheles* mosquitoes. PLoS One 6, e19649.

Kaneko, T., Yano, T., Aggarwal, K., Lim, J.-H., Ueda, K., Oshima, Y., Peach, C., Erturk-Hasdemir, D., Goldman, W.E., Oh, B.-H., et al., 2006. PGRP-LC and PGRP-LE have essential yet distinct functions in the drosophila immune response to monomeric DAP-type peptidoglycan. Nat. Immunol. 7, 715–723.

Kaufman, M.G., Klug, M.J., 1991. The contribution of hindgut bacteria to dietary carbohydrate utilization by crickets (Orthoptera: Gryllidae). Comp. Biochem. Physiol. Part A Physiol. 98, 117–123.

Kim, W., Koo, H., Richman, A.M., Seeley, D., Vizioli, J., Klocko, A.D., O'Brochta, D.A., 2004. Ectopic expression of a cecropin transgene in the human malaria vector mosquito *Anopheles gambiae* (Diptera: Culicidae): effects on susceptibility to *Plasmodium*. J. Med. Entomol. 41, 447–455.

Kim, M., Lee, J.H., Lee, S.Y., Kim, E., Chung, J., 2006. Caspar, a suppressor of antibacterial immunity in *Drosophila*. Proc. Natl. Acad. Sci. U.S.A. 103, 16358–16363.

Kleino, A., Myllymäki, H., Kallio, J., Vanha-Aho, L.-M., Oksanen, K., Ulvila, J., Hultmark, D., Valanne, S., Rämet, M., 2008. Pirk is a negative regulator of the *Drosophila* Imd pathway. J. Immunol. 180, 5413–5422.

Kumar, S., Christophides, G.K., Cantera, R., Charles, B., Han, Y.S., Meister, S., Dimopoulos, G., Kafatos, F.C., Barillas-Mury, C., 2003. The role of reactive oxygen species on *Plasmodium* melanotic encapsulation in *Anopheles gambiae*. Proc. Natl. Acad. Sci. U.S.A. 100, 14139–14144.

Kumar, S., Molina-Cruz, A., Gupta, L., Rodrigues, J., Barillas-Mury, C., 2010. A peroxidase/dual oxidase system modulates midgut epithelial immunity in *Anopheles gambiae*. Science 327, 1644–1648.

Lee, K.-A., Kim, S.-H., Kim, E.-K., Ha, E.-M., You, H., Kim, B., Kim, M.-J., Kwon, Y., Ryu, J.-H., Lee, W.-J., 2013. Bacterial-derived uracil as a modulator of mucosal immunity and gut-microbe homeostasis in *Drosophila*. Cell 153, 797–811.

Lee, K.-A., Kim, B., Bhin, J., Kim, D.H., You, H., Kim, E.-K., Kim, S.-H., Ryu, J.-H., Hwang, D., Lee, W.-J., 2015. Bacterial uracil modulates *Drosophila* DUOX-dependent gut immunity via Hedgehog-induced signaling endosomes. Cell Host Microbe 17, 191–204.

Lemaitre, B., Hoffmann, J., 2007. The host defense of *Drosophila melanogaster*. Annu. Rev. Immunol. 25, 697–743.

Lemaitre, B., Nicolas, E., Michaut, L., Reichhart, J.M., Hoffmann, J.A., 1996. The dorsoventral regulatory gene cassette spätzle/Toll/cactus controls the potent antifungal response in *Drosophila* adults. Cell 86, 973–983.

Levashina, E.A., Moita, L.F., Blandin, S., Vriend, G., Lagueux, M., Kafatos, F.C., 2001. Conserved role of a complement-like protein in phagocytosis revealed by dsRNA knockout in cultured cells of the mosquito, *Anopheles gambiae*. Cell 104, 709–718.

Lhocine, N., Ribeiro, P.S., Buchon, N., Wepf, A., Wilson, R., Tenev, T., Lemaitre, B., Gstaiger, M., Meier, P., Leulier, F., 2008. PIMS modulates immune tolerance by negatively regulating *Drosophila* innate immune signaling. Cell Host Microbe 4, 147–158.

Lim, J., Gowda, D.C., Krishnegowda, G., Luckhart, S., 2005. Induction of nitric oxide synthase in *Anopheles stephensi* by *Plasmodium falciparum*: mechanism of signaling and the role of parasite glycosylphosphatidylinositols. Infect. Immun. 73, 2778–2789.

Lim, J.-H., Kim, M.-S., Kim, H.-E., Yano, T., Oshima, Y., Aggarwal, K., Goldman, W.E., Silverman, N., Kurata, S., Oh, B.-H., 2006. Structural basis for preferential recognition of diaminopimelic acid-type peptidoglycan by a subset of peptidoglycan recognition proteins. J. Biol. Chem. 281, 8286–8295.

Lindh, J.M., Borg-Karlson, A.-K., Faye, I., 2008. Transstadial and horizontal transfer of bacteria within a colony of *Anopheles gambiae* (Diptera: Culicidae) and oviposition response to bacteria-containing water. Acta Trop. 107, 242–250.

Luckhart, S., Vodovotz, Y., Cui, L., Rosenberg, R., 1998. The mosquito *Anopheles stephensi* limits malaria parasite development with inducible synthesis of nitric oxide. Proc. Natl. Acad. Sci. U.S.A. 95, 5700–5705.

Luplertlop, N., Surasombatpattana, P., Patramool, S., Dumas, E., Wasinpiyamongkol, L., Saune, L., Hamel, R., Bernard, E., Sereno, D., Thomas, F., et al., 2011. Induction of a peptide with activity against a broad spectrum of pathogens in the *Aedes aegypti* salivary gland, following infection with Dengue Virus. PLoS Pathog. 7, e1001252.

Lyke, K.E., Laurens, M., Adams, M., Billingsley, P.F., Richman, A., Loyevsky, M., Chakravarty, S., Plowe, C.V., Sim, B.K.L., Edelman, R., et al., 2010. Plasmodium falciparum malaria challenge by the bite of aseptic *Anopheles stephensi* mosquitoes: results of a randomized infectivity trial. PLoS One 5, e13490.

Magalhaes, T., Leandro, D.C., Ayres, C.F.J., 2010. Knockdown of REL2, but not defensin A, augments *Aedes aegypti* susceptibility to *Bacillus subtilis* and *Escherichia coli*. Acta Trop. 113, 167–173.

Maillet, F., Bischoff, V., Vignal, C., Hoffmann, J., Royet, J., 2008. The *Drosophila* peptidoglycan recognition protein PGRP-LF blocks PGRP-LC and IMD/JNK pathway activation. Cell Host Microbe 3, 293–303.

Manda, H., Gouagna, L.C., Nyandat, E., Kabiru, E.W., Jackson, R.R., Foster, W.A., Githure, J.I., Beier, J.C., Hassanali, A., 2007. Discriminative feeding behaviour of *Anopheles gambiae* s.s. on endemic plants in western Kenya. Med. Vet. Entomol. 21, 103–111.

Marchesi, J.R., Adams, D.H., Fava, F., Hermes, G.D.A., Hirschfield, G.M., Hold, G., Quraishi, M.N., Kinross, J., Smidt, H., Tuohy, K.M., et al., 2015. The gut microbiota and host health: a new clinical frontier. Gut 65 (2), 330–339.

Mathieu, A., Delmont, T.O., Vogel, T.M., Robe, P., Nalin, R., Simonet, P., 2013. Life on human surfaces: skin metagenomics. PLoS One 8, e65288.

McMeniman, C.J., Lane, R.V., Cass, B.N., Fong, A.W.C., Sidhu, M., Wang, Y.-F., O'Neill, S.L., 2009. Stable introduction of a life-shortening *Wolbachia* infection into the mosquito *Aedes aegypti*. Science 323, 141–144.

Meister, S., Kanzok, S.M., Zheng, X.-L., Luna, C., Li, T.-R., Hoa, N.T., Clayton, J.R., White, K.P., Kafatos, F.C., Christophides, G.K., et al., 2005. Immune signaling pathways regulating bacterial and malaria parasite infection of the mosquito *Anopheles gambiae*. Proc. Natl. Acad. Sci. U.S.A. 102, 11420–11425.

Meister, S., Agianian, B., Turlure, F., Relógio, A., Morlais, I., Kafatos, F.C., Christophides, G.K., 2009. *Anopheles gambiae* PGRPLC-mediated defense against bacteria modulates infections with malaria parasites. PLoS Pathog. 5, e1000542.

Michel, K., Suwanchaichinda, C., Morlais, I., Lambrechts, L., Cohuet, A., Awono-Ambene, P.H., Simard, F., Fontenille, D., Kanost, M.R., Kafatos, F.C., 2006. Increased melanizing activity in *Anopheles gambiae* does not affect development of *Plasmodium falciparum*. Proc. Natl. Acad. Sci. U.S.A. 103, 16858–16863.

Minard, G., Tran, F.H., Raharimalala, F.N., Hellard, E., Ravelonandro, P., Mavingui, P., Valiente Moro, C., 2013. Prevalence, genomic and metabolic profiles of *Acinetobacter* and *Asaia* associated with field-caught *Aedes albopictus* from Madagascar. FEMS Microbiol. Ecol. 83, 63–73.

Mitraka, E., Stathopoulos, S., Siden-Kiamos, I., Christophides, G.K., Louis, C., 2013. *Asaia* accelerates larval development of *Anopheles gambiae*. Pathog. Glob. Health 107, 305–311.

Mitri, C., Jacques, J.-C., Thiery, I., Riehle, M.M., Xu, J., Bischoff, E., Morlais, I., Nsango, S.E., Vernick, K.D., Bourgouin, C., 2009. Fine pathogen discrimination within the APL1 gene family protects *Anopheles gambiae* against human and rodent malaria species. PLoS Pathog. 5, e1000576.

Moita, L.F., Wang-Sattler, R., Michel, K., Zimmermann, T., Blandin, S., Levashina, E.A., Kafatos, F.C., 2005. In vivo identification of novel regulators and conserved pathways of phagocytosis in *A. gambiae*. Immunity 23, 65–73.

Molina-Cruz, A., DeJong, R.J., Charles, B., Gupta, L., Kumar, S., Jaramillo-Gutierrez, G., Barillas-Mury, C., 2008. Reactive oxygen species modulate *Anopheles gambiae* immunity against bacteria and *Plasmodium*. J. Biol. Chem. 283, 3217–3223.

Molina-Cruz, A., Garver, L.S., Alabaster, A., Bangiolo, L., Haile, A., Winikor, J., Ortega, C., van Schaijk, B.C.L., Sauerwein, R.W., Taylor-Salmon, E., et al., 2013. The human malaria parasite Pfs47 gene mediates evasion of the mosquito immune system. Science 340, 984–987.

Molina-Cruz, A., Canepa, G.E., Kamath, N., Pavlovic, N.V., Mu, J., Ramphul, U.N., Ramirez, J.L., Barillas-Mury, C., 2015. Plasmodium evasion of mosquito immunity and global malaria transmission: the lock-and-key theory. Proc. Natl. Acad. Sci. U.S.A. 112, 15178–15183.

Moll, R.M., Romoser, W.S., Modrzakowski, M.C., Moncayo, A.C., Lerdthusnee, K., 2001. Meconial peritrophic membranes and the fate of midgut bacteria during mosquito (Diptera: Culicidae) metamorphosis. J. Med. Entomol. 38, 29–32.

Myllymäki, H., Rämet, M., 2014. JAK/STAT pathway in *Drosophila* immunity. Scand. J. Immunol. 79, 377–385.

Ngwa, C.J., Glöckner, V., Abdelmohsen, U.R., Scheuermayer, M., Fischer, R., Hentschel, U., Pradel, G., 2013. 16S rRNA gene-based identification of *Elizabethkingia meningoseptica* (Flavobacteriales: Flavobacteriaceae) as a dominant midgut bacterium of the Asian malaria vector *Anopheles stephensi* (Dipteria: Culicidae) with antimicrobial activities. J. Med. Entomol. 50, 404–414.

Nikoh, N., Hosokawa, T., Oshima, K., Hattori, M., Fukatsu, T., 2011. Reductive evolution of bacterial genome in insect gut environment. Genome Biol. Evol. 3, 702–714.

Oliveira, J.H.M., Gonçalves, R.L.S., Lara, F.A., Dias, F.A., Gandara, A.C.P., Menna-Barreto, R.F.S., Edwards, M.C., Laurindo, F.R.M., Silva-Neto, M.A.C., Sorgine, M.H.F., et al., 2011. Blood meal-derived heme decreases ROS levels in the midgut of *Aedes aegypti* and allows proliferation of intestinal microbiota. PLoS Pathog. 7, e1001320.

Osei-Poku, J., Mbogo, C.M., Palmer, W.J., Jiggins, F.M., 2012. Deep sequencing reveals extensive variation in the gut microbiota of wild mosquitoes from Kenya. Mol. Ecol. 21 (20), 5138–5150 n/a-n/a.

Osta, M.A., Christophides, G.K., Kafatos, F.C., 2004. Effects of mosquito genes on *Plasmodium* development. Science 303, 2030–2032.

Pan, X., Zhou, G., Wu, J., Bian, G., Lu, P., Raikhel, A.S., Xi, Z., 2012. Wolbachia induces reactive oxygen species (ROS)-dependent activation of the Toll pathway to control dengue virus in the mosquito *Aedes aegypti*. Proc. Natl. Acad. Sci. U.S.A. 109, E23–E31.

Paredes, J.C., Welchman, D.P., Poidevin, M., Lemaitre, B., 2011. Negative regulation by amidase PGRPs shapes the *Drosophila* antibacterial response and protects the fly from *Innocuous* infection. Immunity 35, 770–779.

Persson, C., Oldenvi, S., Steiner, H., 2007. Peptidoglycan recognition protein LF: a negative regulator of *Drosophila* immunity. Insect Biochem. Mol. Biol. 37, 1309–1316.

Petra, A.I., Panagiotidou, S., Hatziagelaki, E., Stewart, J.M., Conti, P., Theoharides, T.C., 2015. Gut-microbiota-brain axis and its effect on neuropsychiatric disorders with suspected immune dysregulation. Clin. Ther. 37, 984–995.

Pope, P.B., Mackenzie, A.K., Gregor, I., Smith, W., Sundset, M.A., McHardy, A.C., Morrison, M., Eijsink, V.G.H., 2012. Metagenomics of the Svalbard reindeer rumen microbiome reveals abundance of polysaccharide utilization loci. PLoS One 7, e38571.

Povelones, M., Waterhouse, R.M., Kafatos, F.C., Christophides, G.K., 2009. Leucine-rich repeat protein complex activates mosquito complement in defense against *Plasmodium* parasites. Science 324, 258–261.

Povelones, M., Bhagavatula, L., Yassine, H., Tan, L.A., Upton, L.M., Osta, M.A., Christophides, G.K., 2013. The CLIP-domain serine protease homolog SPCLIP1 regulates complement recruitment to microbial surfaces in the malaria mosquito *Anopheles gambiae*. PLoS Pathog. 9, e1003623.

Pumpuni, C.B., Demaio, J., Kent, M., Davis, J.R., Beier, J.C., 1996. Bacterial population dynamics in three anopheline species: the impact on *Plasmodium* sporogonic development. Am. J. Trop. Med. Hyg. 54, 214–218.

Ramirez, J.L., Dimopoulos, G., 2010. The Toll immune signaling pathway control conserved anti-dengue defenses across diverse *Ae. aegypti* strains and against multiple dengue virus serotypes. Dev. Comp. Immunol. 34, 625–629.

Ramirez, J.L., Short, S.M., Bahia, A.C., Saraiva, R.G., Dong, Y., Kang, S., Tripathi, A., Mlambo, G., Dimopoulos, G., 2014a. Chromobacterium Csp_P reduces malaria and dengue infection in vector mosquitoes and has entomopathogenic and in vitro anti-pathogen activities. PLoS Pathog. 10, e1004398.

Ramirez, J.L., Garver, L.S., Brayner, F.A., Alves, L.C., Rodrigues, J., Molina-Cruz, A., Barillas-Mury, C., 2014b. The role of hemocytes in *Anopheles gambiae* antiplasmodial immunity. J. Innate Immun. 6, 119–128.

Ramos-Castañeda, J., González, C., Jiménez, M.A., Duran, J., Hernández-Martínez, S., Rodríguez, M.H., Lanz-Mendoza, H., 2008. Effect of nitric oxide on Dengue virus replication in *Aedes aegypti* and *Anopheles albimanus*. Intervirology 51, 335–341.

Ramphul, U.N., Garver, L.S., Molina-Cruz, A., Canepa, G.E., Barillas-Mury, C., 2015. Plasmodium falciparum evades mosquito immunity by disrupting JNK-mediated apoptosis of invaded midgut cells. Proc. Natl. Acad. Sci. 112, 1273–1280.

Read, A.F., Lynch, P.A., Thomas, M.B., 2009. How to make evolution-proof insecticides for malaria control. PLoS Biol. 7, e1000058.

Riehle, M.M., Markianos, K., Niaré, O., Xu, J., Li, J., Touré, A.M., Podiougou, B., Oduol, F., Diawara, S., Diallo, M., et al., 2006. Natural malaria infection in *Anopheles gambiae* is regulated by a single genomic control region. Science 312, 577–579.

Riehle, M.M., Xu, J., Lazzaro, B.P., Rottschaefer, S.M., Coulibaly, B., Sacko, M., Niare, O., Morlais, I., Traore, S.F., Vernick, K.D., 2008. *Anopheles gambiae* APL1 is a family of variable LRR proteins required for Rel1-mediated protection from the malaria parasite, *Plasmodium berghei*. PLoS One 3, e3672.

Rosenberg, E., Zilber-Rosenberg, I., 2011. Symbiosis and development: the hologenome concept. Birth Defects Res. C. Embryo Today 93, 56–66.

Ryu, J.-H., Kim, S.-H., Lee, H.-Y., Bai, J.Y., Nam, Y.-D., Bae, J.-W., Lee, D.G., Shin, S.C., Ha, E.-M., Lee, W.-J., 2008. Innate immune homeostasis by the homeobox gene caudal and commensal-gut mutualism in *Drosophila*. Science 319, 777–782.

Schaub, G.A., Eichler, S., 1998. The effects of aposymbiosis and of an infection with *Blastocrithidia triatomae* (Trypanosomatidae) on the tracheal system of the reduviid bugs *Rhodnius prolixus* and *Triatoma infestans*. J. Insect Physiol. 44, 131–140.

Schnitger, A.K.D., Kafatos, F.C., Osta, M.A., 2007. The melanization reaction is not required for survival of *Anopheles gambiae* mosquitoes after bacterial infections. J. Biol. Chem. 282, 21884–21888.

Schwartz, A., Koella, J.C., 2002. Melanization of plasmodium falciparum and C-25 sephadex beads by field-caught *Anopheles gambiae* (Diptera: Culicidae) from southern Tanzania. J. Med. Entomol. 39, 84–88.

Sharma, A., Dhayal, D., Singh, O.P., Adak, T., Bhatnagar, R.K., 2013. Gut microbes influence fitness and malaria transmission potential of Asian malaria vector *Anopheles stephensi*. Acta Trop. 128, 41–47.

Shin, S.W., Kokoza, V., Bian, G., Cheon, H.-M., Kim, Y.J., Raikhel, A.S., 2005. REL1, a homologue of *Drosophila* dorsal, regulates toll antifungal immune pathway in the female mosquito *Aedes aegypti*. J. Biol. Chem. 280, 16499–16507.

Shin, S.W., Bian, G., Raikhel, A.S., 2006. A toll receptor and a cytokine, Toll5A and Spz1C, are involved in toll antifungal immune signaling in the mosquito *Aedes aegypti*. J. Biol. Chem. 281, 39388–39395.

Shin, S.C., Kim, S.-H., You, H., Kim, B., Kim, A.C., Lee, K.-A., Yoon, J.-H., Ryu, J.-H., Lee, W.-J., 2011. Drosophila microbiome modulates host developmental and metabolic homeostasis via insulin signaling. Science 334, 670–674.

Sommer, F., Bäckhed, F., 2013. The gut microbiota–masters of host development and physiology. Nat. Rev. Microbiol. 11, 227–238.

Souza-Neto, J.A., Sim, S., Dimopoulos, G., 2009. An evolutionary conserved function of the JAK-STAT pathway in anti-dengue defense. Proc. Natl. Acad. Sci. U.S.A. 106, 17841–17846.

Stathopoulos, S., Neafsey, D.E., Lawniczak, M.K.N., Muskavitch, M.A.T., Christophides, G.K., 2014. Genetic dissection of *Anopheles gambiae* gut epithelial responses to *Serratia marcescens*. PLoS Pathog. 10, e1003897.

Storelli, G., Defaye, A., Erkosar, B., Hols, P., Royet, J., Leulier, F., 2011. *Lactobacillus plantarum* promotes *Drosophila* systemic growth by modulating hormonal signals through TOR-dependent nutrient sensing. Cell Metab. 14, 403–414.

Tchioffo, M.T., Boissière, A., Churcher, T.S., Abate, L., Gimonneau, G., Nsango, S.E., Awono-Ambéné, P.H., Christen, R., Berry, A., Morlais, I., 2013. Modulation of malaria infection in *Anopheles gambiae* mosquitoes exposed to natural midgut bacteria. PLoS One 8, e81663.

Villarino, A.V., Kanno, Y., Ferdinand, J.R., O'Shea, J.J., 2015. Mechanisms of Jak/STAT signaling in immunity and disease. J. Immunol. 194, 21–27.

Villegas, L.M., Pimenta, P.F.P., 2014. Metagenomics, paratransgenesis and the *Anopheles* microbiome: a portrait of the geographical distribution of the anopheline microbiota based on a meta-analysis of reported taxa. Memórias Do Inst. Oswaldo Cruz 109, 672–684.

Vizioli, J., Bulet, P., Charlet, M., Lowenberger, C., Blass, C., Müller, H.M., Dimopoulos, G., Hoffmann, J., Kafatos, F.C., Richman, A., 2000. Cloning and analysis of a cecropin gene from the malaria vector mosquito, *Anopheles gambiae*. Insect Mol. Biol. 9, 75–84.

Vizioli, J., Bulet, P., Hoffmann, J.A., Kafatos, F.C., Müller, H.M., Dimopoulos, G., 2001a. Gambicin: a novel immune responsive antimicrobial peptide from the malaria vector *Anopheles gambiae*. Proc. Natl. Acad. Sci. U.S.A. 98, 12630–12635.

Vizioli, J., Richman, A.M., Uttenweiler-Joseph, S., Blass, C., Bulet, P., 2001b. The defensin peptide of the malaria vector mosquito *Anopheles gambiae*: antimicrobial activities and expression in adult mosquitoes. Insect Biochem. Mol. Biol. 31, 241–248.

Waldock, J., Olson, K.E., Christophides, G.K., 2012. *Anopheles gambiae* antiviral immune response to systemic O'nyong-nyong infection. PLoS Negl. Trop. Dis. 6, e1565.

Wang, Y., Gilbreath, T.M., Kukutla, P., Yan, G., Xu, J., 2011. Dynamic gut microbiome across life history of the malaria mosquito *Anopheles gambiae* in Kenya. PLoS One 6, e24767.

Wang, B., Pakpour, N., Napoli, E., Drexler, A., Glennon, E.K.K., Surachetpong, W., Cheung, K., Aguirre, A., Klyver, J.M., Lewis, E.E., et al., 2015. *Anopheles stephensi* p38 MAPK signaling regulates innate immunity and bioenergetics during *Plasmodium falciparum* infection. Parasit. Vectors 8, 424.

Warr, E., Das, S., Dong, Y., Dimopoulos, G., 2008. The Gram-negative bacteria-binding protein gene family: its role in the innate immune system of *Anopheles gambiae* and in anti-Plasmodium defence. Insect Mol. Biol. 17, 39–51.

Waterhouse, R.M., Kriventseva, E.V., Meister, S., Xi, Z., Alvarez, K.S., Bartholomay, L.C., Barillas-Mury, C., Bian, G., Blandin, S., Christensen, B.M., et al., 2007. Evolutionary dynamics of immune-related genes and pathways in disease-vector mosquitoes. Science 316, 1738–1743.

Weber, A.N.R., Tauszig-Delamasure, S., Hoffmann, J.A., Lelièvre, E., Gascan, H., Ray, K.P., Morse, M.A., Imler, J.-L., Gay, N.J., 2003. Binding of the *Drosophila* cytokine Spätzle to Toll is direct and establishes signaling. Nat. Immunol. 4, 794–800.

WHO, World Malaria Report, 2015.

Wilke, A.B.B., Marrelli, M.T., 2015. Paratransgenesis: a promising new strategy for mosquito vector control. Parasit. Vectors 8, 342.

Xi, Z., Ramirez, J.L., Dimopoulos, G., 2008. The *Aedes aegypti* toll pathway controls dengue virus infection. PLoS Pathog. 4, e1000098.

Yassine, H., Osta, M.A., 2010. *Anopheles gambiae* innate immunity. Cell. Microbiol. 12, 1–9.

Yassine, H., Kamareddine, L., Osta, M.A., 2012. The mosquito melanization response is implicated in defense against the entomopathogenic fungus *Beauveria bassiana*. PLoS Pathog. 8, e1003029.

Yassine, H., Kamareddine, L., Chamat, S., Christophides, G.K., Osta, M.A., 2014. A serine protease homolog negatively regulates TEP1 consumption in systemic infections of the malaria vector *Anopheles gambiae*. J. Innate Immun. 6, 806–818.

Zaidman-Rémy, A., Hervé, M., Poidevin, M., Pili-Floury, S., Kim, M.-S., Blanot, D., Oh, B.-H., Ueda, R., Mengin-Lecreulx, D., Lemaitre, B., 2006. The *Drosophila* amidase PGRP-LB modulates the immune response to bacterial infection. Immunity 24, 463–473.

Zaidman-Rémy, A., Poidevin, M., Hervé, M., Welchman, D.P., Paredes, J.C., Fahlander, C., Steiner, H., Mengin-Lecreulx, D., Lemaitre, B., 2011. Drosophila immunity: analysis of PGRP-SB1 expression, enzymatic activity and function. PLoS One 6, e17231.

Zou, Z., Shin, S.W., Alvarez, K.S., Kokoza, V., Raikhel, A.S., 2010. Distinct melanization pathways in the mosquito *Aedes aegypti*. Immunity 32, 41–53.

Zug, R., Hammerstein, P., 2012. Still a host of hosts for *Wolbachia*: analysis of recent data suggests that 40% of terrestrial arthropod species are infected. PLoS One 7, e38544.

CHAPTER

7

Using an Endosymbiont to Control Mosquito-Transmitted Disease

Eric P. Caragata, Luciano A. Moreira

Centro de Pesquisas René Rachou – Fiocruz, Belo Horizonte, Brazil

The bacterial endosymbiont *Wolbachia pipientis*, more commonly referred to as *Wolbachia*, represents a promising agent of mosquito biological control that could potentially reduce mortality and morbidity associated with mosquito-transmitted pathogens including the parasites that cause malaria and the viruses that cause dengue, chikungunya, and West Nile fever. Recent evidence also suggests *Wolbachia* could be an effective means to limit transmission of Zika virus—a fact that will likely lead to more widespread use of the bacterium in mosquito control. In this chapter we discuss potential approaches and issues associated with using *Wolbachia* as a disease-control agent in mosquitoes.

THE BIOLOGY OF WOLBACHIA *PIPIENTIS*

Wolbachia is a Gram-negative, obligate bacterial endosymbiont of insects. The bacterium was first described in 1924 by Wolbach and Hertig who observed the granular, circular bacterium in *Culex quinquefasciatus* mosquito eggs, larvae, and adults, with the greatest bacterial density observed in the gonads.

Indeed, infection in host reproductive tissues is a common feature of all *Wolbachia* strains, most prominently in the ovaries where the bacterium heavily infects individual oocytes and nearby nurse cells. Consequently, *Wolbachia* infections are most commonly spread from mother to progeny, and most strains are associated with high maternal transmission rates.

Wolbachia infection is highly prevalent among arthropods, where historical estimates suggested that between 60% and 80% of taxa are infected. More recent estimates based on an expanding pool of molecular data suggest that roughly 40% of all terrestrial insect species harbor the bacterium (Zug and Hammerstein, 2012). This would still make *Wolbachia* the most prevalent bacterial endosymbiont on the planet, a fact that has led some to label it a pandemic. *Wolbachia* infect a diverse range of host taxa including spiders, isopods, flies, ants, beetles, and worms. Yet, genetic phylogeny of infected hosts, and of their infecting *Wolbachia* strains, suggests that

Arthropod Vector: Controller of Disease Transmission, Volume 1
http://dx.doi.org/10.1016/B978-0-12-805350-8.00007-6

WOLBACHIA: PROS, CONS, AND LESSONS FROM THE FIELD

Wolbachia-based control

Mosquitoes infected with the endosymbiotic bacterium *Wolbachia* have been deployed in the field to reduce threats to human health caused by diseases such as dengue. These releases take two main forms. *Wolbachia*-based population suppression is used to eradicate target mosquito populations. It offers short-term control and compares favorably to other suppression techniques such as the sterile insect technique (SIT), because *Wolbachia* infection has only little effect on male competitiveness. *Wolbachia*-based population replacement is used to supplant existing mosquito populations with those less capable of transmitting target pathogens. It can be slow to implement but provides enduring control because of the self-sustaining nature of the bacterium in insect populations. In the future, these approaches could potentially be applied against a wide range of pathogens, but the difficulties associated with generating *Wolbachia* infections in new vector species must first be overcome.

Wolbachia-infected mosquitoes

Successful *Wolbachia* releases depend on the nature of the *Wolbachia*–host relationship. Field trials have demonstrated that highly virulent *Wolbachia* strains disastrously impact host competitiveness. More suitable strains induce few fitness effects but also high levels of desirable traits such as cytoplasmic incompatibility (CI) or antipathogen effects. Data from established field populations indicate that these critical antipathogen effects have been stable since initial releases, suggesting that this form of control could be persistent over the long term.

The release area

Understanding and accounting for the impact of environmental factors such as climate, local mosquito population characteristics and dynamics, and local mosquito control programs has been critical to the success of *Wolbachia* releases. The same is true for social issues, where extensive community engagement programs have proven vital to obtaining high levels of stakeholder buy-in. A further concern is regulatory risk, which can be an impediment to any biocontrol program. Yet, the advent of releases of *Wolbachia*-infected mosquitoes in several countries provides an established regulatory pathway that might make this approach a comparatively appealing option.

the bacterium has frequently moved between host species. In comparison to maternal transmission, horizontal transmission appears to be relatively rare but could potentially result from predation during parasitism or by consumption of a resource containing *Wolbachia*.

Several *Wolbachia* genomes have been sequenced and annotated, and they are generally small in size, containing only around 1000 coding sequences. Analysis of these genes indicates that the bacteria do not possess all of the metabolic pathways required to survive without a host. They are consequently highly reliant on their host, with infections characterized by high levels of host–symbiont interaction. *Wolbachia* bind host cytoskeletal proteins, alter host gene expression, and coopt host signaling pathways to manipulate the cellular environment and promote their own replication and transmission. The scope of these host–symbiont interactions differs substantially among infected hosts, and this can result in bizarre manipulations of host physiology.

The majority of *Wolbachia* infections induce some form of reproductive parasitism,

TABLE 7.1 Types of Reproductive Parasitism Induced by *Wolbachia*

Manipulation	Effect on Progeny	Sex Ratio Distortion?	Example Strain and Host Species
Cytoplasmic incompatibility	*Wolbachia*-infected embryos develop normally. Uninfected embryos only develop if both parents were uninfected.	No	*w*Ri infection in *Drosophila simulans*
Feminization	Genetic male embryos become female. Genetic female embryos develop normally.	Yes	*w*VulM in *Armadillidium vulgare*
Male killing	Male embryos do not develop. Females develop normally.	Yes	*w*Bol1 in *Hypolimnas bolina*
Parthenogenesis	Unfertilized and/or fertilized eggs develop into female progeny.	Yes	*w*Uni in *Muscidifurax uniraptor*

wherein they alter their host's reproductive process in a manner that promotes their own replication and the chance of passing to the next generation but also typically decreases host fitness (Table 7.1). Ordinarily, such a bacterial endosymbiont would be quickly eliminated from insect populations because of selection pressure; however, the opposite is true for *Wolbachia*, both because of high rates of maternal transmission and the biological quirks associated with the reproductive parasitism the bacterium induces.

The most widespread of the reproductive manipulations is CI. This occurs when a *Wolbachia*-infected male mates with an uninfected female. The female lays eggs as normal; however, the hatch rate of these eggs is either greatly decreased (known as incomplete CI), or they all fail to hatch (known as complete CI). In contrast, *Wolbachia*-infected females are capable of producing viable progeny by mating with both *Wolbachia*-infected and *Wolbachia*-uninfected male insects. Thus in a given population, they have a larger proportion of viable potential mates. CI can cause an increase in a population's *Wolbachia* infection frequency with each generation, as an increasing proportion of the eggs that hatch contain *Wolbachia*-infected insects.

Other examples of reproductive parasitism include male killing, where all genetic male eggs fail to hatch, feminization, where male eggs become female, or parthenogenesis, where only female offspring are produced. These methods may differ in format, but the effect is the same—an increase in the chance of survival of female, *Wolbachia*-transmitting insects, and increased likelihood of propagation for the bacterium. In nature, these phenotypes reduce the effect of selection pressures against *Wolbachia* infection, even in the face of significant *Wolbachia*-induced fitness costs. They are also a means by which *Wolbachia* can invade and spread to fixation in uninfected insect populations or to spread quickly across large geographic areas, as was the case for *Drosophila simulans* fruit flies in California during the late 1980s (Turelli and Hoffmann, 1991), and it is this feature of infection that gives *Wolbachia* the potential to be used as a form of vector control.

THE USE OF *WOLBACHIA* IN MOSQUITO CONTROL PROGRAMS

Wolbachia is an important part of many biological control programs used to reduce the incidence of disease transmission by mosquitoes. Although these programs take a variety of forms, all involve the release of *Wolbachia*-infected mosquitoes into the field (Table 7.2).

TABLE 7.2 Mosquito Control Techniques Involving *Wolbachia*

	Technique	Released Mosquitoes	Status	Advantages	Disadvantages	Example Target Mosquito
Population suppression	SIT[a] (no *Wolbachia*)	Sterilized males	Widespread use in field	• Fast • Effective • Easily halted	• Male competitiveness • Accidental female release • Leaves a vacant niche	*Anopheles arabiensis*
	CI/IIT	*Wolbachia*-infected males	Field trials	• Effective • Easily halted	• Generating infected line • Accidental female release • Leaves a vacant niche	*Culex quinquefasciatus*
	IIT/SIT	Sterilized *Wolbachia*-infected females and males	Laboratory trials	• Effective • Easily halted	• Potential male competitiveness effects • Generating infected line • Leaves a vacant niche	*Aedes albopictus*
Population replacement	Life shortening	*w*MelPop-infected females and males	Field releases	• Reduced disease transmission	• Generating infected line • Requires continuous release • *w*MelPop fitness costs	*Aedes aegypti*
	Egg viability	*w*MelPop-infected females and males	Semifield trials	• Effective	• Generating infected line • Difficult to spread *w*MelPop • Leaves a vacant niche	*A. aegypti*
	Pathogen interference	*Wolbachia*-infected females and males	Field releases	• Self-sustaining • Reduced disease transmission	• Generating infected line • Potential competi tiveness issues • Difficult to halt • Slow to implement	*A. aegypti*

[a] *SIT included for comparative purposes.*

Population Suppression

Wolbachia control programs designed to crash or suppress mosquito populations are closely related to a more established form of insect biological control, the SIT. The SIT has been utilized since the 1950s and involves the release of male mosquitoes sterilized by exposure to chemicals or mild doses of radiation. In the field, these males mate with wild females, which are then unable to produce viable eggs. The consequence is rapid reduction in the size of the wild population, because large numbers of females produce no progeny. And with lower mosquito numbers in the area, there is then, theoretically, a reduced incidence of disease transmission. The SIT has been used effectively to target a range of vector and pest insects, including screwworms, fruit flies, and many mosquito species. However, there are some associated limitations, such as the requirement for mass rearing of mosquitoes and the fact that the sterilization process typically reduces the competitiveness of released males. There are issues with regard to accidental female release, which would hinder suppression efforts, because no current sexing technique is both 100% effective and timely. And finally there is an ecological question of the vacant niche left by the loss of the mosquito population, which could easily be filled by new mosquito immigrants—a problem common to all population suppression techniques.

The net result of the SIT is similar to the effects of releasing male mosquitoes infected with a CI-causing *Wolbachia* strain into the field. These males mate with wild, uninfected females and crash the population. Nominally, such a strategy would avoid the pitfalls associated with the SIT in terms of released male competiveness. However, it does not address the issue of accidental female release, because this would help to establish a *Wolbachia* infection in the target mosquito population.

CI can also occur between different *Wolbachia* strains, known as bidirectional incompatibility (BI). The result of mating in a mixed population with two *Wolbachia* strains that cause BI is that infected females can only successfully mate with males infected with the same *Wolbachia* strain. This forms the basis of a further form of mosquito control using *Wolbachia*—the incompatible insect technique (IIT). This technique is targeted at mosquito populations that are already infected with *Wolbachia*, and releasing males again serves to crash the target population. The IIT works very well in mosquito species that are naturally infected with a range of *Wolbachia* strains that cause BI—as is the case for the *Culex pipiens* species complex where there are many BI-causing genetic variants of the native *Wolbachia* strain, *w*Pip (Chen et al., 2013).

In target mosquito species where there is only one native strain or strain genetic variant, then new *Wolbachia*–host associations must be generated artificially by transferring the bacterium from one host to another through the process of transinfection (Generating and Evaluating a *Wolbachia*-Infected Line section). An example of this is seen in *Aedes polynesiensis*, a common mosquito vector of the filarial nematodes that cause lymphatic filariasis across the pacific islands. These mosquitoes are naturally infected with a native *Wolbachia* strain, and current control efforts utilize transinfections with one of the native strains from *Aedes albopictus*, *w*AlbB.

The effects of accidentally released females are less severe with the IIT than with the SIT; however, it is still possible that an infection with the released *Wolbachia* strain could be established in the field, which would detrimentally affect future releases. To overcome this, it is possible to combine the IIT with the SIT—treat all released insects with a low dose of radiation rendering accidentally released females sterile, without affecting male competitiveness, or to utilize one of the newly developed sexing techniques to reduce the chance of female release (Gilles et al., 2014). IIT field trials show that the technique is very effective; however, no mosquito control method is without its flaws, and the IIT may still be subject to issues of male competitiveness because many *Wolbachia* infections also have implicit fitness costs (Effects of Wolbachia in Different Host Species section). Regardless, *Wolbachia*-based population suppression techniques could prove to be an important part of Zika virus control efforts in the near future (Bourtzis et al., 2014).

Population Replacement

Other forms of *Wolbachia*-based mosquito control depend on the fact that CI can spread a specific *Wolbachia* strain through uninfected mosquito populations. The crux of these strategies is not population suppression but population replacement. To date, all population replacement strategies involve *Wolbachia* transinfections, and the difficulty surrounding the generation of these infected lines is the greatest impediment to using *Wolbachia* to target a greater range of mosquito vectors. The aim of population replacement strategies is that, postrelease, all mosquitoes in the target population, or at least a high proportion of individuals, carry a *Wolbachia* infection at the end of the release period. Given the reproductive advantage associated with CI, this spread can be rapid, depending on (1) the size of the mosquito population in the release area; (2) optimal environmental conditions including temperature, rainfall, and access to quality breeding sites; and (3) the relative competitiveness and fitness effects associated with the particular *Wolbachia* strain being used (Hoffmann et al., 2014). Population replacement again requires mass rearing but often without the laborious sexing process—as released females spread the bacterium and released males reduce uninfected larvae production through CI. In contrast to population suppression approaches, population replacement typically requires a longer rollout period and therefore is more suitable as a long-term control option, rather than a quick response to an ongoing disease outbreak.

Once a high level of *Wolbachia* infection is established in the target population, the importance then shifts to other physiological manipulations of host biology caused by *Wolbachia*, and here the population replacement techniques differ quite substantially. Most of the current strategies in this category are focused on the key dengue and Zika vector *Aedes aegypti*; however, there have been recent advancements, which suggest that an expansion to vectors of human malaria may be possible in the near future. Initial release experiments suggest that *Wolbachia*-infected mosquitoes do not spread beyond release areas at a high frequency, and given that it is difficult to establish an infection from a low starting frequency, there appears to be little risk of accidental invasion occurring in nontarget areas. *Wolbachia* infections in these programs can be self-sustaining in the field, depending on host competitiveness. However, once established in the field, they would be difficult to clear, and such a process would involve the use of insecticides or the release of mosquitoes with a different strain that causes BI.

Initially, population replacement was considered with a single *Wolbachia* strain in mind, *w*MelPop. This strain does not occur naturally but rather in a laboratory line of *Drosophila melanogaster* fruit flies. *w*MelPop is unique among the *Wolbachia* strains for its virulence and the deleterious physiological changes it causes. Infected *Drosophila* experience abnormally high bacterial density in many tissues, including the nervous system, and the density increases rapidly with age and greatly reduces life span.

*w*MelPop infection in mosquitoes also cuts life span in half, which is beneficial because most pathogens (including malaria-causing *Plasmodium* and dengue and Zika viruses) require a period of time to cross the mosquito midgut epithelium. Before this extrinsic incubation period (EIP), the mosquito is unable to transmit the pathogen to a new host during feeding.

Post-EIP, the pathogen would have invaded a range of mosquito tissues, including the salivary glands, which serve as a staging platform for invasion of subsequent hosts. Consequently, the vast majority of pathogen transmission comes from the older individuals in a mosquito population, and the introduction of a life-shortening agent–such as *w*MelPop–shifts the population structure toward a younger mean age, removing a large proportion of the potential vectors and theoretically reducing pathogen transmission (Brownstein et al., 2003).

As *w*MelPop does not occur naturally in mosquitoes, infections were generated via transinfection. Initial attempts involving direct transfer from *D. melanogaster* to *A. aegypti* proved unsuccessful; however, further attempts proved successful after a lengthy period of adaptation in mosquito cell culture before transfer to live mosquitoes. Infection in *A. aegypti* resulted in a less extreme life-shortening effect but induced a range of other fitness costs that reduced host competitiveness (see The Wolbachia Density Trade-Off section). Although preliminary attempts were made to characterize potential changes to mosquito population age structure resulting from *w*MelPop infection, interest in this technique has died because of the severity of the strains' associated fitness costs and the discovery of *Wolbachia*'s antipathogenic effects.

A further *w*MelPop fitness effect has been utilized as the basis for a novel form of population replacement, referred to as a crash and burn strategy. *w*MelPop infection is associated with a decrease in the long-term viability of mosquito eggs; the longer it has been since they were laid, the less likely the eggs are to hatch, with less than 5% of eggs hatching 1 month after they were laid (Ritchie et al., 2015). Frequent release of mosquitoes can lead to high levels of *w*MelPop infection in wild mosquito populations. Each *w*MelPop-infected egg that is laid has a reduced chance of hatching, particularly during dry conditions when uninfected eggs would otherwise remain viable during quiescence, waiting for rainfall to stimulate hatching. The net result of this technique is population suppression but through population replacement, which makes it unique among the *Wolbachia* strategies. While this technique could effectively reduce mosquito population sizes, it could also prove difficult to implement given the problems associated with the invasion of *w*MelPop-infected mosquitoes (see The Wolbachia Density Trade-Off section).

The most frequently employed example of *Wolbachia*-based population replacement takes advantage of the pathogen interference phenotype, the inhibition of infection and replication of pathogens in *Wolbachia*-infected insects. In mosquitoes, this effect applies to arboviruses such as dengue, chikungunya, Zika, West Nile fever and yellow fever, and nonviral pathogens including human malaria *Plasmodium* parasites, and filarial nematodes such as *Brugia pahangi* (Bian et al., 2013; Kambris et al., 2009; Moreira et al., 2009). As a consequence, the chance of pathogen transmission by *Wolbachia*-infected mosquitoes is greatly decreased. The interference mechanism is unclear, but the phenotype occurs most frequently and is more effective among *Wolbachia* infections with high bacterial density and in transinfected mosquitoes. Interference has also been associated with changes in immune gene expression, with changes in oxidative stress response and increased levels of H_2O_2, and competition for nutritional resources, although none of these effects are universal to all insects where pathogen interference is observed (Caragata et al., 2016). The classic pathogen interference example is for *A. aegypti* and dengue virus. When mosquitoes are infected with the *w*MelPop strain, viral prevalence is reduced almost to zero, as is viral presence in the saliva. Using this technique, an established *Wolbachia* infection in a mosquito population would theoretically lead to a significant decrease in the chance of pathogen transmission, and given the effects of CI, would be self-sustaining after an initial release period. This strategy has been deployed against dengue in Australia, Brazil, Colombia, Indonesia, and

TABLE 7.3 Undertaking the Release of *Wolbachia*-Infected Mosquitoes

Stage	Tasks
I. Choose type of control	• Compare desired outcome against the different *Wolbachia* techniques.
II. Test *Wolbachia*-infected line	• Test for the presence of key *Wolbachia* traits. • Compare competitiveness with wild mosquitoes.
III. Characterize field site	• Mosquito population size, composition, and distribution. • Expected variance in environmental conditions. • Site geography and layout. • Other mosquito control efforts that might impact the release.
IV. Engage stakeholders	• Politicians. • Regulators. • Health officials. • Local community members.
V. Design release parameters	• Model potential release outcomes. • Determine optimal release cohort size. • Determine where releases will occur within the field site. • Estimate duration of releases. • Develop a rearing plan to generate sufficient mosquitoes.
VI. Release	• Conduct regular monitoring of local mosquito population size, *Wolbachia* infection frequency, and local disease transmission. • Use data from monitoring to alter release design or halt releases, if necessary. • Terminate releases when local mosquito population has been eliminated, or local *Wolbachia* infection frequency is very high.
VII. Postrelease	• Maintain regular monitoring. • For population suppression techniques, recommence releases if mosquito numbers increase. • For population replacement, monitor *Wolbachia* infection frequency and measure drift in key *Wolbachia* traits. • Monitor disease transmission in the area.

Vietnam and will likely be an integral part of future Zika control efforts.

PRERELEASE CONSIDERATIONS

While *Wolbachia* infections in mosquitoes have demonstrated great potential as a control agent under laboratory conditions, taking these techniques to the field is not a simple process and involves the consideration of scientific, social, regulatory, and logistical issues (Table 7.3). The latter three are critical to the success of a control strategy but are not areas that are typically considered by scientists.

Generating and Evaluating a *Wolbachia*-Infected Line

Scientific concerns associated with releasing *Wolbachia*-infected mosquitoes center primarily on the relationship between the host mosquito and its infecting *Wolbachia* strain, and there is a clear need to pair the right strain with the right host. Many mosquitoes possess native *Wolbachia* infections, but a large proportion of key vector species do not. In these cases, or if the native strain is unsuitable for control purposes, the desired strain must be introduced artificially via transinfection (Hughes and Rasgon, 2014). This involves the injection of *Wolbachia*-infected material (purified from the cytoplasm of infected insects or from infected cell culture) into the eggs of the target species. The process is difficult in that most injected eggs fail to hatch, and many surviving adult females will blood feed or produce eggs. Obtaining a single, stably infected line typically requires 5000—10,000 injections, and even then *Wolbachia* is often lost during the initial generations.

Once the infected line has been established, the effects of *Wolbachia* infection must then be characterized to determine if key *Wolbachia* effects occur and to compare competitiveness with mosquitoes from the release area population. These studies should focus on traits such as fecundity, egg viability, mating competitiveness, host seeking behavior, and longevity, as all of these are critical to the success in the field. All of the aforementioned control strategies require a strain that has a high rate of maternal transmission and cause CI with mosquitoes in the target release area. Low maternal transmission or incomplete CI will make it harder for *Wolbachia* to spread, and to that end it is also important to stage invasion experiments, potentially including large-scale field or semifield cage trials. If pathogen interference is required, there is merit to conducting experimental infections using pathogens likely to be present in the release area—for instance, currently or recently circulating dengue virus isolates. Perhaps, the most important aspect of these assays is to look at the effect on transmission by measuring prevalence and intensity of infection in the saliva.

Planning, Modeling, and Community Engagement

In many respects, planning before release is as important as establishing the release line, and perhaps the most important step in this process is obtaining approval to release from all relevant stakeholders. The population replacement *Wolbachia* approaches involve releasing female mosquitoes into areas where people live. This will augment the number of mosquitoes in the area, increase the chance that local residents will be bitten, and could theoretically increase the chance of disease transmission (unless the release strain causes pathogen interference). As such, there is a clear need to seek approval for release from the local community, in addition to the appropriate regulatory and government bodies. As *Wolbachia* occurs in nature, and is not a genetically modified organism, this may lead community members to view the bacterium more favorably but might also cause issues of classification for regulatory agencies. Balancing the risks associated with regulatory delays and low community buy-in are a key factor in selecting a release site. Experience from previous releases of *Wolbachia*-infected mosquitoes suggests that there are benefits to targeting multiple levels in the community, holding meetings where the public can have questions answered by scientists, and maintaining these relationships during the course of releases, as this helps the community to feel a sense of ownership over the project.

Field site characterization is another critical factor in a successful release, and the first step of this process is to characterize the size, species composition, immigration rate, and spatial distribution of the local mosquito population. It is also necessary to determine how these factors might change in response to short-term environmental factors such as heavy rainfall, and longer term factors such as seasonal changes in mosquito population sizes, as these will all influence the ability of *Wolbachia* to spread.

It is also important to understand what other mosquito control measures occur in the release area. For example, if there is regular insecticide spraying conducted by local health officials, this could help to suppress the local population prior to release and could facilitate invasion. The use of insecticides during releases is a complicated issue. A difference in insecticide resistance levels between released and wild mosquitoes could affect the ability of *Wolbachia* to spread. If resistance levels were higher among wild mosquitoes, *Wolbachia*-infected mosquitoes would be less competitive and less likely to spread, whereas conversely, the bacterium would spread more easily if the wild population were more susceptible. Local insecticide spraying programs typically continue, whenever needed, during *Wolbachia* releases, as part of municipality mosquito control measures. *Wolbachia*-infected mosquitoes have to have the same chance of survival as the wild mosquitoes, and it is expected that there will be no change in the behavior of the residents in terms of mosquito breeding site eradication.

Data from one *Wolbachia* control effort are not perfectly applicable to future releases in different locations. All *Wolbachia*–host associations have different implicit fitness effects—the same strain in two different mosquito genetic backgrounds can produce different outcomes. Similarly, there can be variation in release area conditions between countries. To that end, it is also important to obtain a wide range of mosquito physiological data and to use these data in conjunction with data on the local mosquito population, and geographical and environmental variables associated with the release site. Together, these data can be used to create mathematical models that estimate the likelihood of a successful population suppression or replacement release. These models can consider a variety of scenarios accounting for variation in key factors such as (1) local mosquito population size, (2) release cohort size, (3) distribution of local population and release sites across the release area, and (4) the duration of the release period. Once ideal conditions have been estimated, release strategies can be optimized accordingly.

FIELD DEPLOYMENT

The two most important aspects of a successful field release are logistics and monitoring. Releases have typically been conducted on a weekly or fortnightly basis, usually involving release cohorts of between 10,000 and 20,000 mosquitoes for a release area encompassing 600–700 houses. Preparation and release of material on this scale necessitates effective mosquito-rearing systems be in place. To reduce the space and time costs associated with releases, *Wolbachia*-infected mosquitoes can be released as eggs rather than adults. However, maintaining a laboratory facility close to the release site can reduce logistical issues associated with mosquito distribution.

Mass rearing of mosquitoes is an issue that is more complicated than it appears, as generating sufficient eggs requires large-scale blood feeding, and this relies on human or animal blood or potentially an artificial diet. Utilizing blood from any source has implicit ethical issues, because it must be fed to mosquitoes using an artificial feeding system or fed directly from human volunteers or laboratory animals. In either case, blood must be screened for the presence of pathogens that could potentially infect the release material. Obtaining sufficient volumes of human blood may also be problematic in some countries, where there are regulatory difficulties associated with feeding mosquitoes or high levels of endemic disease transmission. Feeding animal blood is a viable alternative for mosquitoes that are uninfected by *Wolbachia*; however, some *Wolbachia*-infected mosquitoes exhibit reduced egg production, potentially because of metabolic costs associated with infection and the nutritional composition of animal blood (McMeniman et al., 2011). Using an artificial diet might overcome both of these issues; however, this too comes with disadvantages, as such diets are often expensive and laborious to prepare and would first need to be tested in conjunction with *Wolbachia*-infected mosquitoes.

Measuring the success of *Wolbachia*-based control strategies is facilitated in large part by in-release monitoring. For population replacement strategies, this includes monitoring of changes in the *Wolbachia* infection frequency during and postrelease. For population suppression strategies, this typically includes monitoring the size of the mosquito population in the release area, although similar monitoring is beneficial for replacement as well, because it can help to detect immigration or unforeseen population crashes, which might affect spread. Monitoring is typically conducted on mosquito material captured in sentinel traps deployed across the release area. Molecular analysis of this material involves screening for the presence of *Wolbachia*, or a specific pathogen, likely via quantitative PCR or a similar assay. With these data, the progress of the invasion can be tracked, and if necessary the release conditions can be altered. For example, the size of the release cohort could be increased if *Wolbachia* was struggling to gain a foothold. Monitoring for the presence of pathogens in this material might provide an indicator of an incipient outbreak, which could be used as a cue to temporarily halt releases. The same is true for monitoring vector-borne disease transmission among the human residents of the release area. However, given that they can freely move beyond the release area, cases of infection among the residents would not necessarily indicate that the control strategy is failing.

Long-term postrelease monitoring is an important means of determining the stability of a *Wolbachia* strategy overtime. For population suppression, this process can be as simple as maintaining traps in the area and waiting to see if the number of mosquitoes rises. For population replacement strategies, this is a bit more complicated. It can involve monitoring of *Wolbachia* infection frequencies but can also the capture of live mosquitoes to test for the presence and strength of the critical phenotypes. Traits such as pathogen interference in transinfected mosquitoes are not hypothesized to be stable over the long term, given the potential for resistance to infection to evolve either on the part of the host or pathogen, or the potential loss of the interference mechanism through coadaptation with the host. Native *Wolbachia* infections in mosquitoes have likely undergone long periods of coadaptation with their hosts, and there are many examples of these strains producing incomplete CI. If CI in transinfected mosquitoes became less effective, the invasive and suppressive potential of *Wolbachia* would be diminished. Given the current lack of evidence, it is not possible to determine the time frame over which these changes might occur, or whether they would occur at all. Nevertheless, long-term monitoring of *Wolbachia*-infected *A. aegypti* in Australia revealed that key phenotypes such as CI and pathogen interference had not dissipated over the 2 years postrelease, which suggests that the phenotypic effects of transinfections are likely to be stable at least in the short term, and this is likely sufficient to make a significant impact on disease transmission.

SELECTING THE RIGHT *WOLBACHIA* STRAIN

Successful control depends on selecting the right *Wolbachia* strain for the target mosquito species, and this involves finding the right balance between beneficial and detrimental *Wolbachia*-induced changes to host biology (Caragata et al., 2016). The choice of strains is currently limited by the fact that there are still only a handful of *Wolbachia* transinfections in a small number of mosquito species. However, there are a wide range of *Wolbachia* strains that naturally infect *Drosophila* species, which grow to high density, induce CI, and inhibit viral infection and consequently could prove to be good candidates for transinfection. A wider range of infections in mosquitoes would enable a greater degree of optimization of infected mosquitoes to suit release conditions and goals. For example, when using the IIT, it may be beneficial to pick a lower density strain that only causes CI, as any further phenotypes would be unnecessary and might actually limit competitiveness.

Effects of *Wolbachia* in Different Host Species

Wolbachia interact with their hosts on the molecular, physiological, and behavioral levels. The scope of these interactions varies among infected hosts, and understanding these differences is a key part of deciding on the right strain.

At the physiological level, the most common consequence of infection is reproductive parasitism (CI, male killing, feminization, and parthenogenesis). As *Wolbachia* are typically present at high density in the ovaries, infection can also cause a range of other effects on reproduction, including naturally increasing or decreasing fecundity, which is seen in the parasitoid wasps *Leptopilina heterotoma* and *Trichogramma pretiosum*. In other species, such as parasitoid wasps from the *Asobara* genus, oogenesis is impossible without *Wolbachia* infection, likely because the bacterium contributes key nutrients to this process. The bacterium can also counteract environmental stresses that detrimentally affect reproduction in uninfected insects. For example, *Wolbachia*-infected *D. melanogaster* produce more viable eggs than uninfected flies, under conditions of iron toxicity or scarcity. Physiological effects associated with infection also extend beyond reproductive biology. For instance, some strains alter host longevity, while others alter the rate of host locomotor activity (Fleury et al., 2000). These differences likely result because of unique interactions between different strains and hosts, and while the bacterium is primarily concerned with its own propagation, this does not mean that there will always be negative consequences for the host.

Consequently, it is not surprising that there is similar diversity among the behavioral changes associated with *Wolbachia* infection, although these effects are less well characterized than changes to host physiology. *Wolbachia* can affect host learning and memory. Infected *Armadillidium vulgare* isopods were less likely to remember the correct directional response to positive stimulus (Temple and Richard, 2015), suggesting that there may be some impairment of cognitive function due to *Wolbachia*. In *Drosophila*, different *Wolbachia* strains can increase or decrease host responsiveness to olfactory cues, indicating altered chemosensory response. However, infected male *D. melanogaster* are less likely to engage in aggressive, competitive behaviors because of reduced levels of the hormone octopamine and decreased expression levels of associated genes (Rohrscheib et al., 2015). Such changes might occur because of the presence of *Wolbachia* in host neural tissues.

Physiological and behavioral changes induced by *Wolbachia* likely have a molecular basis, although the mechanisms underlying the majority of these changes are poorly understood. The scope of these molecular interactions is fairly broad but often includes some manner of change to host metabolism. Here also, the scope of effects differs between associations. Nutritional mutualism, where *Wolbachia* provides metabolites that the host cannot synthesize, is commonly associated with the *Wolbachia* strains infecting nematodes. These strains do not cause reproductive manipulations such as CI and so cannot afford to engage in parasitic manipulations lest they be subject to increased host resistance. The *Wolbachia* of dipterans tend to fall more on the parasitic side of the spectrum. They typically have metabolically limited genomes and display a reliance on host metabolic process for their own propagation. However, there are also dipteran *Wolbachia* that display aspects of both traits, known as Jekyll and Hyde infections. One example occurs in the mosquito *Aedes fluviatilis*, where the native strain causes CI but also increases levels of the key sugar reserve glycogen in developing eggs, which likely provides an energy advantage to infected larvae.

Many strains affect host immunity, although here there are obvious differences between native and artificial associations. Transinfections, particularly among mosquitoes, tend to stimulate a wide range of host immune pathways, resulting in the upregulation of large swathes of immune genes including the antimicrobial peptides of the Toll and IMD immune pathways, which are responsible for pathogen recognition. In contrast, native associations typically do not have a large effect on host immunity, which is hypothesized to be due to the evolution of tolerance on the part of the host. However, some native *Wolbachia* infections increase the rate of apoptosis in host cells, indicating that some measure of resistance still occurs.

Many *Wolbachia* strains cause a perturbation of the host oxidative stress response, which can manifest as an increase in levels of reactive oxygen species or through changes in the expression of genes encoding enzymes involved in oxidative stress response, including oxidases and oxidoreductases. This effect occurs in both native and transinfections, although primarily among those where bacterial density is highest. In transinfection, these traits might indicate host resistance, whereas the fact that an oxidative stress imbalance persists among native associations has been ascribed to a switch to tolerance (Zug and Hammerstein, 2015). *Wolbachia* appear to have a strong need for host iron and heme, as many strains alter the expression of genes involved in heme biosynthesis, iron transfer, and sequestration (Kremer et al., 2009). These effects are not always beneficial to the host, as high levels of the protein ferritin, which is involved in iron sequestration, causes neurodegeneration in the brains of *Wolbachia*-infected *Drosophila* (Kosmidis et al., 2014).

The *Wolbachia* Density Trade-Off

Transinfection with *Wolbachia* can be a driver of extreme change in host biology, an effect that is perhaps best illustrated by the *w*AlbB strain. In its native host, the mosquito *A. albopictus*, *w*AlbB is associated with strong CI but appears to have few detrimental effects on its host. In fact, some *w*AlbB lineages have actually been associated with a fitness benefit in the form of increased host fecundity in comparison to uninfected *A. albopictus*. The strain has also been used in a range of transinfections, and these are associated with more extreme examples of parasitism and fitness costs. In *A. aegypti*, infection substantially increased bacterial density and increased activation of immune genes. Similar phenotypes were observed in *A. polynesiensis*, but in that species *w*AlbB also induced a high rate of mortality after blood feeding. This strain was also used to make the first transinfection of a human malaria vector, *Anopheles stephensi*, where it increased longevity (if the mosquitoes did not feed on blood) but also reduced egg viability and male competitiveness. What this suggests is that it can be difficult to predict the potential physiological consequences resulting from transinfection, even if the strain is relatively benign in its original host.

When it comes to choice of release strain there is a further consideration in the trade-off between fitness costs and desired manipulations, which is primarily driven by bacterial density. For instance, there are a wide range of *Wolbachia* strains that are present in *D. simulans*, some naturally and others that have been established via transinfection. When these lines are challenged with a pathogenic virus such as *Drosophila* C virus, or Flock house virus, there is a clear correlation between high *Wolbachia* density and increased survival (Martinez et al., 2014). However, the same strains that induce this stronger pathogen interference are also more likely to have higher fitness costs such as reduced egg viability, male fertility, and fecundity (Martinez et al., 2015). Finding the right balance in this trade-off is critical for control strategies that rely on pathogen interference, as stronger interference is a very desirable trait, but higher fitness costs could prevent successful establishment of the strain in the field.

Similar effects occur among *Wolbachia*-transinfected mosquitoes, as exhibited by the *w*MelPop strain in *A. aegypti*, which produces near complete inhibition of dengue virus (DENV), chikungunya virus (CHIKV), yellow fever virus (YFV), West Nile virus (WNV), and a range of other pathogens. However, *w*MelPop has extremely high bacterial density in a range of nonreproductive tissues and alters many aspects of host physiology inducing a serious reduction in life span, decreased egg viability, altered locomotory behavior, altered feeding rates, altered feeding behavior, differential immune gene expression, altered gene methylation, altered dopamine levels, and altered metabolic needs.

In contrast, *w*Mel-infected *A. aegypti* have approximately 6–10 times lower bacterial density. *w*Mel induces a high level of DENV interference but is slightly less efficacious than *w*MelPop in that some level of infection persists in both mosquito tissues and saliva. *w*Mel also strongly inhibits infection with Zika virus in both mosquito tissues and saliva (Dutra et al., 2016). No effects of *w*Mel infection on fecundity or egg viability have been characterized. Although initial experiments revealed that there was a life span reduction of approximately 10% in comparison to uninfected mosquitoes (Walker et al., 2011), this appears to have abated in the years since the establishment of the line, at least for female mosquitoes (Joubert et al., 2016). Interestingly, *w*Mel infection actually decreased larval development time under conditions of high nutrition, indicating that infection might actually improve some aspects of host fitness.

Both *w*MelPop- and *w*Mel-infected *A. aegypti* have been released into the field as part of population replacement control strategies. In northeastern Australia, the *w*Mel strain reached high infection frequencies in local mosquito populations after a release period of 10 weeks. Infection rates continued to climb postrelease, indicating that CI was facilitating spread and that the population was not just becoming saturated with infected mosquitoes. Regular monitoring over the course of 2–3 years revealed that although the infection frequency fluctuated, it did not drop below 0.80. Mosquitoes infected with *w*MelPop strain have been released in both Australia and Vietnam. Releases in both areas followed a similar pattern; during the release period, it took longer for infection frequencies to reach high levels, and once the releases ceased, the *Wolbachia* infection frequency gradually declined until eventually the infection was lost. This effect was likely due to decreased poor competitiveness of infected mosquitoes. This failure highlights the necessity of balancing fitness costs with levels of pathogen interference, and this can be difficult to do, given that high bacterial density drives both of these facets of infection.

PATHOGEN INTERFERENCE VERSUS PATHOGEN ENHANCEMENT

One issue of concern that has arisen in the last few years is the tendency of some *Wolbachia* strains to enhance infection with certain pathogens, with this enhancement either taking the form of an increase in prevalence or infectivity of infection (Table 7.4). But how important is this enhancement to the use of *Wolbachia* to control mosquito-transmitted disease?

Enhancement was first observed in *Anopheles gambiae* mosquitoes transiently infected with the *w*AlbB strain. Transient infection occurs when purified *Wolbachia* is injected into adult mosquitoes. Afterward, *Wolbachia* can establish infections in host somatic tissues, and although low-level infection can occur in reproductive tissues, the bacterium is not typically maternally transmitted. Transient infections have been used as a model to understand the potential effects of *Wolbachia* on *Plasmodium* infection, given the dearth of stable *Wolbachia* infections in human malaria vectors. Upon experimental infection with *Plasmodium berghei*, *w*AlbB-infected *A. gambiae* had a significantly higher number of oocysts per midgut than *Wolbachia*-uninfected mosquitoes. However, experiments conducted

TABLE 7.4 Incidences of Pathogen Enhancement by *Wolbachia*

Mosquito Species	Pathogen	Transmitted by this Mosquito in Nature?	Pathogen Infects	*Wolbachia* Strain	Type of *Wolbachia* Infection	Infection Enhancement Effect	References
Aedes fluviatilis	*Plasmodium gallinaceum*	No	Birds	*w*Flu	Native	Increased intensity in midguts	Baton et al. (2013)
Anopheles gambiae	*Plasmodium berghei*	No	Rodents	*w*AlbB	Transient	Increased intensity in midguts	Hughes et al. (2012)
Anopheles stephensi	*Plasmodium yoelii*	No	Rodents	*w*AlbB	Transient	Increased intensity in midguts at 24°C	Murdock et al. (2014)
Culex pipiens	*Plasmodium relictum*	Yes	Birds	*w*Pip	Native	Increased prevalence and intensity in midguts, increased prevalence in salivary glands	Zele et al. (2014)
Culex tarsalis	West Nile virus	Yes	Humans, birds	*w*AlbB	Transient	Increased prevalence in whole mosquitoes	Dodson et al. (2014)

in parallel with *A. gambiae* transiently infected with the *w*MelPop strain resulted in a consistent but nonsignificant decrease in *P. berghei* oocyst numbers. This result, and previous experiments displaying interference of *Plasmodium* infection among transiently infected *A. gambiae*, demonstrate that pathogen enhancement is not universal among all transient infections.

Interestingly, among the mosquitoes in that study, *w*AlbB density was 2–3 times higher than that of the typically high-density *w*MelPop. This effect appears to be the result of the somatic injection process, as the *Wolbachia* density among transiently infected mosquitoes appears to be highly variable in that it can differ between experiments, it is dependent on ambient temperature and appears to decline over the initial week postinfection, before increasing quite significantly over the rest of the mosquito life span (Hughes et al., 2011; Murdock et al., 2014).

Transient infection also appears to have a variable effect on host immune gene expression, as studies have shown both induction and suppression of immune gene activity (Hughes et al., 2011; Kambris et al., 2010). This effect varies both over the time postinfection and between *Wolbachia* strains, which indicates that the host immune response is somewhat unstable. Critically, it is unclear if the distinction between enhancement and interference of a pathogen during transient infection is due to immune expression, because transcriptional changes associated with transient infection have typically been conducted for associations that produce interference and not enhancement. The one exception to this is a study in *Culex tarsalis*, where transient infection with *w*AlbB increased the prevalence of WNV infection. In that study, levels of the immune genes Rel2, Cactus, Defensin, and Diptericin were not significantly different between *Wolbachia*-infected and uninfected mosquitoes, whereas levels of Rel1 were significantly lower among *Wolbachia*-infected mosquitoes (Dodson et al., 2014). This might suggest that there is an immune effect that underlies enhancement, and although given the complicated and variable biological changes that appear to result from transient *Wolbachia* infection, it may prove difficult to determine the causal mechanism of enhancement.

The majority of native mosquito *Wolbachia* infections that have been tested have no effect on pathogen infection. However, there have been two cases of enhancement observed for avian malaria. The first example was in *w*Flu-infected *A. fluviatilis*, where higher *Plasmodium gallinaceum* oocyst load was observed among *Wolbachia*-infected mosquitoes in two of the four experiments, with no effect of *Wolbachia* observed in the other two (Baton et al., 2013). The other example occurred in *w*Pip-infected *C. pipiens* mosquitoes challenged with *Plasmodium relictum*, where *Wolbachia* infection increased the prevalence and intensity of oocyst stage infection in midguts and prevalence of sporozoite stage infection in salivary glands (Zele et al., 2014). While *A. fluviatilis* is not a natural vector of *P. gallinaceum*, *C. pipiens* does naturally transmit *P. relictum*, indicating that enhancement in native associations is not simply a product of putting a novel pathogen into an unfamiliar system.

Although there is currently no clear mechanistic pathway to enhancement or interference, comparison of the host–strain–pathogen combinations that result in enhancement and those that do not, reveals a few interesting observations. The first and most important of these is that, to date, enhancement has not been observed in a transinfection, and while the phenotype remains something that must be studied in more detail, it is unclear whether enhancement could actually occur for a *Wolbachia*-host association that will be utilized in a vector control program. Some authors have questioned whether population replacement programs contain the implicit risk of enhancement of nontarget pathogens, for example, enhancement of Zika virus for mosquitoes deployed to combat DENV. This should be a consideration in the future, but in addition to the fact that enhancement has not been observed for a transinfection, there has been no observed case of an association interfering with one viral infection and enhancing another. Likewise, enhancement has not been observed for a *Plasmodium* species that causes human malaria, only for rodent or avian malaria, and the only case of enhancement of a human pathogen (WNV) was in a transiently infected mosquito, and that association is unlikely to ever be used for vector control purposes. All cases of enhancement in transient infections also involve the *w*AlbB strain, which may be due to the fact that it is also one of the most common strains used for transient infection or perhaps due to biological quirks associated with that strain.

The severity of enhancement as an issue to *Wolbachia*-based vector control is still questionable. The effect has been observed in a laboratory setting, but it is also currently unclear if increased prevalence or intensity of infection translates to increased pathogen transmission or if the effect will ever occur in the field, with a circulating pathogen, and in a *Wolbachia*-infected mosquito used for vector control. An important point to consider is whether the biology of a transient *Wolbachia* infection is a realistic representation of the biology of the transinfections that will be utilized in vector control programs and if these differences are the result of many of the cases of enhancement that have been observed. The development of further transinfections in malaria vectors will provide the answers to these questions and a better idea of the actual risk associated with enhancement.

THE FUTURE

The next five years could see the extension of *Wolbachia*-based control programs to other mosquito species and pathogens, in a greater range of countries. In light of the recent outbreak in Latin America, Zika virus is likely to be the first target. Local authorities are considering more widespread releases of *Wolbachia*-infected mosquitoes using the pathogen interference approach. This will be made easier by the fact that *Wolbachia* infections are already present in local genetic backgrounds and the fact that *Wolbachia* strongly inhibits infection with recently circulating Zika viruses in these mosquitoes. Broader releases would likely have the additional benefit of reducing dengue and chikungunya transmission, which remain problematic across the continent.

Such releases will likely represent a single facet of an integrated control program that will also involve population suppression techniques such as the IIT or RIDL (release of insects carrying a dominant lethal gene) (Harris et al., 2011). Either of these techniques would complement a pathogen interference–based release, because of their differing modes of action. They could be used to reduce the size of existing mosquito populations before releases for population replacement, which would probably facilitate *Wolbachia* invasion. Likewise, they could be used to remove populations of competing mosquito species or to eliminate populations of secondary or nontarget vectors.

Recent advances in genetic sexing of mosquitoes will likely provide the ability to kill genetic female larvae or convert them into fertile males (Adelman and Tu, 2016). Either would increase the efficiency of existing population suppression techniques that require sexing of pupae. It is unclear if these techniques will function for *Wolbachia*-infected mosquitoes, but they could potentially lead to a more effective version of the IIT, where there was no risk of accidental female release.

Antipathogen GM approaches to mosquito control have been restricted by the lack of a suitable drive mechanism. Considerable advances are likely to occur in this field over the coming years, including those based on homing endonuclease genes and CRISPR/Cas9. Yet, this will not remove the need for *Wolbachia*-based control, as multiple antipathogen mechanisms will provide additional impediments to the development of resistance on the part of the target pathogen. To that end, it is possible to envision future control programs where GM and *Wolbachia*-infected mosquitoes are deployed sequentially or simultaneously, which could be beneficial if neither approach is 100% effective at preventing transmission.

Likewise, *Wolbachia*-based control will likely be compatible with future vaccines of both the traditional and transmission-blocking varieties, as vaccine-derived immunity rarely occurs in all those inoculated. A transmission-blocking vaccine would likely have a greater degree of compatibility with *Wolbachia*-based control, as it would prevent transmission from people who become infected with a pathogen, whereas the *Wolbachia* approaches would serve to reduce the likelihood of that infection occurring in the first place.

Extending the use of *Wolbachia* to target malaria transmission is a more long-term goal, and efforts are liable to be complicated by the increased difficulty of generating transinfections in anophelines and the high number of vectors of human malaria. However, initial laboratory studies demonstrating interference against human malaria suggest that the *Wolbachia* approach is worth persisting with (Bian et al., 2013). Furthermore, the native *Wolbachia* strains discovered in *A. gambiae* (Baldini et al., 2014) could prove to be a better option to generate transinfections and also suggest that a stable transinfection in that key vector species is feasible.

References

Adelman, Z.N., Tu, Z., 2016. Control of mosquito-borne infectious diseases: sex and gene drive. Trends Parasitol. 32 (3), 219–229. http://dx.doi.org/10.1016/j.pt.2015.12.003. PubMed PMID: 26897660; PubMed Central PMCID: PMCPMC4767671.

Baldini, F., Segata, N., Pompon, J., Marcenac, P., Robert Shaw, W., Dabire, R.K., et al., 2014. Evidence of natural *Wolbachia* infections in field populations of *Anopheles gambiae*. Epub 2014/06/07 Nat. Commun. 5, 3985. http://dx.doi.org/10.1038/ncomms4985. PubMed PMID: 24905191; PubMed Central PMCID: PMC4059924.

Baton, L.A., Pacidonio, E.C., Goncalves, D.S., Moreira, L.A., 2013. wFlu: characterization and evaluation of a native *Wolbachia* from the mosquito *Aedes fluviatilis* as a potential vector control agent. PLoS One 8 (3), e59619. http://dx.doi.org/10.1371/journal.pone.0059619. PubMed PMID: 23555728; PubMed Central PMCID: PMCPMC3608659.

Bian, G., Joshi, D., Dong, Y., Lu, P., Zhou, G., Pan, X., et al., 2013. *Wolbachia* invades *Anopheles stephensi* populations and induces refractoriness to *Plasmodium* infection. Science 340 (6133), 748–751. http://dx.doi.org/10.1126/science.1236192. PubMed PMID: 23661760.

Bourtzis, K., Dobson, S.L., Xi, Z., Rasgon, J.L., Calvitti, M., Moreira, L.A., et al., 2014. Harnessing mosquito-*Wolbachia* symbiosis for vector and disease control. Acta Trop. 132 (Suppl), S150–S163. http://dx.doi.org/10.1016/j.actatropica.2013.11.004. PubMed PMID: 24252486.

Brownstein, J.S., Hett, E., O'Neill, S.L., 2003. The potential of virulent *Wolbachia* to modulate disease transmission by insects. J. Invertebr. Pathol. 84 (1), 24–29. PubMed PMID: 13678709.

Caragata, E.P., Dutra, H.L., Moreira, L.A., 2016. Exploiting intimate relationships: controlling mosquito-transmitted disease with *Wolbachia*. Trends Parasitol. 32 (3), 207–218. http://dx.doi.org/10.1016/j.pt.2015.10.011. PubMed PMID: 26776329.

Chen, L., Zhu, C., Zhang, D., 2013. Naturally occurring incompatibilities between different *Culex pipiens* pallens populations as the basis of potential mosquito control measures. PLoS Negl. Trop. Dis. 7 (1), e2030. http://dx.doi.org/10.1371/journal.pntd.0002030. PubMed PMID: 23383354; PubMed Central PMCID: PMCPMC3561155.

Dodson, B.L., Hughes, G.L., Paul, O., Matacchiero, A.C., Kramer, L.D., Rasgon, J.L., 2014. *Wolbachia* enhances West Nile virus (WNV) infection in the mosquito *Culex tarsalis*. PLoS Negl. Trop. Dis. 8 (7), e2965. http://dx.doi.org/10.1371/journal.pntd.0002965. PubMed PMID: 25010200; PubMed Central PMCID: PMCPMC4091933.

Dutra, H.L., Rocha, M.N., Dias, F.B., Mansur, S.B., Caragata, E.P., Moreira, L.A., 2016. *Wolbachia* blocks currently circulating Zika virus isolates in Brazilian *Aedes aegypti* mosquitoes. Cell Host Microbe 19 (6), 771–774. http://dx.doi.org/10.1016/j.chom.2016.04.021. PubMed PMID: 27156023.

Fleury, F., Vavre, F., Ris, N., Fouillet, P., Bouletreau, M., 2000. Physiological cost induced by the maternally-transmitted endosymbiont *Wolbachia* in the *Drosophila* parasitoid *Leptopilina* heterotoma. Parasitology 121 (Pt. 5), 493–500. PubMed PMID: 11128800.

Gilles, J.R., Schetelig, M.F., Scolari, F., Marec, F., Capurro, M.L., Franz, G., et al., 2014. Towards mosquito sterile insect technique programmes: exploring genetic, molecular, mechanical and behavioural methods of sex separation in mosquitoes. Acta Trop. 132 (Suppl), S178–S187. http://dx.doi.org/10.1016/j.actatropica.2013.08.015. PubMed PMID: 23994521.

Harris, A.F., Nimmo, D., McKemey, A.R., Kelly, N., Scaife, S., Donnelly, C.A., et al., 2011. Field performance of engineered male mosquitoes. Nat. Biotechnol. 29 (11), 1034–1037. http://dx.doi.org/10.1038/nbt. 2019. PubMed PMID: 22037376.

Hoffmann, A.A., Goundar, A.A., Long, S.A., Johnson, P.H., Ritchie, S.A., 2014. Invasion of *Wolbachia* at the residential block level is associated with local abundance of *Stegomyia aegypti*, yellow fever mosquito, populations and property attributes. Med. Vet. Entomol. 28 (Suppl. 1), 90–97. http://dx.doi.org/10.1111/mve.12077. PubMed PMID: 25171611.

Hughes, G.L., Rasgon, J.L., 2014. Transinfection: a method to investigate *Wolbachia*-host interactions and control arthropod-borne disease. Insect Mol. Biol. 23 (2), 141–151. http://dx.doi.org/10.1111/imb.12066. PubMed PMID: 24329998; PubMed Central PMCID: PMCPMC3949162.

Hughes, G.L., Koga, R., Xue, P., Fukatsu, T., Rasgon, J.L., 2011. *Wolbachia* infections are virulent and inhibit the human malaria parasite *Plasmodium falciparum* in *Anopheles gambiae*. PLoS Pathog. 7 (5), e1002043. http://dx.doi.org/10.1371/journal.ppat.1002043. PubMed PMID: 21625582; PubMed Central PMCID: PMCPMC3098226.

Hughes, G.L., Vega-Rodriguez, J., Xue, P., Rasgon, J.L., 2012. *Wolbachia* strain wAlbB enhances infection by the rodent malaria parasite *Plasmodium berghei* in *Anopheles gambiae* mosquitoes. Appl. Environ. Microbiol. 78 (5), 1491–1495. http://dx.doi.org/10.1128/AEM.06751-11. PubMed PMID: 22210220; PubMed Central PMCID: PMCPMC3294472.

Joubert, D.A., Walker, T., Carrington, L.B., De Bruyne, J.T., Kien, D.H., Hoang Nle, T., et al., 2016. Establishment of a *Wolbachia* superinfection in *Aedes aegypti* mosquitoes as a potential approach for future resistance management. PLoS Pathog. 12 (2), e1005434. http://dx.doi.org/10.1371/journal.ppat.1005434. PubMed PMID: 26891349; PubMed Central PMCID: PMCPMC4758728.

Kambris, Z., Cook, P.E., Phuc, H.K., Sinkins, S.P., 2009. Immune activation by life-shortening *Wolbachia* and reduced filarial competence in mosquitoes. Science 326 (5949), 134–136. http://dx.doi.org/10.1126/Science.1177531. PubMed PMID: ISI:000270355600053.

Kambris, Z., Blagborough, A.M., Pinto, S.B., Blagrove, M.S., Godfray, H.C., Sinden, R.E., et al., 2010. *Wolbachia* stimulates immune gene expression and inhibits *Plasmodium* development in *Anopheles gambiae*. PLoS Pathog. 6 (10), e1001143. http://dx.doi.org/10.1371/journal.ppat.1001143. PubMed PMID: 20949079; PubMed Central PMCID: PMCPMC2951381.

Kosmidis, S., Missirlis, F., Botella, J.A., Schneuwly, S., Rouault, T.A., Skoulakis, E.M., 2014. Behavioral decline and premature lethality upon pan-neuronal ferritin overexpression in *Drosophila* infected with a virulent form of *Wolbachia*. Front Pharmacol. 5, 66. http://dx.doi.org/10.3389/fphar.2014.00066. PubMed PMID: 24772084; PubMed Central PMCID: PMCPMC3983519.

Kremer, N., Voronin, D.A., Charif, D., Mavingui, P., Mollereau, B., Vavre, F., 2009. *Wolbachia* interferes with ferritin expression and iron metabolism in insects. PLoS Pathog. 5 (10), e1000630.

Martinez, J., Longdon, B., Bauer, S., Chan, Y.S., Miller, W.J., Bourtzis, K., et al., 2014. Symbionts commonly provide broad spectrum resistance to viruses in insects: a comparative analysis of *Wolbachia* strains. PLoS Pathog. 10 (9), e1004369. http://dx.doi.org/10.1371/journal.ppat.1004369. PubMed PMID: 25233341; PubMed Central PMCID: PMCPMC4169468.

Martinez, J., Ok, S., Smith, S., Snoeck, K., Day, J.P., Jiggins, F.M., 2015. Should symbionts be nice or selfish? Antiviral effects of *Wolbachia* are costly but reproductive parasitism is not. PLoS Pathog. 11 (7), e1005021. http://dx.doi.org/10.1371/journal.ppat.1005021. PubMed PMID: 26132467; PubMed Central PMCID: PMCPMC4488530.

McMeniman, C.J., Hughes, G.L., O'Neill, S.L., 2011. A *Wolbachia* symbiont in *Aedes aegypti* disrupts mosquito egg development to a greater extent when mosquitoes feed on nonhuman versus human blood. J. Med. Entomol. 48 (1), 76–84. http://dx.doi.org/10.1603/Me09188. PubMed PMID: ISI:000287009400011.

Moreira, L.A., Iturbe-Ormaetxe, I., Jeffery, J.A., Lu, G.J., Pyke, A.T., Hedges, L.M., et al., 2009. A *Wolbachia* symbiont in *Aedes aegypti* limits infection with dengue, chikungunya, and *Plasmodium*. Cell 139 (7), 1268–1278. http://dx.doi.org/10.1016/J.Cell.2009.11.042. PubMed PMID: ISI:000273048700015.

Murdock, C.C., Blanford, S., Hughes, G.L., Rasgon, J.L., Thomas, M.B., 2014. Temperature alters *Plasmodium* blocking by *Wolbachia*. Sci. Rep. 4, 3932. http://dx.doi.org/10.1038/srep03932. PubMed PMID: 24488176; PubMed Central PMCID: PMCPMC3909897.

Ritchie, S.A., Townsend, M., Paton, C.J., Callahan, A.G., Hoffmann, A.A., 2015. Application of wMelPop *Wolbachia* strain to crash local populations of *Aedes aegypti*. PLoS Negl. Trop. Dis. 9 (7), e0003930. http://dx.doi.org/10.1371/journal.pntd.0003930. PubMed PMID: 26204449; PubMed Central PMCID: PMCPMC4512704.

Rohrscheib, C.E., Bondy, E., Josh, P., Riegler, M., Eyles, D., van Swinderen, B., et al., 2015. *Wolbachia* influences the production of octopamine and affects *Drosophila* male aggression. Appl. Environ. Microbiol. 81 (14), 4573–4580. http://dx.doi.org/10.1128/AEM.00573-15. PubMed PMID: 25934616; PubMed Central PMCID: PMCPMC4551182.

Temple, N., Richard, F.J., 2015. Intra-cellular bacterial infections affect learning and memory capacities of an invertebrate. Front Zool. 12, 36. http://dx.doi.org/10.1186/s12983-015-0129-6. PubMed PMID: 26675213; PubMed Central PMCID: PMCPMC4678612.

Turelli, M., Hoffmann, A.A., 1991. Rapid spread of an inherited incompatibility factor in California *Drosophila*. Nature 353 (6343), 440–442. http://dx.doi.org/10.1038/353440a0. PubMed PMID: 1896086.

Walker, T., Johnson, P.H., Moreira, L.A., Iturbe-Ormaetxe, I., Frentiu, F.D., McMeniman, C.J., et al., 2011. A non-virulent *Wolbachia* infection blocks dengue transmission and rapidly invades *Aedes aegypti* populations. Nature 476, 450–455.

Zele, F., Nicot, A., Berthomieu, A., Weill, M., Duron, O., Rivero, A., 2014. *Wolbachia* increases susceptibility to *Plasmodium* infection in a natural system. Proc. Biol. Sci. 281 (1779), 20132837. http://dx.doi.org/10.1098/rspb.2013.2837. PubMed PMID: 24500167; PubMed Central PMCID: PMCPMC3924077.

Zug, R., Hammerstein, P., 2012. Still a host of hosts for *Wolbachia*: analysis of recent data suggests that 40% of terrestrial arthropod species are infected. PLoS One 7 (6), e38544. http://dx.doi.org/10.1371/journal.pone.0038544. PubMed PMID: 22685581; PubMed Central PMCID: PMCPMC3369835.

Zug, R., Hammerstein, P., 2015. *Wolbachia* and the insect immune system: what reactive oxygen species can tell us about the mechanisms of *Wolbachia*-host interactions. Front Microbiol. 6, 1201. http://dx.doi.org/10.3389/fmicb.2015.01201. PubMed PMID: 26579107; PubMed Central PMCID: PMCPMC4621438.

CHAPTER

8

Effect of Host Blood–Derived Antibodies Targeting Critical Mosquito Neuronal Receptors and Other Proteins: Disruption of Vector Physiology and Potential for Disease Control

Jacob I. Meyers[1], *Brian D. Foy*[2]

[1]Texas A&M University, College Station, TX, United States; [2]Colorado State University, Fort Collins, CO, United States

KEY LEARNING POINTS

- The negative effects of host-derived anti-vector antibodies on vector fitness have been recognized in a diverse range of disease vectors since the 1930s.
- Antibodies against both salivary and concealed antigens found in vector tissues can reduce vector competence.
- Antibodies against the midgut antigen Bm86 of *Rhipicephalus* (formerly *Boophilus*) *microplus* reduce tick survivorship and Bm86 was developed into an antigen for a commercialized vaccine for cattle.
- The antivector approach has not been as successful for insect vectors, potentially due to differences in vector feeding behavior and blood meal digestion physiology combined with a focus on gut antigens, rather antigens located in the hemocoel.
- Recently, blood meal–derived antivector antibodies against extracellular domains of neuronal channels have been shown to reduce mosquito survivorship.
- Antibodies against neuronal channels appear to inhibit channel activity, potentially through obstructing conformational changes necessary for channel opening.
- Antagonistic effects of blood meal–derived antibodies on protein function may represent a novel laboratory tool to study the physiological role of target proteins.

Arthropod Vector: Controller of Disease Transmission, Volume 1
http://dx.doi.org/10.1016/B978-0-12-805350-8.00008-8

- Antibodies targeting antigens expressed exclusively in the hemocoel must translocate across the midgut, a process that is yet to be determined, but varies widely across mosquito species.
- To be successful, most antivector vaccine strategies should be based on reducing disease transmission by targeting vectorial capacity through decreasing vector fitness following blood feeding on a vaccinated host.

BACKGROUND

In 1939, William Trager was the first person to demonstrate that antivector immunity could be developed in animals (Trager, 1939). In his first reported experiments, he blood-fed larvae or nymphs of *Dermacentor variabilis* on guinea pigs (and a few rabbits), and then subsequently performed blood-feeding challenges on the same animals with *D. variabilis* larvae or nymphs weeks to months later. The proportion of engorged ticks in the second challenge was usually considerably lower relative to that on previously unexposed animals; many ticks dropped off prematurely, failed to initiate feeding, were often an unusual brown or pale color, failed to molt into the next stadia normally, and some died. Further, their average postengorgement weight and the amount of excretia they produced post blood meal were reduced. He went on to show that previous infestation with *D. variabilis* provided some cross-protection against other tick species. Trager also explored pathology and the cellular reaction at the tick attachment site using histological sections in previously exposed guinea pigs, which showed large edematous inflammatory reactions and cellular infiltrates in the dermis under the tick mouthpart attachment site.

In separate experiments of this pioneering study, Trager also made crude extracts of *D. variabilis* larvae by grinding hundreds of them in buffer, clarifying the homogenate by centrifugation, and injecting the supernatant into guinea pigs in serial immunizations. Nymphs challenged on these animals failed to engorge and those that did were an unusual pale brown color rather than black. While he had trouble demonstrating acquired immunity in the serum via antibody presence and activity using the rudimentary assays of that time, he was successful in demonstrating that passively transferred sera into naïve guinea pigs induced some of the same effects. Trager discussed that the "bright red" ticks, only found on immune guinea pigs, always died soon after detaching. He stated that, "At a stage when the gut contents of normal larvae, examined in the fresh state, contained large hemoglobin crystals and many black granules, the gut contents of the red larvae contained intact red blood cells and no black granules. It would seem that something had interfered with the normal digestive processes of the tick."

Trager's prescient study 77 years ago launched two parallel fields of research. The first field focused on vector salivary antigens and how repeated exposure or immunization against them could sometimes induce immunity against subsequent vector infestation or bites. Most of Trager's experiments in that seminal paper were focused in this research field, as he more often exposed his animals to biting ticks alone to stimulate their immunity. Paradoxically, as secreted salivary gland proteins are introduced into the host, it has been subsequently discovered that many are often very effective at modulating host immunity. This field is very well developed and explored (Abdeladhim et al., 2014; McDowell, 2015). The entire saliomes of many vectors are now known, many antigens are well described for their ability to modulate the physiology and immunity of the host (Ribeiro et al., 2010; Francischetti et al., 2009), and researchers have demonstrated how vectorborne pathogens have co-opted the mechanism of these salivary proteins to gain a foothold in the host (Ramamoorthi et al., 2005).

The other research field arises from Trager's fewer experiments where he injected tick extracts into the animals to stimulate their immunity. In those experiments, the immunogens would not necessarily be tick salivary proteins but include other antigens in the tick body that likely were targeted *inside the tick* by antibodies or other immune factors ingested with the blood meal. This field is much less developed, particularly in mosquitoes and other insect vectors, but no less promising, and it will be the focus of this chapter. While no single antigen was identified from Trager's studies, they nonetheless suggested that these so-called "concealed antigens" in vectors could be important targets and could be bound by immune factors present in the blood of hosts that the vectors feed upon. Many of such antigens are expected to be critical for vector homeostasis and to be circulating in hemolymph or exposed on tissues and the surface of cells. These are called "concealed" because they are distinguished from antigens in vector saliva that are introduced into the host upon biting, which would naturally foster immune responses in the host. Despite this distinction, it is likely that certain proteins or immunogenic epitopes on proteins are present in both vector saliva and in the body of the vector, and so in some instances this distinction may be misleading. Trager's experiments set in motion later studies in ticks that eventually culminated in the development of a commercial vaccine in the 1990s (Willadsen et al., 1995) and set the stage for vaccine studies to target many different arthropod vectors. However, despite these advances, one could argue that the field has developed surprisingly slow and has received too little attention relative to the potential it has to reduce vectorborne disease transmission. Also underdeveloped is our understanding of how antibodies and other blood molecules bind and affect their targets in the vector, and how they transit from the midgut and affect targets in the vector hemocoel. Given the diversity and ancient evolutionary divergence of arthropod vectors, of which most taxonomic families independently developed the propensity to blood feed (Ribeiro et al., 2010), we can almost certainly expect that efficacy of antibody binding, transit, and activity will vary among vector species. The knowledge to be gained from this research will not only help our ability to make antivector vaccines, but to perform basic biological function studies in vectors to define how extracellular proteins and other macromolecules function in vector homeostasis.

Vaccine Studies Against Concealed Antigens to Reduce the Survival of Ticks and Blood-Feeding Insects

As alluded to above, the work against ticks is by far the most advanced. Many reviews have highlighted these efforts and many candidate vaccine antigens have been discovered (Willadsen, 2001, 2006). Similar to Trager's experiments, the most advanced studies from the 1980s used homogenates of ticks or tick organs; they serially immunized semi-purified fractions, performed more rounds of biochemical antigen purification and immunizations, and eventually identified the antigens from fractions that stimulated the most killing activity in immunized animals, such as the Bm86 antigen, through peptide sequencing (Willadsen et al., 1989). The Bm86 antigen was ultimately developed into the TickGard vaccine in Australia and Gavac vaccine in Latin America, which successfully reduced tick infestations on cattle and reduced acarcide usage (de la Fuente et al., 2007). When anti-Bm86 antibody enters the tick midgut, endocytosis of the blood meal contents by tick midgut cells is inhibited, followed by cell lysis, which disrupts the integrity of the gut (Willadsen and Kemp, 1988). Beyond this knowledge, the normal role of the Bm86 antigen in tick homeostasis is still mostly undefined, as are many other candidate concealed antigens in ticks. While the achievement of commercialized anti-tick vaccines proves the method of biochemical fractionation, serial vaccination and purification as successful, the efforts to identify critical antigens remain very difficult and slow. For similar studies targeting mosquitoes and other blood-feeding insects, most of the studies never proceeded past immunizations with crude organ homogenates, and so the results between immunized animals and repeated experiments were highly variable and confounded specific antigen identification (Jacobs-Lorena and Lemos, 1995; Foy et al., 2002; Billingsley et al., 2008). Now, with the availability of many vector genomes on VectorBase and NCBI, more targeted molecular and bioinformatics approaches have recently predominated. Researchers can directly identify targets based on their potential availability to be easily targeted with the largest quantity of antibody, such as proteins expressed on the luminal surface of the midgut at the time of blood feeding or most proteins with a secretion signal on the N-terminus localizing them to the cell membrane. Expression library immunization screens by immunizing partial or whole cDNA sequences in expression plasmids has had some success (Foy et al., 2003; Almazan et al., 2003) and RNA interference assays have been used to prove that a selected target or antigen class may be critical to the vector's survival (de la Fuente et al., 2005; Marr et al., 2014).

Even with these more molecular-based methods to identify targets, only a handful of antigens have been developed successfully against blood-feeding insects and particularly against mosquitoes. It is tempting to speculate that some inherent biological differences between these two types of blood-feeding arthropods favors ticks for antivector vaccine success. Willadsen (2001) described three characteristics of tick feeding that may make them more amenable to immunological control relative to blood-feeding insects: (1) Hard ticks have slow feeding habits, (2) the gut surface is exposed to antibody while the insect gut is partially or entirely protected by the peritrophic matrix, and (3) digestion in ticks is mostly intracellular via pinocytosis of gut contents and intravacuole protein digestion and adsorption. This also means that the tick gut has a neutral pH and is mostly free of the proteases that can degrade host immune factors. Similar points were also highlighted by Kay and Kemp (1994) in their review. To this we might add that the gut of blood-feeding insects is usually highly glycosylated and very few specific antigens have been identified from the luminal surface that can be targeted by antibodies to induce rapid and unequivocal mortality. The exceptional few are the peritrophins from the myiasis-causing *Lucilia cuprina*, which are bound by incoming antibody that acts to block transit of biomolecules through the peritrophic matrix, starving the maggot (Wijffels et al., 1999). Overall, it would seem that the vector insect gut is very well equipped to deal with the potential physiological insult of ingested host factors and hemoglobin that, when degraded, release toxic heme among other things. Perhaps attempts to develop antigens from insect luminal gut proteins will continue to be confounded. However, we know that antibodies can translocate the gut in many different insect vectors, and these can bind targets in the hemocoel.

Antibody Translocation Across the Insect Midgut and Binding of Concealed Antigens

The ingestion of a blood meal follows a well-documented series of events in the posterior midgut where the majority of digestion, nutrient absorption, and ion/water regulation takes place. By analyzing the hemolymph of vectors following ingestion of a blood meal, blood meal-proteins have been shown to cross the midgut and gain access to the hemolymph of blood-feeding arthropods. In ixodid ticks, the mechanism for this phenomenon is quite complex, as blood immunoglobulins such as IgG undergoes selective uptake into the hemocoel, whereby IgG-binding proteins transport it to the salivary glands and it is eventually secreted into the host with the tick saliva during subsequent blood feeding (Ackerman et al., 1981; Jasinskas et al., 2000; Jasinskas and Barbour, 2005). This mechanism seems to serve a specific physiological purpose that has yet to be fully defined, but it has been proposed that the proteins involved in this would be good candidate antigens for anti-tick vaccines (Wang and Nuttall, 1999).

In contrast, the passage of IgG and other proteins across the gut of insect vectors is less well studied and not highly conserved across taxonomic groups. Mostly antibodies, but also albumin, other serum proteins, and hemoglobin, have all been found to occur in the hemolymph of certain Dipteran, Hemipteran, and Siphonapteran species (Jeffers and Michael Roe, 2008). Hatfield (1988) discovered mouse IgG closely associated with the membranes and within the cytosol of microvilli of *Aedes aegypti* midgut epithelial cells using immunogold labeling. However, this is the only study looking at potential mechanisms for how antibodies traverse the midgut in blood-feeding insect vectors, and conflicting data have been published about whether *A. aegypti* even successfully allows IgG to pass across the gut (Hatfield, 1988; Ramasamy et al., 1988; Vaughan and Azad, 1988). Our own published study supports other investigations suggesting that IgG is not capable of traversing the gut of this species (Meyers et al., 2015). However, in several studies of multiple *Anopheles* species, IgG was consistently detected in the hemolymph shortly after a blood meal (Vaughan and Azad, 1988; Meyers et al., 2015; Vaughan et al., 1990; Vaughan et al., 1988; Beier et al., 1989; Lackie and Gavin, 1989). Vaughan and Azad (1988) reported that antibody was detectable in Anopheline hemolymph between 3 and 24h post blood meal. In a different study with *Anopheles stephensi*, hemolymph IgG levels only reached ~1 μg/mL at 3h post blood feeding (Vaughan et al., 1990). Using fleas, Vaughan et al. (1998), estimated that host IgG concentration 1h post blood meal in *Ctenocephalides felis* was ~35 μg/mL, and consistent antibody concentrations were maintained in the hemolymph of fleas allowed to feed ad libitum for 3 days. Finer resolutions of the time course and location of antibody passage have not been reported and the dependence on energetic processes as opposed to simple diffusion has also not been addressed. Importantly, at least some, if not most, of the antibody detected in insect vector hemolymph remains capable of binding to its selected target (Vaughan et al., 1998). Brennan et al. (2000) demonstrated that antisalivary gland antibodies fed to the malaria vector mosquito *Anopheles gambiae* were bound to the salivary glands when they were dissected after the blood meal. In total, these studies show that concealed antigens in the hemocoel can be targeted by translocated host IgG.

There are likely limitations in the quantity and activity of passaged IgG; however, the antigens do not necessarily need to be in high abundance, but they likely should be highly accessible to IgG by exposure on the surface of cells and tissues that are bathed in hemolymph. Also, they necessarily should be very sensitive to antibody binding that would cause disruption of their normal activity and which would directly lead to increased vector mortality (see Fig. 8.1). Obvious target antigens that fulfill these requirements are receptors on insect neuronal tissue, many of which stimulate activity in vital cells and tissues through ligand binding or voltage-gated molecular movement and steric effects. Chemoreceptors and ion channels are the primary target of most chemical insecticides, mainly because of this sensitivity and accessibility. More than 20 years ago, two separate review papers in the same issue of the *International Journal of Parasitology* suggested that such antigen targets should be the focus of future antivector vaccine studies (Elvin and Kemp, 1994; Sauer et al., 1994). Elvin and Kemp (1994) focused much of their discussion on the ion channels that might be disrupted by antibody binding to interfere with chemosensory nerve cells, including vector gustatory and olfactory sensilla. It might be difficult to see how targeting these

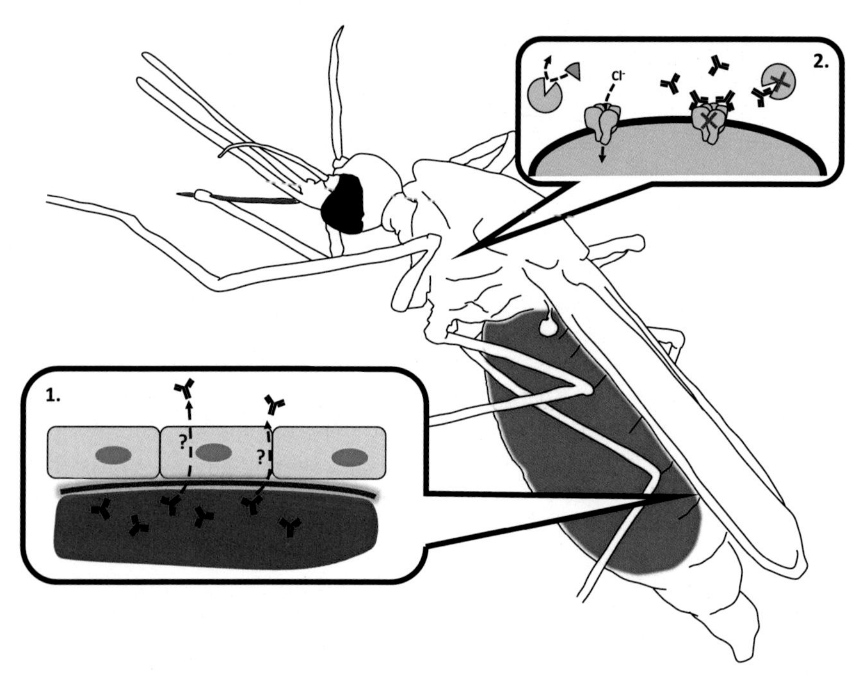

FIGURE 8.1 (1) In certain vectors such as *Anopheles gambiae* mosquitoes, antibodies present in the host blood are able to quickly transit from the mosquito's blood meal, across the midgut epithelium, and into the hemolymph, but the mechanisms and route of this transit are currently unknown. (2) Once in the hemolymph, antibodies produced by immunization of the host with vector proteins can bind the native protein (e.g., hemolymph enzymes or extracellular regions of ion channels) and disrupt their function in vivo, which may act to kill or impair the mosquito.

sensory systems would quickly lead to vector death without needing to perform detailed behavioral bioassays of the blood-fed insects when screening for activity. However, they also discussed voltage-gated sodium and potassium channels as potential targets, and mentioned calcium-activated channels, ion antiporters, and ATPases. Sauer et al. (1994) centered their broad review on cell membrane receptors of ticks, specifically gated ion channels, agonist-stimulated tyrosine kinases, and G-coupled receptors. As for blood-feeding insects, they believed that targeting receptors (as well as neuropeptide hormones) involved in vitellogenesis and blood meal diuresis might be effective in disrupting the physiology of both *A. aegypti* and *Rhodnius prolixus*.

Surprisingly, subsequent studies that acted on the advice of these two reviews have not been published for ticks as far as we are aware (Nuttall et al., 2006), and none from blood-feeding insects until our demonstration that antibodies targeting an *A. gambiae* ligand-gated chloride ion channel could affect mosquito survival (Meyers et al., 2015). Parallel work in this field, however, regarding peptides toxins made by various invertebrate predators and autoimmune channelopathies that occur in vertebrates can inform this research.

Insights From the Activity and Targets of Invertebrate Peptide Toxins

Some venomous animals have evolved to produce bioactive peptide toxins to deter predators and kill prey. There is remarkable diversity in peptide toxins, which vary in their protein structure, target specificity, and mode of action (Smith et al., 2013). The targets of these peptide

TABLE 8.1 Target and Species of Insect-Specific Toxins

Target	Species	Peptide Toxin	Reference
Voltage-gated sodium channel	Spider	μ-Agatoxins	Adams (2004)
		Curtatoxins	Stapleton et al. (1990)
		DTX9.2,	Bloomquist et al. (1996)
		Hainantoxin-I	Li et al. (2003)
		Magi 2	Corzo et al. (2003)
		δ-Palutoxins	Corzo et al. (2000)
		Tx4(6-1)	Figueiredo et al. (1995)
	Scorpion	Scorpion α-insect toxins	Gordon et al. (2007)
		Excitatory scorpion β-toxins	Leipold et al. (2012)
		Depressant scorpion β-toxins	Leipold et al. (2012)
	Sea anemone	Av3	Moran et al. (2007)
		CgII	Salgado and Kem (1992)
		Sh-I	Salgado and Kem (1992)
Voltage-gated calcium channel	Spider	ω-ACTX-Hv1a	Fletcher et al. (1997)
		Omega-ACTX-Hv2a	Wang et al. (2001)
		Huwentoxin-V	Zhang et al. (2003)

toxins may also make efficacious targets for antivector antibodies. Several families of peptide toxins have evolved in spiders, scorpions, and sea anemones that specifically act on insect voltage-gated sodium (VGSC) and calcium channels (VGCC) and have no bioactivity against their mammalian orthologs (Table 8.1). Their specificity is a result of high binding affinity to insect-specific channel sequences and low sequence homology between mammals and insects (King et al., 2008).

Insects carry a single VGSC gene and obtain functional VGSC diversity through high rates of alternative splicing. This VGSC gene is highly conserved making insect-specific toxins active across diverse orders of insects (Davies et al., 2007). The allosteric binding sites for numerous insect-specific peptide toxins of VGSC are well described. These binding sites are primarily found on specific extracellular loops on the VGSC, and to a lesser degree, on transmembrane domains (Stevens et al., 2011). There is a wide range of toxin-induced VGSC activity and modulation including the blockage of ion conductance, shifts in voltage activation, repetitive activity, and inhibition of inactivation (Klint et al., 2012). Since VGSCs play a critical role in neuronal excitability and action potential firing, small perturbations in VGSC activity have strong neurotoxic effects resulting in paralysis and death.

The specificity and binding activity of these insect-specific toxins lend support to the approach of developing antivector antibodies directed against these same binding sites to similarly disrupt normal channel activity causing neurotoxic effects. Importantly, the majority of toxin binding sites are located on extracellular domains, which are also accessible to antibodies in the hemolymph (Stevens et al., 2011). Additionally, these insect-specific toxins act simply through binding to their target and do not cleave their target like tetanus neurotoxin and other toxins, suggesting that homologous binding by host-derived antibodies could elicit similar toxic effects.

Insights From Autoimmune Channelopathies in Vertebrates

Autoimmune channelopathies can occur in mammals against nearly all classes of metabotropic and ionotropic receptors, which cause diseases like myasthenia gravis that often result in either muscle weakness or hyperactivity. Autoantibody binding to self-receptors can result in a shifted conformation of the receptor channels that physically agonize *or* antagonize ion flow into the cells (Vincent et al., 2006). Antibody binding can also block access to the ligand binding site, thus antagonizing the channel (Gomez et al., 2010). These scenarios should be possible to replicate in blood-feeding insects when they bite a vertebrate host immunized against arthropod vector epitopes of metabotropic and ionotropic receptors that are exposed on the cell surface. The result is expected to be the death of the arthropod. In mammals, autoantibody binding can also stimulate complement or T-cell attack on one's own cells, eventually leading to physiologic dysregulation and/or receptor loss on the cell (Gomez et al., 2010; Kordas et al., 2014). This latter scenario is a hallmark of autoimmune channelopathies that are associated with an inflammatory process in the host. However, assuming arthropod receptor channel targets are in the hemocoel, active complement or T cells from the blood meal would need to co-translocate into the vector hemocoel along with the antibodies to cause this effect, making such a scenario highly unlikely. It is important to note that myasthenia gravis first became recognized as an autoimmune disease when researchers immunized rabbits with the acetylcholine receptors (AChRs) purified from an electric eel and the rabbits developed flaccid paralysis and irregular electromyographs (Patrick and Lindstrom, 1973). Clearly there was enough homology between the AChRs of the electric eel and the rabbit that an autoimmune channelopathy was stimulated in the immunized host. Thus, for antivector vaccines, it will be necessary to focus on receptor channels or epitopes that are unique to invertebrates or that have a very low possibility of cross-reactivity to related vertebrate channels. Fortunately, the preponderance of annotated genomes among both vertebrates and invertebrates, and especially well-developed phylogenies of receptor channels, allows for easy analysis of orthologous proteins and epitopes across taxa.

CURRENT ADVANCES IN ANTIMOSQUITO ANTIBODY DEVELOPMENT

Over the last 50 years, in contrast to anti-tick vaccine research, there have been only intermittent research testing candidate antigens for a potential antimosquito vaccine. Few studies have tested the mosquito homologs of efficacious antitick antigens. Antibodies against the transcription factor subolesin/akirin have been shown to affect survivorship and fecundity of a wide range of hard and soft ticks (Shakya et al., 2014). However, blood feeding of *A. gambiae*, *A. aegypti*, and *Aedes albopictus* on subolesin/akirin (derived from *A. albopictus*)-immunized mice had no effect on survivorship and caused a minor reduction in the number of eggs per ovary in *A. albopictus* alone (da Costa et al., 2014; de la Fuente et al., 2013). Furthermore, this reduction in eggs per ovary had no effect on the overall number of eggs laid per female. The failure of anti-subolesin/akirin antibodies to affect mosquito fitness is likely due to the differences in blood feeding and blood meal-digestion between mosquitoes and ticks described previously. The limitation of blood meal–derived antibodies to extracellular spaces pares the potential list of candidate antimosquito antigens to only secreted proteins or extracellular domains of transmembrane proteins that are accessible to free-floating antibodies in the hemolymph. Given that subolesin/akirin functions as a transcription factor in the cytosol and nucleus (da Costa et al., 2014), it is unlikely that this protein is ever exposed to the extracellular fluid where ingested antibodies are likely restricted in the mosquito.

Recently, the *A. gambiae* glutamate-gated chloride channel (AgGluCl) was tested as a candidate antigen for an antimosquito vaccine against the mosquito disease vectors *A. gambiae*, *A. aegypti*, and *Culex tarsalis* (Meyers et al., 2015). AgGluCl is member of the cys-loop family of ligand-gated ion channels (Meyers et al., 2015). GluCl is an ionotropic neurotransmitter receptor which gates an inhibitory chloride current at postsynaptic terminals. It has unique characteristics which make it a strong candidate antigen for an antimosquito vaccine: (1) AgGluCl has a large extracellular domain that would potentially be exposed to antibodies found in the hemolymph. (2) AgGluCl activity requires conformational changes in the extracellular domain. If these conformational changes are obstructed, the channel will not open and allow ion passage. (3) GluCls are only expressed in invertebrates, where it is highly conserved. The closest mammalian homology is the glycine-gated chloride channel, which shares minimal sequence identity to AgGluCl, specifically in the extracellular region. (4) AgGluCl is the target of the insecticidal drug ivermectin (IVM) which is toxic to *A. gambiae* at low concentrations found in human blood for several days following a standard IVM dose (Sylla et al., 2010). IVM is remarkably nontoxic to humans and is widely distributed in mass drug administrations for the control of filarial diseases such as onchocerciasis and lymphatic filariasis (Campbell, 2016).

To create antibodies targeting AgGluCl (anti-AgGluCl IgG), rabbits were immunized against the large, N-terminal extracellular domain of AgGluCl and total IgG was isolated and purified. The toxicity of anti-AgGluCl IgG on *A. gambiae* was tested by blood-feeding anti-AgGluCl IgG-containing blood to *A. gambiae* through an artificial membrane feeder (Meyers et al., 2015). Notably, anti-AgGluCl IgG caused an increase in *A. gambiae* mortality over 5 days following blood feeding in a dose-dependent manner and the LD_{50} was calculated to be 2.82 mg/mL. Unfortunately, this IgG concentration is 10 times greater than what would be expected in human blood following immunization, suggesting that a single blood meal on an AgGluCl-immunized host would have little effect on *A. gambiae*. However, when *A. gambiae* were serially fed every 4 days over a 20-day period on a physiological concentration of anti-AgGluCl IgG (282 μg/mL), there was a significant reduction in survivorship when compared to controls. The basis for this serial feeding effect is currently unknown, though it could be due to antibody accumulation in the mosquito hemolymph or age-dependent anti-AgGluCl IgG susceptibility.

The original hypothesis for anti-AgGluCl IgG's mode of action is that it antagonizes channel activity by binding dynamic extracellular regions that must undergo conformational changes for AgGluCl to open. To begin to test this hypothesis, anti-AgGluCl IgG was co-administered with a known AgGluCl agonist, IVM, to *A. gambiae* through a blood meal. Interestingly, anti-AgGluCl IgG attenuated the toxic effects of the AgGluCl agonist IVM suggesting that the antibody was able to block or antagonize the agonistic effects of IVM (Meyers et al., 2015). However, this is not conclusive evidence on anti-AgGluCl IgGs mode of action. Future work is needed to show exactly how the antibodies are affecting AgGluCl activity which could include arresting the channel in a specific conformational state, reducing agonist binding of IVM or the natural ligand glutamate, or inducing AgGluCl endocytosis making the channel physically inaccessible.

GluCl is highly conserved in mosquitoes making it likely that anti-AgGluCl IgG's toxicity to *A. gambiae* might translate to other medically important mosquito disease vectors. However, when blood meals containing anti-AgGluCl IgG were fed to *A. aegypti* and *C. tarsalis* there were no effects on mortality (Meyers et al., 2015). There are two explanations for their resistance; anti-AgGluCl IgG might be unable to bind to GluCl expressed by *A. aegypti* and *C. tarsalis* due to differences in GluCl protein sequence; or anti-AgGluCl IgG was not able to access its target when administered through a blood meal. The capacity for intact antibodies to translocate across the midgut after blood-feeding has been tested in many disease vectors. However, there have been conflicting results on antibody passage in *A. aegypti* and limited work with *Culex* vectors (Vaughan and Azad, 1988). To test if anti-AgGluCl IgG had toxicity effects in the mosquito hemocoel, anti-AgGluCl IgG was intrathoracically injected directly into the hemolymph of all three mosquito disease vectors. This resulted in an equal reduction in survivorship for all three mosquitoes suggesting that anti-AgGluCl IgG has similar toxicity effects on all three. To confirm that resistance to blood meals containing anti-AgGluCl IgG was due to a midgut barrier, all three mosquito species were fed a blood meal containing anti-AgGluCl IgG and then had their hemolymph extracted and blotted for antibodies. As expected, only *A. gambiae* had antibodies present in the hemolymph. Importantly, immunohistochemical staining of fixed adult mosquitoes with anti-AgGluCl IgG indicated AgGluCl expression does not occur in the midgut lumen and was confined to tissues in the hemocoel (Meyers et al., 2015). This confirmed that antibody translocation across the midgut is necessary for anti-AgGluCl IgG to reach its target and cause mosquito toxicity.

FUTURE RESEARCH DIRECTIONS

Understanding Basic Vector Biology by Disrupting Protein Function In Vivo

In addition to being a prospective novel direction for vector control, antivector antibodies are also a promising new tool for studying basic vector biology. Blood meals containing anti-AgGluCl IgG have been shown to block the toxic effects of the AgGluCl agonist IVM, suggesting that antibodies in a blood meal can antagonize target protein activity. Current research tools to inhibit protein function in insect vectors rely on reducing or eliminating target protein expression. These methods include RNA interference (RNAi) and transgenic approaches including CRISPR/Cas9 and zinc-finger nucleases (Hannon, 2002; Aryan et al., 2014; Kistler et al., 2015). Both transgenic approaches utilize directed nuclease activity to cause double-stranded DNA breaks at a target site within the genome. This activates the highly error prone nonhomologous end joining pathway which often results in indels, causing frame shifts and dysfunctional protein products (Kistler et al., 2015). These transgenic approaches are difficult to accomplish, require multiple generations to create homozygous knockouts, and have limited flexibility because target genes (some of which are necessary for growth and development) are permanently knocked out soon after the egg is deposited. RNAi-based knockdown of a target gene is accomplished through injection of double-stranded RNA into the vector, which is processed by the protein Dicer and loaded into the RNA-induced silencing complex to degrade target mRNA transcripts. This approach also has many limitations including limited, transient knockdown effects, high variability between treated individuals, and a dependence on high rates of target protein turnover (Hannon, 2002).

Antibody-based antagonism of target protein activity should have some advantages over current molecular techniques. Antibodies can be fed through a blood meal to *Anopheles* mosquitoes, and other hematophagous insects that allow antibody translocation across the midgut, providing a high throughput method to antagonizing protein function in a large number of individuals. The antibody effects will occur within hours following blood feeding unlike RNAi, which takes multiple days to take effect and is highly dependent on gene expression and protein turnover (Vaughan and Azad, 1988; Meyers et al., 2015). However, the length of time after antibody administration for which the effects will persist is currently unknown. Previous results suggest antibodies in a blood meal are able to bind to targets within the head and thorax, though the efficacy against these differentially expressed targets requires further study. Unlike the previously described genetic approaches, antibodies provide flexibility in the timing of when a target protein is inhibited and can be used to transiently or permanently inhibit protein function, thus distinguishing protein effects from transcription or translational effects.

There are also clear limitations to using antibodies to experimentally inhibit protein. It may be unlikely that target proteins in the head and head appendages will be efficiently inhibited if the antibodies are administered through a blood meal. This may be improved by intrathoracic injection of the antibody or by ingestion through a sugar meal. The use of smaller single chain variable fragments (scFVs) or fragment antigen binding domains (Fabs) might have greater success at translocating into the hemocoel while maintaining the ability to bind to its target. Lastly, only secreted proteins or extracellular domains of transmembrane proteins can be targeted by antibodies. Previous work has targeted extracellular domains of ligand-gated ion channels that undergo conformational changes during protein activity. It seems reasonable that similar targeting of metabotropic ligand-gated receptors and of enzyme active sites might occlude ligand or substrate binding respectively, though this is yet to be tested.

Translational Opportunities for Disease Control and Prevention

The first-order effect of a successful antivector vaccine is that arthropod vectors will die prematurely by imbibing blood from a vaccinated host. For nidicolous ectoparasites that reside on the host, they will blood feed upon the host for all, or at least a long portion, of their lifespan (e.g., many ticks, mites, fleas, and lice), and so the immediate effect is reduction or clearance of ectoparasite vectors infesting the vaccinated animal. Antivector vaccine immunity studies with these arthropods have tended to focus on this immediate vector-killing effect, with the simple goal of reducing infestation of individual, vaccinated animals or of vaccinated herds. Heavy infestation can weaken farm animals, decrease their production (milk, weight gain), damage their hides, and greatly disturb individual animals; so, they rub and itch their hides on objects that can cause lesions that become septic (Byford et al., 1992). Reduction of infestation is paramount for the owners of companion animals, who often have strong personal connections to their pets. However, there is a high bar to achieve for antivector vaccines that must compete with the numerous, relatively cheap, and fast-acting endectocides and acaracides/insecticides that are used in and on farm and companion animals. The primary factors in favor of a vaccine over these chemicals is the prospect of not contaminating animals, people (via the meat or milk they consume), and the environment with potentially toxic pesticides. Insecticide resistance against many chemical classes is also widespread, and so a vaccine might circumvent this or even integrate with pesticide use to counterbalance selection pressures in the arthropod (de la Fuente et al., 2007). While it must be acknowledged that selection against antibody binding and target perturbation may occur just as easily as has resistance against chemical insecticides, mutated antigens can be cloned and an updated vaccine could be relatively easily created compared to developing new chemical insecticides. Finally, vaccination might offer a longer period of protection because antibody-titers can be long lasting, especially if there is some natural boosting from similar epitopes within vector salivary gland antigens injected into the host. Most chemical insecticides must be regularly re-administered to, or reapplied, on the host.

The link between infestation reduction in individual animals or herds by vaccination and the reduction of vectorborne diseases that may be spread by these nidicolous vectors, however, is not linear. Most vectorborne pathogens are very efficiently transmitted, so that even a large reduction of vector numbers, or infectious bites by vectors, can fail to significantly reduce the spread of disease (Beier et al., 1999; Smith et al., 2001). Some manuscripts suggest the Gavac vaccine may reduce the transmission of *Babesia* spp. and *Anaplasma* spp., but the data are weak and the studies not well controlled (one measured reduction of disease from integrated control using the vaccine and acaricides, so it is impossible to tell which treatment led to disease control) (de la Fuente et al., 1998; Valle et al., 2004).

For insect vectors like mosquitoes that do not live on the host, and only momentarily blood-feed on it, the goal of a vaccine should not be reduction of infestation but reduction of disease transmission. The vaccinated host is acting as a conduit to kill the blood-feeding insect prematurely, after it has ingested the host's blood. The killing effect stemming from the host reduces the chance of pathogen transmission at a later time, to the host itself, its kin, or its neighbors. In this concept, mosquitocidal and other insect-killing vaccines offer the promise of a type of herd immunity and so rest on a premise of high effective coverage rates (Foy et al., 2002), similar to parasite transmission-blocking vaccines that are well into development (Carter, 2002). While it is clear that the vaccination will not protect the vacciné from infection if the mosquito is infectious at the time it first bites the host, the indirect personal benefit afforded by the vaccine via this type of herd immunity is real, and so mosquitocidal vaccines should not be called "altruistic" as they sometimes have been. Not surprisingly, the parallel field of IVM development for malaria parasite transmission control, which first lead to our research discovery of using the AgGluCl as a mosquitocidal vaccine antigen (described above), can also serve as a model for how to consider, develop, and deploy an insecticidal/mosquitocidal vaccine.

Similar to how we understand that antichannel antibodies work, the IVM molecule targets invertebrate-specific GluCl by altering its conformation. The differences are that while antibody can only access the extracellular domain, the lipophilic IVM molecule integrates into the cell membrane and intercalates into a binding pocket among the transmembrane domains between different subunits in the assembled channel. IVM binding agonizes the channel, forcing it to open and allows chloride ions to flow into the cell aberrantly. This action hyperpolarizes mosquito neurons, and causes flaccid paralysis and death (Campbell, 1989), although there may be other homeostatic pathways in the mosquito that are also disrupted by the drug (Seaman et al., 2015). Anti-AgGluCl antibody is much less researched, but we currently hypothesize that the binding activity also changes the kinetics of the channel in vivo, however, in the opposite direction, as an antagonist.

When IVM was first being researched as an agent that effectively killed *Anopheles* malaria vectors, most groups were simply focused on how it may kill mosquitoes that bite treated hosts (Jones et al., 1992; Bockarie et al., 1999), similar to infestation control strategies described above. But it was clear that if the goal is reduction of *Plasmodium* transmission, it needed to be developed differently. The first works that attempted to change this thinking were by Wilson (1993) in his review of IVM, where he espoused thinking about the drug within the concept of changing vectorial capacity, and Foley et al. (2000), who developed models that utilized concepts of vectorial capacity to show how IVM treatments might reduce malaria in a community. In our own work, we took this further and hypothesized that the effect of a discrete single IVM mass drug administration (MDA) in West African villages would lead to a high probability that any mosquito will ingest a lethal blood meal before it can become infectious in late adult life. Subsequently, the local adult mosquito population structure would be shifted to younger age classes that cannot support pathogen transmission, but the effects on the total vector population would be modest. This hypothesis was supported by our natural field experiments. Single MDAs achieving ≥75% drug coverage reduced the daily probability of *A. gambiae* s.l. (indoor-resting, blood-fed mosquitoes) survival by ~11% for approximately one week. This mortality effect resulted in ~25% reduction in parity rate of mosquitoes collected host-seeking outdoors for approximately 2 weeks, which demonstrated that the age structure of the vector population significantly shifts to younger age classes around the village. As young mosquitoes have not lived long enough to become infectious, this results in at least a ≥78% reduction in vectorial capacity, and significant reductions (>77%) in sporozoite rates for two weeks following the MDA (Sylla et al., 2010; Kobylinski et al., 2011; Alout et al., 2014). These field data have been supported by modeling efforts (Sylla et al., 2010; Foy et al., 2011; Slater et al., 2014). Likely supplementing the observed reductions are the abundant effects on mosquitoes that we and others have documented from sub-lethal IVM doses that would be present in humans 2–7 days after they ingest the drug in an MDA, including slowed digestion (Kobylinski et al., 2010), inhibited flight (Butters et al., 2012), reduced re-blood-feeding frequency (Kobylinski et al., 2010), inhibited sporogony (Kobylinski et al., 2012), and dramatically inhibited egg production (Fritz et al., 2012). It is not clear if all the added effects seen from IVM could be expected of mosquitocidal antibodies ingested by the mosquito, but the general concept is similar: a mosquitocidal vaccine should be designed and deployed for high effective coverage in a community that would limit transmission of a mosquito-borne pathogen, primarily by affecting the daily probability of mosquito survivorship that would most influence the vector population's vectorial capacity (Black and Moore, 2005).

Some insectborne diseases and localized disease scenarios more easily lend themselves to be controlled by such antiinsect vector vaccines, due the life cycles of the vector and pathogen, the transmission cycles and the practicality associated with vaccination. For example, vectorborne diseases in which the causative agents are commonly found in abundant, wild, zoonotic reservoirs might be very difficult to control with an insecticidal vaccine unless there is a simple way to vaccinate the reservoir host. The behavior of specific vector species or populations is likely to be relatively important too. Repeated blood feeding on a defined set of hosts that can be easily vaccinated in high proportions is likely to be advantageous.

We have already been developing IVM for control of malaria parasite transmission, and our preliminary data suggest that the extracellular domain of *Anopheles* GluCl, and other ion channels, can be effective mosquitocidal vaccine target antigens. Malaria is a disease for which mosquitocidal vaccines should be developed and serve as an addition to the growing arsenal of control tools. There is clear evidence that *Anopheles* vectors permit host antibody to translocate from the gut to the hemolymph (Vaughan and Azad, 1988; Meyers et al., 2015), and ion channel sequences among different *Anopheles* vectors are generally highly conserved (Seaman et al., 2015; Jones et al., 2005). In places where vector species often blood-feed on cattle, such as *Anopheles arabiensis* in Africa, *A. stephensi* in India and Pakistan, and *Anopheles minimus* and *Anopheles dirus* in southeast Asia, it may be relatively simple to vaccinate the cattle herds living in and around humans afflicted by the disease and where the vaccine would affect residual malaria transmission scenarios (Killeen, 2014). A vaccine in this scenario should also be safe for the humans who rely on the meat and milk from their animals. Additionally, human vaccinations might be targeted to key demographic groups who are critical to maintaining localized transmission cycles, such as rubber plantation workers in southeast Asia (Tangena et al., 2016).

Other than malaria, lymphatic filariasis may be very amenable to control by mosquitocidal vaccines in areas where it is spread by susceptible *Anopheles* vectors. These tend to be pockets of disease in rural areas where the few hosts that are infecting mosquitoes are hard to find, asymptomatic, or not taking part in MDA campaigns aimed at eliminating the *Wuchereria* and *Brugia* spp. helminth parasites. Similarly, antibody has been shown to cross into the hemocoel of blood-fed tsetse fly vectors (Nogge and Giannetti, 1979). Both sexes of adult flies take large blood meals that are an absolute necessity for their survival, and both human and livestock trypanosomiases are confined to certain pockets of mostly rural areas. It should be possible to vaccinate livestock with an antitsetse vaccine that would work to limit transmission of both forms of the disease. Antiflea vaccines should be able to control infestation on companion animals (Nisbet and Huntley, 2006), and may be able to control the enzootics and epizootics of *Yersinia pestis* from endemic rodent communities. However, the vaccine would likely need to be an oral bait for the rodents, and thus it would need to effectively compete against endectocide baits that are designed to do the same thing.

For many vector insects, such as *Pediculus humanus*, the ability of antibody to cross into the hemolymph following a blood meal is currently unknown. This is similarly the case for the blood-feeding *Ceratopogonidae* that can transmit many arboviruses, including Oropouche virus to humans and Bluetongue and Schmallenberg viruses to ruminants. Likewise, we do not know if antibody can cross into the hemolymph of blood-feeding members of the *Psychodidae* that transmit *Leishmania* parasites and arboviruses. While Wigglesworth (1943) demonstrated many years ago that the kissing bug *R. prolixus* can transport hemoglobin across its midgut, we are not aware of similar data demonstrating transit of antibody. Even if these vectors of Chagas disease were able to be affected by a vaccine against a concealed antigen, it is not clear how one could vaccinate the many wild vertebrate reservoirs that initially infect the insects with *Trypanasoma cruzi*.

Lastly, it may be very difficult to target the many mosquito arbovirus vectors with a concealed antigen vaccine. As described above, while there are reports to the contrary, most data suggest an inability of *A. aegypti* and other Culicine vectors to allow antibody to cross into the vector hemolymph. Even if specific arbovirus vector species were permissive to antibody attack, most arboviruses circulate in diverse, wild reservoir hosts and so the vertebrate reservoirs or the blood meal hosts of the primary mosquito bridge vectors would need to be efficiently targeted with the vaccine. It might be possible, for example, to vaccinate pigs with a mosquitocidal vaccine in an effort to prevent the spread of Japanese encephalitis virus (van den Hurk et al., 2009), but it would also likely be easier and more effective to simply vaccinate them with the Japanese encephalitis vaccine. Arboviruses that circulate in urban and semi-urban cycles between humans and mosquitoes in the absence of an obvious zoonotic reservoir (such as dengue viruses, chikungunya virus, and Zika virus) occur in highly punctuated epidemics, and their transmission events are closely linked to individual human movement (Stoddard et al., 2013), both factors that would resist a mosquitocidal vaccine strategy. Further, it is unlikely that a high enough effective coverage for a mosquitocidal vaccine could be achieved in an urban setting and sexual transmission of Zika virus would be an additional confounder (Foy et al., 2011; D'Ortenzio et al., 2016).

CONCLUSION

The road from Trager's pioneering study to the present state of antivector vaccine research has been long and circuitous. Success in ticks against midgut luminal proteins is owed to their unique digestion processes relative to that of blood-feeding insects. These successes, and those of transmission-blocking vaccine studies in Anopheline vectors of *Plasmodium* spp. and Phlebotomine vectors of *Leishmania* spp., have likely biased research toward searching for "critical" gut proteins of blood-feeding insects that could be targeted by host antibodies. As such critical gut proteins have not been commonly found in insect vectors, research into mosquitocidal/insecticidal vaccine discovery and development has been stifled. We have hope that the clear evidence demonstrating dose-dependent effects of blood meal antibodies targeting hemocoelic, extracellular ion channel domains can change this picture. These extracellular concealed antigen targets were first pushed forward more than 20 years ago, and so researchers have much ground to make up to live up to the foresight of their promoters. The data also suggest that antibodies targeting extracellular epitopes in arthropod vectors will be very useful in dissecting the mechanisms of protein function in vivo. The larger body of work shows that development of mosquitocidal and insecticidal vaccines can follow the research path of developing endectocides for vectorborne disease control. And this research should focus on studying vectors that allow translocation of antibody into their hemolymph, on diseases that could be practically controlled given the vector life cycle and the dynamics of the disease, and the need to vaccinate specific blood mealhosts.

References

Abdeladhim, M., Kamhawi, S., Valenzuela, J.G., 2014. What's behind a sand fly bite? The profound effect of sand fly saliva on host hemostasis, inflammation and immunity. Infect. Genet. Evol. 28, 691–703.

Ackerman, S., et al., 1981. Passage of host serum components, including antibody, across the digestive tract of *Dermacentor variabilis* (Say). J. Parasitol. 67 (5), 737–740.

Adams, M.E., 2004. Agatoxins: ion channel specific toxins from the American funnel web spider, *Agelenopsis aperta*. Toxicon 43 (5), 509–525.

Almazan, C., et al., 2003. Identification of protective antigens for the control of Ixodes scapularis infestations using cDNA expression library immunization. Vaccine 21 (13–14), 1492–1501.

Alout, H., et al., 2014. Evaluation of ivermectin mass drug administration for malaria transmission control across different West African environments. Malar. J. 13, 417.

Aryan, A., Myles, K.M., Adelman, Z.N., 2014. Targeted genome editing in *Aedes aegypti* using TALENs. Methods 69 (1), 38–45.

Beier, J.C., et al., 1989. Effect of human circumsporozoite antibodies in Plasmodium-infected *Anopheles* (Diptera: Culicidae). J. Med. Entomol. 26 (6), 547–553.

Beier, J.C., Killeen, G.F., Githure, J.I., 1999. Short report: entomologic inoculation rates and *Plasmodium falciparum* malaria prevalence in Africa. Am. J. Trop. Med. Hyg. 61 (1), 109–113.

Billingsley, P.F., Foy, B., Rasgon, J.L., 2008. Mosquitocidal vaccines: a neglected addition to malaria and dengue control strategies. Trends Parasitol. 24 (9), 396–400.

Black, W.C.I., Moore, C.G., 2005. Population biology as a tool to study vector-borne diseases. In: Marquardt, W.C. (Ed.), Biology of Disease Vectors. Elsevier Academic Press, San Diego, CA, pp. 187–206.

Bloomquist, J.R., et al., 1996. Mode of action of an insecticidal peptide toxin from the venom of a weaving spider (Diguetia canities). Toxicon 34 (9), 1072–1075.

Bockarie, M.J., et al., 1999. Mass treatment with ivermectin for filariasis control in Papua New Guinea: impact on mosquito survival. Med. Vet. Entomol. 13 (2), 120–123.

Brennan, J.D., et al., 2000. *Anopheles gambiae* salivary gland proteins as putative targets for blocking transmission of malaria parasites. Proc. Natl. Acad. Sci. U S A 97 (25), 13859–13864.

Butters, M.P., et al., 2012. Comparative evaluation of systemic drugs for their effects against *Anopheles gambiae*. Acta Trop. 121 (1), 34–43.

Byford, R.L., Craig, M.E., Crosby, B.L., 1992. A review of ectoparasites and their effect on cattle production. J. Anim. Sci. 70 (2), 597–602.

Campbell, W.C., 1989. Ivermectin and Abamectin. Springer Verlag, New York, p. 363.

Campbell, W.C., 2016. Lessons from the history of ivermectin and other antiparasitic agents. Annu. Rev. Anim. Biosci. 4, 1–14.

Carter, R., 2002. Spatial simulation of malaria transmission and its control by malaria transmission blocking vaccination. Int. J. Parasitol. 32 (13), 1617–1624.

Corzo, G., et al., 2000. Isolation, synthesis and pharmacological characterization of delta-palutoxins IT, novel insecticidal toxins from the spider Paracoelotes luctuosus (Amaurobiidae). Eur. J. Biochem. 267 (18), 5783–5795.

Corzo, G., et al., 2003. Distinct primary structures of the major peptide toxins from the venom of the spider Macrothele gigas that bind to sites 3 and 4 in the sodium channel. FEBS Lett. 547 (1–3), 43–50.

da Costa, M., et al., 2014. Mosquito Akirin as a potential antigen for malaria control. Malar. J. 13, 470.

Davies, T.G., et al., 2007. A comparative study of voltage-gated sodium channels in the Insecta: implications for pyrethroid resistance in Anopheline and other Neopteran species. Insect Mol. Biol. 16 (3), 361–375.

de la Fuente, J., et al., 1998. Field studies and cost-effectiveness analysis of vaccination with Gavac against the cattle tick *Boophilus microplus*. Vaccine 16 (4), 366–373.

de la Fuente, J., et al., 2005. RNA interference screening in ticks for identification of protective antigens. Parasitol. Res. 96 (3), 137–141.

de la Fuente, J., et al., 2007. A ten-year review of commercial vaccine performance for control of tick infestations on cattle. Anim. Health Res. Rev. 8 (1), 23–28.

de la Fuente, J., et al., 2013. Subolesin/Akirin vaccines for the control of arthropod vectors and vectorborne pathogens. Transbound. Emerg. Dis. 60 (Suppl. 2), 172–178.

D'Ortenzio, E., et al., 2016. Evidence of sexual transmission of Zika virus. N. Engl. J. Med. 374 (22), 2195–2198.

Elvin, C.M., Kemp, D.H., 1994. Generic approaches to obtaining efficacious antigens from vector arthropods. Int. J. Parasitol. 24 (1), 67–79.

Figueiredo, S.G., et al., 1995. Purification and amino acid sequence of the insecticidal neurotoxin Tx4(6-1) from the venom of the 'armed' spider Phoneutria nigriventer (Keys). Toxicon 33 (1), 83–93.

Fletcher, J.I., et al., 1997. The structure of a novel insecticidal neurotoxin, omega-atracotoxin-HV1, from the venom of an Australian funnel web spider. Nat. Struct. Biol. 4 (7), 559–566.

Foley, D.H., Bryan, J.H., Lawrence, G.W., 2000. The potential of ivermectin to control the malaria vector *Anopheles farauti*. Trans. R Soc. Trop. Med. Hyg. 94 (6), 625–628.

Foy, B.D., et al., 2002. Immunological targeting of critical insect antigens. Am. Entomol. 48 (3), 150–163.

Foy, B.D., et al., 2003. Induction of mosquitocidal activity in mice immunized with *Anopheles gambiae* midgut cDNA. Infect. Immun. 71 (4), 2032–2040.

Foy, B.D., et al., 2011a. Endectocides for malaria control. Trends Parasitol. 27 (10), 423–428.

Foy, B.D., et al., 2011b. Probable non-vector-borne transmission of Zika virus, Colorado, USA. Emerg. Infect. Dis. 17 (5), 880–882.

Francischetti, I.M., et al., 2009. The role of saliva in tick feeding. Front Biosci. (Landmark Ed) 14, 2051–2088.

Fritz, M.L., Walker, E.D., Miller, J.R., 2012. Lethal and sublethal effects of avermectin/milbemycin parasiticides on the African malaria vector, *Anopheles arabiensis*. J. Med. Entomol. 49 (2), 326–331.

Gomez, A.M., et al., 2010. Antibody effector mechanisms in myasthenia gravis-pathogenesis at the neuromuscular junction. Autoimmunity 43 (5–6), 353–370.

Gordon, D., et al., 2007. The differential preference of scorpion alpha-toxins for insect or mammalian sodium channels: implications for improved insect control. Toxicon 49 (4), 452–472.

Hannon, G.J., 2002. RNA interference. Nature 418 (6894), 244–251.

Hatfield, P.R., 1988. Anti-mosquito antibodies and their effects on feeding, fecundity and mortality of *Aedes aegypti*. Med. Vet. Entomol. 2, 331–338.

Jacobs-Lorena, M., Lemos, F.J.A., 1995. Immunologic strategies for control of insect disease vectors: a critical assessment. Parasitol. Today 11, 144–147.

Jasinskas, A., Barbour, A.G., 2005. The Fc fragment mediates the uptake of immunoglobulin C from the midgut to hemolymph in the ixodid tick *Amblyomma americanum* (Acari: Ixodidae). J. Med. Entomol. 42 (3), 359–366.

Jasinskas, A., Jaworski, D.C., Barbour, A.G., 2000. *Amblyomma americanum*: specific uptake of immunoglobulins into tick hemolymph during feeding. Exp. Parasitol. 96 (4), 213–221.

Jeffers, L.A., Michael Roe, R., 2008. The movement of proteins across the insect and tick digestive system. J. Insect Physiol. 54 (2), 319–332.

Jones, J.W., et al., 1992. Lethal effects of ivermectin on *Anopheles quadrimaculatus*. J. Am. Mosq. Control. Assoc. 8 (3), 278–280.

Jones, A.K., Grauso, M., Sattelle, D.B., 2005. The nicotinic acetylcholine receptor gene family of the malaria mosquito, *Anopheles gambiae*. Genomics 85 (2), 176–187.

Kay, B.H., Kemp, D.H., 1994. Vaccines against arthropods. Am. J. Trop. Med. Hyg. 50 (6), 87–96.

Killeen, G.F., 2014. Characterizing, controlling and eliminating residual malaria transmission. Malar. J. 13, 330.

King, G.F., Escoubas, P., Nicholson, G.M., 2008. Peptide toxins that selectively target insect Na(V) and Ca(V) channels. Channels (Austin) 2 (2), 100–116.

Kistler, K.E., Vosshall, L.B., Matthews, B.J., 2015. Genome engineering with CRISPR-Cas9 in the mosquito *Aedes aegypti*. Cell Rep. 11 (1), 51–60.

Klint, J.K., et al., 2012. Spider-venom peptides that target voltage-gated sodium channels: pharmacological tools and potential therapeutic leads. Toxicon 60 (4), 478–491.

Kobylinski, K.C., et al., 2010. The effect of oral anthelmintics on the survivorship and re-feeding frequency of anthropophilic mosquito disease vectors. Acta Trop. 116 (2), 119–126.

Kobylinski, K.C., et al., 2011. Ivermectin mass drug administration to humans disrupts malaria parasite transmission in Senegalese villages. Am. J. Trop. Med. Hyg. 85 (1), 3–5.

Kobylinski, K.C., Foy, B.D., Richardson, J.H., 2012. Ivermectin inhibits the sporogony of *Plasmodium falciparum* in *Anopheles gambiae*. Malar. J. 11, 381.

Kordas, G., et al., 2014. Direct proof of the in vivo pathogenic role of the AChR autoantibodies from myasthenia gravis patients. PLoS One 9 (9), e108327.

Lackie, A.M., Gavin, S., 1989. Uptake and persistence of ingested antibody in the mosquito *Anopheles stephensi*. Med. Vet. Entomol. 3 (3), 225–230.

Leipold, E., Borges, A., Heinemann, S.H., 2012. Scorpion beta-toxin interference with NaV channel voltage sensor gives rise to excitatory and depressant modes. J. Gen. Physiol. 139 (4), 305–319.

Li, D., et al., 2003. Function and solution structure of hainantoxin-I, a novel insect sodium channel inhibitor from the Chinese bird spider Selenocosmia hainana. FEBS Lett. 555 (3), 616–622.

Marr, E.J., et al., 2014. RNA interference for the identification of ectoparasite vaccine candidates. Parasite Immunol. 36 (11), 616–626.

McDowell, M.A., 2015. Vector-transmitted disease vaccines: targeting salivary proteins in transmission (SPIT). Trends Parasitol. 31 (8), 363–372.

Meyers, J.I., et al., 2015. Characterization of the target of ivermectin, the glutamate-gated chloride channel, from *Anopheles gambiae*. J. Exp. Biol. 218 (Pt 10), 1478–1486.

Meyers, J.I., Gray, M., Foy, B.D., 2015. Mosquitocidal properties of IgG targeting the glutamate-gated chloride channel in three mosquito disease vectors (Diptera: Culicidae). J. Exp. Biol. 218 (Pt 10), 1487–1495.

Moran, Y., et al., 2007. Molecular analysis of the sea anemone toxin Av3 reveals selectivity to insects and demonstrates the heterogeneity of receptor site-3 on voltage-gated Na+ channels. Biochem. J. 406 (1), 41–48.

Nisbet, A.J., Huntley, J.F., 2006. Progress and opportunities in the development of vaccines against mites, fleas and myiasis-causing flies of veterinary importance. Parasite Immunol. 28 (4), 165–172.

Nogge, G., Giannetti, M., 1979. Midgut absorption of undigested albumin and other proteins by tsetse, Glossina M. morsitans (Diptera: Glossinidae). J. Med. Entomol. 16 (3), 263.

Nuttall, P.A., et al., 2006. Exposed and concealed antigens as vaccine targets for controlling ticks and tick-borne diseases. Parasite Immunol. 28 (4), 155–163.

Patrick, J., Lindstrom, J., 1973. Autoimmune response to acetylcholine receptor. Science 180 (4088), 871–872.

Ramamoorthi, N., et al., 2005. The Lyme disease agent exploits a tick protein to infect the mammalian host. Nature 436 (7050), 573–577.

Ramasamy, M.S., et al., 1988. Anti-mosquito antibodies decrease the reproductive capacity of *Aedes aegypti*. Med. Vet. Entomol. 2, 87–93.

Ribeiro, J.M., Mans, B.J., Arca, B., 2010. An insight into the sialome of blood-feeding Nematocera. Insect Biochem. Mol. Biol. 40 (11), 767–784.

Salgado, V.L., Kem, W.R., 1992. Actions of three structurally distinct sea anemone toxins on crustacean and insect sodium channels. Toxicon 30 (11), 1365–1381.

Sauer, J.R., McSwain, J.L., Essenberg, R.C., 1994. Cell membrane receptors and regulation of cell function in ticks and blood-sucking insects. Int. J. Parasitol. 24 (1), 33–52.

Seaman, J.A., et al., 2015. Age and prior blood feeding of *Anopheles gambiae* influences their susceptibility and gene expression patterns to ivermectin-containing blood meals. BMC Genomics 16 (1), 797.

Shakya, M., et al., 2014. Subolesin: a candidate vaccine antigen for the control of cattle tick infestations in Indian situation. Vaccine 32 (28), 3488–3494.

Slater, H.C., et al., 2014. The potential impact of adding ivermectin to a mass treatment intervention to reduce malaria transmission: a modelling study. J. Infect. Dis. 210 (12), 1972–1980.

Smith, J.J., et al., 2013. The insecticidal potential of venom peptides. Cell Mol. Life Sci. 70 (19), 3665–3693.

Smith, T.A., Leuenberger, R., Lengeler, C., 2001. Child mortality and malaria transmission intensity in Africa. Trends Parasitol. 17 (3), 145–149.

Stapleton, A., et al., 1990. Curtatoxins. Neurotoxic insecticidal polypeptides isolated from the funnel-web spider Hololena curta. J. Biol. Chem. 265 (4), 2054–2059.

Stevens, M., Peigneur, S., Tytgat, J., 2011. Neurotoxins and their binding areas on voltage-gated sodium channels. Front Pharmacol. 2, 71.

Stoddard, S.T., et al., 2013. House-to-house human movement drives dengue virus transmission. Proc. Natl. Acad. Sci. U S A 110 (3), 994–999.

Sylla, M., et al., 2010. Mass drug administration of ivermectin in south-eastern Senegal reduces the survivorship of wild-caught, blood fed malaria vectors. Malar. J. 9, 365.

Tangena, J.A., et al., 2016. Risk and control of mosquito-borne diseases in southeast Asian rubber plantations. Trends Parasitol. 32 (5), 402–415.

Trager, W., 1939. Acquired immunity to ticks. J. Parasitol. 25, 137–139.

Valle, M.R., et al., 2004. Integrated control of *Boophilus microplus* ticks in Cuba based on vaccination with the anti-tick vaccine Gavac. Exp. Appl. Acarol. 34 (3–4), 375–382.

van den Hurk, A.F., Ritchie, S.A., Mackenzie, J.S., 2009. Ecology and geographical expansion of Japanese encephalitis virus. Annu. Rev. Entomol. 54, 17–35.

Vaughan, J.A., Azad, A.F., 1988. Passage of host immunoglobulin G from blood meal into hemolymph of selected mosquito species (Diptera: Culicidae). J. Med. Entomol. 25 (6), 472–474.

Vaughan, J.A., et al., 1988. *Plasmodium falciparum*: ingested anti-sporozoite antibodies affect sporogony in *Anopheles stephensi* mosquitoes. Exp. Parasitol. 66, 171–182.

Vaughan, J.A., et al., 1990. Quantitation of antisporozoite immunoglobulins in the hemolymph of *Anopheles stephensi* after bloodfeeding. Am. J. Trop. Med. Hyg. 42 (1), 10–16.

Vaughan, J.A., et al., 1998. Quantitation of cat immunoglobulins in the hemolymph of cat fleas (Siphonaptera: Pulicidae) after feeding on blood. J. Med. Entomol. 35 (4), 404–409.

Vincent, A., Lang, B., Kleopa, K.A., 2006. Autoimmune channelopathies and related neurological disorders. Neuron 52 (1), 123–138.

Wang, X.H., et al., 2001. Discovery and structure of a potent and highly specific blocker of insect calcium channels. J. Biol. Chem. 276 (43), 40306–40312.

Wang, H., Nuttall, P.A., 1999. Immunoglobulin-binding proteins in ticks: new target for vaccine development against a blood-feeding parasite. Cell Mol. Life Sci. 56 (3–4), 286–295.

Wigglesworth, V.B., 1943. The fate of haemoglobin in *Rhodnius prolixus* (Hemiptera) and other blood-sucking arthropods. Proc. Roy. Soc. B 131, 313–339.

Wijffels, G., et al., 1999. Peritrophins of adult dipteran ectoparasites and their evaluation as vaccine antigens. Int. J. Parasitol. 29 (9), 1363–1377.

Willadsen, P., et al., 1989. Immunologic control of a parasitic arthropod. Identification of a protective antigen from *Boophilus microplus*. J. Immunol. 143 (4), 1346–1351.

Willadsen, P., et al., 1995. Commercialisation of a recombinant vaccine against *Boophilus microplus*. Parasitology 110 (Suppl.), S43–S50.

Willadsen, P., Kemp, D.H., 1988. Vaccination with 'concealed' antigens for tick control. Parasitol. Today 4 (7), 196–198.

Willadsen, P., 2001. The molecular revolution in the development of vaccines against ectoparasites. Vet. Parasitol. 101 (3–4), 353–368.

Willadsen, P., 2006. Vaccination against ectoparasites. Parasitology 133 (Suppl.), S9–s25.

Wilson, M.L., 1993. Avermectins in arthropod vector management - prospects and pitfalls. Parasitol. Today 9 (3), 83–87.

Zhang, P.F., et al., 2003. Huwentoxin-V, a novel insecticidal peptide toxin from the spider Selenocosmia huwena, and a natural mutant of the toxin: indicates the key amino acid residues related to the biological activity. Toxicon 42 (1), 15–20.

CHAPTER

9

Role of the Microbiota During Development of the Arthropod Vector Immune System

Aurélien Vigneron, Brian L. Weiss
Yale School of Public Health, New Haven, CT, United States

SPECTRUM OF VECTOR–MICROBE INTERACTIONS

Arthropod vectors are most commonly associated with the transmission of viral, bacterial, protozoan, and metazoan pathogens that cause disease in vertebrate animals. While only a small percentage of arthropods are infected with disease-causing microorganism, they all house a community of symbiotic bacteria that effect their host by largely known mechanisms. These symbiotic associations can positively or negatively influence their host's physiology and are thus classified along a continuous spectrum that is defined by each partner's physiological impact on the other. As such, outcomes can range from parasitic (i.e., harmful for one partner) to mutualistic (i.e., both partners are beneficial), with commensalism (i.e., neutral interaction) falling in-between. The gut-associated (i.e., gut lumen and/or intracellular within specialized cells attached to the gut) component of the cumulative whole-body microbiota is an important mediator of arthropod host development and growth. These bacteria provide nutriments that allow their hosts to feed exclusively on nutritionally unbalanced sources such as vertebrate blood (Snyder et al., 2010; Rio et al., 2016), promote the maturation of their host from immature to adult stages (Coon et al., 2014), and modulate immune system development and function (Weiss and Aksoy, 2011; Cirimotich et al., 2011a). Finally, some arthropod vectors also harbor bacteria that manipulate their host's reproductive physiology (Shaw et al., 2015; Segata et al., 2016). Information related to the molecular mechanisms that underlie functional relations between arthropod vector hosts and their symbiont communities is beginning to emerge. In this review we summarize the current state of knowledge regarding this topic, including how indigenous microbes regulate the progression of their host's lifecycle, function of their host's immune system during adulthood, and development of their host's immune system during immature stages.

Arthropod Vector: Controller of Disease Transmission, Volume 1
http://dx.doi.org/10.1016/B978-0-12-805350-8.00009-X

ENVIRONMENTALLY ACQUIRED COMMENSAL BACTERIA SUPPORT THEIR HOST'S DEVELOPMENT

A growing body of literature demonstrates a robust functional correlation between gut microbes and host physiological homeostasis (both digestive and systemic) and development (Diaz Heijtz et al., 2011; Nicholson et al., 2012). In mosquitoes, the gut microbiota plays an important role in completion of the insect's lifecycle. In fact, total absence of the microbiota from larval *Aedes aegypti* and *Anopheles gambiae* (axenic individuals) leads to developmental arrest during the first instar (Coon et al., 2014). Interestingly, development can be rescued by adding back members of both the native microbiota as well as exogenous bacteria (*Escherichia coli*). Importantly, the development-rescuing phenotype is only observed when the supplemented bacteria are alive (Coon et al., 2014). Altogether, these findings suggest that conserved bacterial products may provide signaling cues that regulate larval mosquito growth processes and that mosquitoes require their indigenous bacteria presence to achieve their development. The effects of a symbiosis are not always as dramatic as the latter example. Rather, the presence of a symbiont can confer an advantage to its host. For example, although the mosquito *Anopheles stephensi* can survive when it lacks symbiotic *Asaia* spp. (Chouaia et al., 2012), it presents delayed larval development. The mechanism(s) by which *Asaia* bacteria support their host is unknown. However, in the Mediterranean fruit fly (*Ceratitis capitata*), *Asaia* plays a nutritional role by supplementing its host with nitrogen (Behar et al., 2005). This bacterium may perform a similar function in *A. stephensi*.

MICROBIOME INFLUENCES ON ARTHROPOD HOST VECTOR COMPETENCE

Enteric microbes can modulate the ability of their arthropod host to transmit disease-causing microorganisms, and this phenotype can be either direct or indirect (Weiss and Aksoy, 2011; Cirimotich et al., 2011b; Jupatanakul et al., 2014; Johnson, 2015; Narasimhan and Fikrig, 2015). An elegant example of direct microbial mediation of arthropod vector competence can be observed in the African malaria vector, *A. gambiae*. A strain of *Enterobacter* (*Esp_Z*) was isolated from midguts of *Plasmodium*-refractory *A. gambiae* collected in Zambia. This bacterium induced a similar *Plasmodium*-refractory phenotype in normally parasite-susceptible laboratory-reared *A. gambiae*. In this case, reactive oxygen species (ROS) generated by *Esp_Z* interfered with parasite development, thus prohibiting invasion of the mosquito midgut epithelium (Cirimotich et al., 2011a).

Chromobacterium (*Csp_P*), isolated from midguts of field-captured *A. aegypti*, is another microbe that regulates its host's vector competence. This effect is multifaceted and results from both direct and indirect *Csp_P* modulated anti-*Plasmodium* and antidengue activity in laboratory-reared *A. gambiae* and *A. aegypti*, respectively. Specifically, exposure to *Csp_P* dramatically shortens the life span of both larval and adult mosquitoes, thus reducing host fecundity. *Csp_P* also activates mosquito immune responses and produces stable antipathogen molecules in vitro (Ramirez et al., 2014). The combined antipathogen and entomopathogenic properties of *Csp_P* account for this bacterium's vector competency–reducing phenotype in mosquito hosts.

Interestingly, microbiota-induced homeostasis in the arthropod gut can be exploited by pathogens to facilitate transmission through their vector host. This phenomenon was recently described in the blacklegged tick, *Ixodes scapularis*, where gut microbes activate expression of the transcription factor STAT. STAT expression is necessary for production of a structurally robust peritrophic matrix (PM), which lines the tick gut thus separating the intestinal lumen from epithelial cells. The tick-transmitted bacterial pathogen, *Borrelia burgdorferi* (the causative agent of Lyme disease), finds protection from the harsh luminal environment by hiding on the epithelial side (called the ectoperitrophic space) of the tick PM. Dysbiosis of *I. scapularis'* gut microbiota reduces STAT expression, thus resulting in the production of a relatively porous PM. Under such circumstances, luminal contents breach the PM and interfere with *Borrelia* colonization (Narasimhan et al., 2014).

The above-described examples detail the impact of environmentally acquired gut microbes on their host's vector competency. However, maternally transmitted microbes also play an important role in determining this host phenotype. The best example of this type of microbe is the parasitic symbiont *Wolbachia*, which is most commonly associated with manipulation of arthropod host reproduction (Werren et al., 2008). *Wolbachia* infection in mosquitoes primes the vector's innate immune system, thus limiting pathogen colonization (Bian et al., 2010; Pan et al., 2012; Rances et al., 2012). This effect confers mosquitoes with refractoriness to infection with pathogens such as *Plasmodium falciparum* (Bian et al., 2013a,b; Shaw et al., 2016), dengue (Frentiu et al., 2014), chikungunya (van den Hurk et al., 2012; Aliota et al., 2016), yellow fever (van den Hurk et al., 2012), and filarial nematodes (Kambris et al., 2009). Additionally, *Wolbachia* can significantly reduce the life span of some mosquitoes (McMeniman et al., 2009) such that the vector perishes before the pathogen is able to complete its extrinsic incubation period (EID; the interval between vector uptake of an infectious agent and transmission of the agent to a new host). *Wolbachia's* two-pronged vector competency reducing phenotype makes this microbe attractive for use in novel disease control strategies.

MUTUALISTIC ENDOSYMBIONTS SUPPORT THEIR HOST'S DEVELOPMENT

The information above describes how transient, environmentally acquired, and maternally transmitted, parasitic bacteria alter arthropod vector competency. Another type of symbiotic association, obligate mutualism, arises when animal hosts and bacteria closely associate over long evolutionary time frames. Under these circumstances, both organisms are physiologically dependent on one another (neither can survive in the absence of the other), and the bacterial symbiont is designated an "endosymbiont" because it resides within specialized host cells that collectively form a bacteriome organ. Endosymbiosis is widespread and represents a potent driver of insect evolution (Margulis, 1993; Moran, 2006, 2007; Douglas, 2014). The human body louse (*Pediculus humanus humanus*; Perotti et al., 2007), the tsetse fly (*Glossina* spp.; Aksoy, 1995), and several tick species (Sassera et al., 2006; Smith et al., 2015b) also house endosymbiotic bacteria. These mutualistic associations are maintained across generations via maternal transmission of the endosymbionts. The genomes of these bacteria are dramatically reduced to reflect a lifestyle that includes little, if any, exposure to the extracellular environment (Wernegreen, 2002; Moran et al., 2008). Despite this reduction, mutualistic bacteria typically produce essential nutriments that permit their hosts to survive on nutritionally restricted diets (Rio et al., 2016). This support is essential for the arthropod host, as their fitness significantly diminishes in the absence of their endosymbionts (Zhong et al., 2007; Pais et al., 2008; Snyder and Rio, 2015). In addition to nutritional supplementation, little is known about how mutualistic bacteria contribute to the homeostasis of other host physiological systems.

THE TSETSE FLY AS A MODEL SYSTEM FOR STUDYING SYMBIONT CONTRIBUTIONS TO HOST IMMUNE SYSTEM DEVELOPMENT

Tsetse flies serve as a useful model system for studying animal host–symbiont interactions because they house a taxonomically simple population of endogenous symbiotic bacteria. In fact, laboratory-reared flies house only three

TABLE 9.1 Tsetse's Indigenous Symbiotic Bacteria and Their Functional Contributions to Host Physiology

Symbiont	Symbiotic Integration	Functional Contribution to Host Physiology	Host Tissue Tropism	Defining Characteristics	References
Wigglesworthia	Obligate mutualist	Nutrient provisioning; modulation of tsetse host immune system development	Intracellular within gut associated bacteriocytes; extracellular in maternal milk gland	Ancient association with tsetse host (50–80 million years) and found in all flies; highly streamlined genome (700 kb); genome encodes vitamin biosynthesis pathways	Akman et al. (2002), Rio et al. (2012), Michalkova et al. (2014), Balmand et al. (2013), Attardo et al. (2008), Chen et al. (1999), Weiss et al. (2011), and Weiss et al. (2012)
Sodalis	Commensal	Experimentally undetermined	Wide tissue tropism, intra and extracellular within gut, milk gland, salivary glands, muscle, male and female reproductive tract	Recent association with tsetse host (~1 million years) and Found in all colonized and most field-captured individuals; genome similar in size and synteny to those of several free-living bacteria; genome in a state of decay	Cheng and Aksoy (1999), Balmand et al. (2013), De Vooght et al. (2015), Toh et al. (2006), and Aksoy et al. (2014a,b)
Wolbachia	Reproductive parasite	Manipulation of host reproductive biology	Intracellular within female and male germ line cells	Found in all colonized and some field-captured individuals; induces cytoplasmic incompatibility in tsetse	Cheng et al. (2000), Balmand et al. (2013), and Alam et al. (2011)

distinct bacterial taxa, which include obligate *Wigglesworthia*, commensal *Sodalis*, and parasitic *Wolbachia* (Table 9.1; Maltz et al., 2012; Wang et al., 2013). Tsetse reproduce via a process called adenotrophic viviparity, during which all embryonic and larval stages mature within their mom's uterus. While in this position, larval tsetse receive nourishment via milk secretions produced by a modified accessory gland (Benoit et al., 2015). This milk also contains extracellular populations of *Wigglesworthia* and *Sodalis*, and these microbes colonize subsequent generations of tsetse via this maternal route of transmission (Attardo et al., 2008). Unlike *Wigglesworthia and Sodalis*, *Wolbachia* is transmitted transovarially via germ line cells (Cheng et al., 2000). Field-captured tsetse all house *Wigglesworthia*, while *Sodalis* and *Wolbachia* are restricted to specific tsetse species and geographic populations (Wamwiri et al., 2013; Doudoumis et al., 2013; Mbewe et al., 2015). Wild tsetse can also house transient populations of environmentally acquired endogenous bacteria (Lindh and Lehane, 2011; Geiger et al., 2013). However, these microbes make up a diminutive proportion of the total bacterial population present in tsetse (the vast majority are *Wigglesworthia*; Aksoy et al., 2014a), and their functional role within the fly is currently unknown.

Generation of Dysbiotic Tsetse

The functional role of tsetse's maternally transmitted microbiota, as it relates to immune system development and function, was evaluated by generating dysbiotic larvae. To do so, pregnant females were administered blood meals supplemented with either ampicillin, or tetracycline and yeast extract. Treatment with ampicillin clears all milk gland–associated *Wigglesworthia* (but not *Sodalis*) while leaving the intracellular bacteriome-associated population (*Wigglesworthia*-free offspring are designated Gmm^{Wgm-}; Pais et al., 2008) undisturbed. Tetracycline treatment eliminates all bacterial symbionts from pregnant females (hence the yeast extract supplement, which provided essential nutrients normally produced by *Wigglesworthia*) such that offspring mature in the absence of their entire endogenous microbiome (and are thus designated "aposymbiotic," or Gmm^{Apo}; Alam et al., 2011; Weiss et al., 2012). Because these flies undergo their entire developmental program in the absence of indigenous symbiotic bacteria, this experimental system is useful for investigating the functional association between these microbes and the maturation of their host's immune system.

Wigglesworthia's Role in the Development of Tsetse's Cellular Immune Response

Both Gmm^{Wgm-} and Gmm^{Apo} adults are severely immunocompromised when compared to wild-type individuals (Weiss et al., 2011, 2012). Systemic challenge with *E. coli* K12, of individuals from either group, results in unimpeded bacterial replication that in turn results in tsetse host death. Conversely, wild-type tsetse (Gmm^{WT}) are able to control the same infection and survive. Following cuticular wounding with a sterile needle, dysbiotic tsetse adults also fail to produce melanin and thus do not form a clot at the lesion site. Wounded individuals are thus vulnerable to dehydration via unrestricted loss of hemolymph and are exposed to systemic infection with environmental microbes. The immunocompromised phenotypes presented by Gmm^{Wgm-} and Gmm^{Apo} adults result from both cellular and humoral immune deficiencies. Dysbiotic tsetse house a severely depleted population of sessile and circulating hemocytes (Weiss et al., 2011, 2012). In closely related *Drosophila*, hemocytes are classified into three distinct subtypes; lamellocytes, crystal cells, and phagocytes. Lamellocytes and crystal cells are involved in encapsulating foreign objects introduced into the insect's hemocoel (e.g., parasitic wasp eggs) and initiating the melanization cascade, respectively. Phagocytes, which are the functional equivalent of vertebrate macrophages, engulf (or phagocytose) and digest exogenous microbes that have been introduced into the insect hemocoel (Honti et al., 2014).

Phagocytosis serves as a potent first line of defense against systemic infection with pathogenic microbes in several insect models, including *Drosophila*, *Anopheles*, and *Manduca* (Lamiable et al., 2016; Parsons and Foley, 2016; Smith et al., 2015a; Eleftherianos et al., 2009). Additionally, hemocytes, along with the insect fat body, also produce effector molecules, including antimicrobial peptides (AMPs; cecropin and attacin), signal-mediating ROS (inducible nitric oxide synthase and dual oxidase), and pathogen recognition (thioester-containing proteins 2 and 4) and pathogen clearance (prophenoloxidase)–associated molecules (Lemaitre and Hoffmann, 2007). Gmm^{Apo} adults exhibit atypical expression of genes that encode these immune-related molecules (Weiss et al., 2011, 2012), which may reflect the markedly reduced hemocyte population present in these flies. Interestingly, *Wigglesworthia* does not directly enhance immunity in Gmm^{WT} individuals, because elimination of this bacterium from adult flies does not result in an immune-compromised phenotype. Rather, this obligate must be present during intrauterine larval for tsetse's immune system to mature and function properly during adulthood.

Signaling pathways and transcription factors that regulate blood cell differentiation within the hematopoietic niche are functionally conserved across many animal taxa (Hartenstein, 2006). In *Drosophila*'s lymph gland, prohemocytes express the GATA factor *serpent*, and these cells subsequently differentiate into either phagocytes or crystal cell lineages via induction of the transcription factors, *glial cells missing* or *lozenge*, respectively (Lebetsky et al., 2000; Ferjoux et al., 2007). Dysbiotic tsetse larvae express significantly fewer *serpent* and *lozenge* transcripts than their *Gmm*^{WT} counterparts do (Weiss et al., 2011, 2012), and this phenomenon likely accounts for their depleted population of phagocytic hemocytes and crystal cells (Fig. 9.1). The specific symbiont-derived molecules that induce expression of tsetse hematopoietic transcription factors

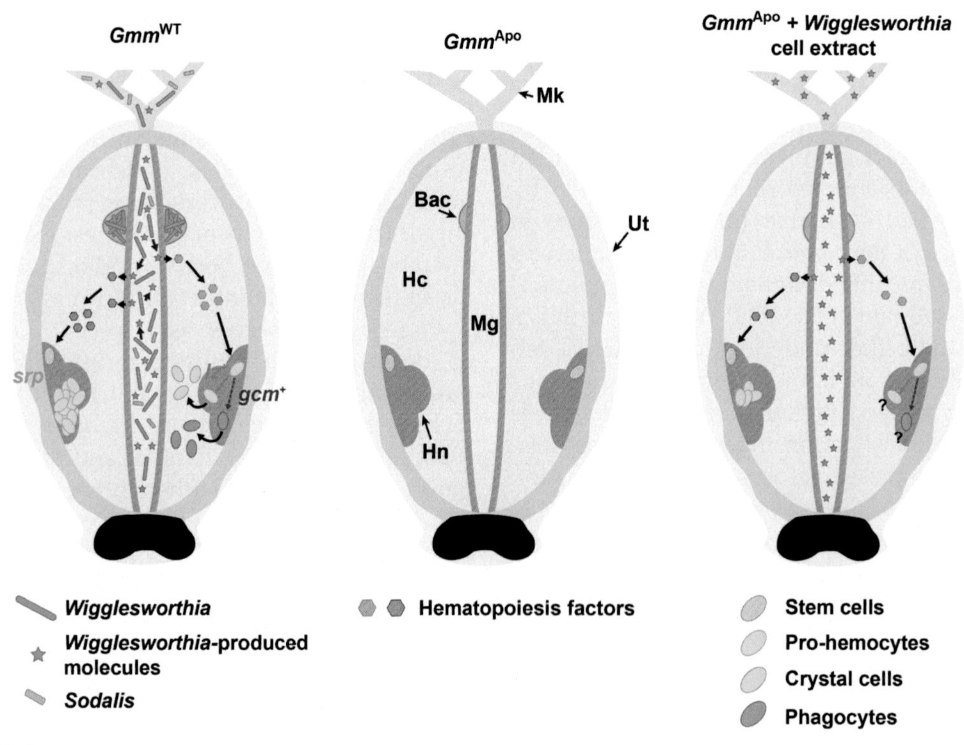

FIGURE 9.1 Model illustrating the functional impact of maternally transmitted symbionts on the immune system development in larval tsetse flies. In Gmm^{WT} tsetse (left diagram), *Wigglesworthia* (*blue rod*) are transmitted to developing intrauterine larvae via milk that is synthesized by the maternal milk gland (Mk). Molecule(s) (*blue stars*) derived from this bacterium stimulates the synthesis of factors (*brown/beige hexagons*) that promote blood cell differentiation, or hematopoiesis. Consequently, in the larval hematopoietic niche, stem cells (*gray ellipse*) differentiate into prohemocytes (*turquoise ellipse*) following expression of the transcription factor serpent (srp). Subsequently, prohemocytes express lozenge (lz) and glial cells missing (gcm), which promotes their further differentiation into crystal cells (*green ellipse*) or phagocytes (*pink ellipses*), respectively. Sodalis, also present in tsetse milk, does not exert a hematopoiesis-stimulating effect. GmmApo tsetse (right diagram) present no hemocytes. This phenotype can be partially rescued by supplementing the diet of symbiont-free moms with *Wigglesworthia* cell extracts (center diagram). However, under this circumstance, hemocyte production does not reach wild-type levels and, whether or not hemocyte differentiation occurs, has yet to be established. *Bac*, bacteriome; *Hc*, hemocoel; *Hn*, hematopoietic niche; *Ut*, uterus.

are currently unknown. However, hematopoiesis is partially restored in *Gmm*^{Apo} larvae when their symbiont-cured females are fed a diet supplemented with *Wigglesworthia* cell extracts. This restorative process does not occur in *Gmm*^{Apo} larvae when their symbiont-cured moms receive *Sodalis* cell extracts, thus indicating that the stimuli are likely *Wigglesworthia* specific. Furthermore, *Wigglesworthia*-produced vitamin metabolites are likely not the stimulus, as fecundity rescuing yeast extracts included in the blood meal of symbiont-cured females fail to actuate immune system development in aposymbiotic intrauterine larvae (Weiss et al., 2011, 2012).

Wigglesworthia and the Development of Tsetse Gut Barriers That Modulate Trypanosome Infection Outcomes

Tsetse are cyclic vectors of African trypanosomes, and as such, these parasites must pass through the fly to complete their lifecycle and be transmitted to a new vertebrate host (Rotureau and Van Den Abbeele, 2013). This fly presents several passive and active gut-associated immune barriers that trypanosomes must overcome to successfully colonize the fly. In *Gmm*^{WT} flies these barriers are formidable, thus resulting in adult flies that are highly resistant to infection (Hu and Aksoy, 2006; Aksoy et al., 2014b). However, from a comparative perspective, *Gmm*$^{Wgm-}$ and *Gmm*^{Apo} adults are unusually susceptible to infection with parasites (Pais et al., 2008; Wang et al., 2009; Weiss et al., 2013). These findings indicate that tsetse's endogenous microbiota modulates development of the fly's antitrypanosomal immune responses (reviewed in Weiss and Aksoy, 2011).

The first prominent barrier that trypanosomes encounter following entry into tsetse's gut is the PM. The PM, which is a chitinous and proteinaceous sleevelike structure that extends the length of the fly's midgut (Lehane, 1997; Rose et al., 2014), modulates pathogen infection outcomes and alters the temporal kinetics of host immune responses (Hegedus et al., 2009; Weiss et al., 2014). Newly enclosed adult tsetse (designated "teneral") are highly susceptible to infection with trypanosomes, whereas mature individuals (≥8 days posteclosion) that have fed multiple times are highly refractory (Walshe et al., 2011). The structural integrity of tsetse's PM increases as a function of adult age posteclosion. The higher parasite susceptibility exhibited by teneral adults is attributed to the absence of a robust PM (Welburn and Maudlin, 1992; Haines, 2013) and an immature immune system (Weiss et al., 2013) at this stage of the fly's lifecycle.

Similar to teneral *Gmm*^{WT} adults, mature *Gmm*$^{Wgm-}$ and *Gmm*^{Apo} individuals present a structurally underdeveloped PM, and this phenomenon significantly alters the dynamics of trypanosome infection establishment in tsetse's gut (Weiss et al., 2013). Specifically, epithelial cells in the gut of mature *Gmm*^{WT} flies do not immunologically detect parasites until they have differentiated from vertebrate-adapted bloodstream forms (BSFs) into insect-adapted procyclics (~12h postingestion) and then circumvented the PM barrier (~3 days postingestion). Conversely, dysbiotic tsetse that house a structurally compromised PM immediately recognize and immunologically respond to BSF trypanosomes present in a newly acquired blood meal. BSF trypanosomes present a dense coat of variant surface glycoproteins and thus may not be detrimentally impacted by antitrypanosomal proteins produced early in the infection process (Rudenko, 2011). Additionally, directly after feeding, effector molecules diluted in the voluminous, potentially pH-unfavorable blood meal could exhibit reduced trypanocidal activity. Finally, tsetse rapidly excretes abundant fluid volumes via diuresis following completion of a blood meal (Gee, 1975), and this process would substantially decrease the quantity of soluble effector molecules present in the resulting trypanosome-containing blood bolus. Cumulatively, these conditions, which result from the presence of a defective PM barrier, likely contribute to the trypanosome-susceptible phenotype presented by *Gmm*^{Apo} flies.

SUMMARY AND CONCLUDING THOUGHTS

Herein we summarize how microbes that reside within arthropod disease vectors contribute to their host's physiological homeostasis (Table 9.2). Specifically, we describe the following:

1. The association between the presence of indigenous bacterial symbionts and arthropod host developmental progression through immature larval stages and into adulthood.
2. How indigenous bacterial symbionts modulate vector competency of adult arthropod hosts.
3. The relationship between the presence of indigenous bacterial symbionts in larval tsetse and induction of host immune system development.

The bulk of the content in this review focuses on Point 3 above. Notably, this work was performed exclusively in the tsetse fly model system, and conspicuously absent from the literature are the following:

1. Information that pertains to symbiont-modulated immune system development in other arthropod model systems.
2. Details related to the specific bacterial contributors, and molecular mechanisms, which underlie symbiont-modulated development of arthropod host immunity. For example, what member(s) of the cumulative microbial community is responsible for inducing development

TABLE 9.2 Arthropod Vectors in Which Symbiotic Bacteria Exhibit an Experimentally Confirmed Role as Regulators of Their Host's Physiology

Arthropod Vector	Functionally Confirmed Microbial Contributors[a]	Symbiont Functional Contribution to Arthropod Host Physiology	References
Aedes aegypti	**1.** *Acinetobacter, Aeromonas, Aquitalea, Chryseobacterium, Microbacterium, Paenibacillus, Escherichia coli* **2.** *Chromobacterium* (Csp_P)	**1.** Promote mosquito lifecycle (development from 1st instar larvae to adulthood) **2.** Mediate mosquito dengue vector competency	Coons et al. (2014) and Ramirez et al. (2014)
Anopheles stephensi	*Asaia*	Promote mosquito larval development	Chouaia et al. (2012)
Anopheles gambiae	*Enterobacter (Esp_Z), Chromobacterium (Csp_P)*	Mediate mosquito plasmodium vector competency	Cirimotich et al. (2011a,b) and Ramirez et al. (2014)
Ixodes scapularis	Wild-type microbiota	Mediate host production of a structurally robust peritrophic matrix	Narasimhan et al. (2014)
Glossina morsitans	*Wigglesworthia*	Promote the development of cellular immunity during immature host stages; mediate host production of a structurally robust peritrophic matrix	Wang et al. (2009), Weiss et al. (2011), Weiss et al. (2012), and Weiss et al. (2013)

[a] *All of the bacteria listed below, with the exception of* E. coli, *are found naturally in their arthropod host.*

of their host's immune system, and what is (are) the microbiota-derived factor(s) responsible? Additionally, do these factors directly actuate immune system development, or do they induce the expression of downstream immunostimulatory compounds?

These questions, and others, must be addressed through future experimentation using the tsetse fly and other arthropod model systems.

Every year, 1 billion people contract a vector-borne disease, and greater than 1 million of these infections will result in host death (World Health Organization, 2016). These epidemiological outcomes indicate that more must be done to reduce the spread of pathogens transmitted by arthropods. Current control strategies, including insecticides, traps, and sterile insect technique, are expensive and difficult to sustain. Recent vector-borne disease control research is aimed at developing novel strategies designed to block pathogen transmission through arthropod vectors by boosting their refractoriness to infection. Basic, translational research on the functional association between symbiotic bacteria and arthropod immunity is necessary to accomplish this goal.

References

Akman, L., Yamashita, A., Watanabe, H., Oshima, K., Shiba, T., Hattori, M., Aksoy, S., 2002. Genome sequence of the endocellular obligate symbiont of tsetse flies, *Wigglesworthia glossinidia*. Nat. Genet. 32, 402–407.

Aksoy, E., Telleria, E.L., Echodu, R., Wu, Y., Okedi, L.M., Weiss, B.L., Aksoy, S., Caccone, A., 2014a. Analysis of multiple tsetse fly populations in Uganda reveals limited diversity and species-specific gut microbiota. Appl. Environ. Microbiol. 80, 4301–4312.

Aksoy, S., Weiss, B.L., Attardo, G.M., 2014b. Trypanosome transmission dynamics in tsetse. Curr. Opin. Insect Sci. 3, 43–49.

Aksoy, S., 1995. *Wigglesworthia* gen. nov. and *Wigglesworthia glossinidia* sp. nov., taxa consisting of the mycetocyte-associated, primary endosymbionts of tsetse flies. Int. J. Syst. Bacteriol. 45, 848–851.

Alam, U., Medlock, J., Brelsfoard, C., Pais, R., Lohs, C., Balmand, S., Carnogursky, J., Heddi, A., Takac, P., Galvani, A., Aksoy, S., 2011. *Wolbachia* symbiont infections induce strong cytoplasmic incompatibility in the tsetse fly *Glossina morsitans*. PLoS Pathog. 7, e1002415.

Aliota, M.T., Walker, E.C., Uribe Yepes, A., Dario Velez, I., Christensen, B.M., Osorio, J.E., 2016. The wMel strain of *Wolbachia* reduces transmission of chikungunya virus in *Aedes aegypti*. PLoS Negl. Trop. Dis. 10, e0004677.

Attardo, G.M., Lohs, C., Heddi, A., Alam, U.H., Yildirim, S., Aksoy, S., 2008. Analysis of milk gland structure and function in *Glossina morsitans*: milk protein production, symbiont populations and fecundity. J. Insect Physiol. 54, 1236–1242.

Balmand, S., Lohs, C., Aksoy, S., Heddi, A., 2013. Tissue distribution and transmission routes for the tsetse fly endosymbionts. J. Invertebr Pathol. (112 Suppl.), S116–S122.

Behar, A., Yuval, B., Jurkevitch, E., 2005. Enterobacteria-mediated nitrogen fixation in natural populations of the fruit fly *Ceratitis capitata*. Mol. Ecol. 14, 2637–2643.

Benoit, J.B., Attardo, G.M., Baumann, A.A., Michalkova, V., Aksoy, S., 2015. Adenotrophic viviparity in tsetse flies: potential for population control and as an insect model for lactation. Annu. Rev. Entomol. 60, 351–371.

Bian, G., Xu, Y., Lu, P., Xie, Y., Xi, Z., 2010. The endosymbiotic bacterium *Wolbachia* induces resistance to dengue virus in *Aedes aegypti*. PLoS Pathog. 6, e1000833.

Bian, G., Joshi, D., Dong, Y., Lu, P., Zhou, G., Pan, X., Xu, Y., Dimopoulos, G., Xi, Z., 2013a. *Wolbachia* invades *Anopheles stephensi* populations and induces refractoriness to *Plasmodium* infection. Science 340, 748–751.

Bian, G., Zhou, G., Lu, P., Xi, Z., 2013b. Replacing a native *Wolbachia* with a novel strain results in an increase in endosymbiont load and resistance to dengue virus in a mosquito vector. PLoS Negl. Trop. Dis. 7, e2250.

Chen, X., Li, S., Aksoy, S., 1999. Concordant evolution of a symbiont with its host insect species: molecular phylogeny of genus *Glossina* and its bacteriome-associated endosymbiont, *Wigglesworthia glossinidia*. J. Mol. Evol. 48, 49–58.

Cheng, Q., Aksoy, S., 1999. Tissue tropism, transmission and expression of foreign genes in vivo in midgut symbionts of tsetse flies. Insect Mol. Biol. 8, 125–132.

Cheng, Q., Ruel, T.D., Zhou, W., Moloo, S.K., Majiwa, P., O'Neill, S.L., Aksoy, S., 2000. Tissue distribution and prevalence of *Wolbachia* infections in tsetse flies, *Glossina* spp. Med. Vet. Entomol. 14, 44–50.

Chouaia, B., Rossi, P., Epis, S., Mosca, M., Ricci, I., Damiani, C., Ulissi, U., Crotti, E., Daffonchio, D., Bandi, C., Favia, G., 2012. Delayed larval development in *Anopheles* mosquitoes deprived of *Asaia* bacterial symbionts. BMC Microbiol. (12 Suppl. 1), S2.

Cirimotich, C.M., Dong, Y., Clayton, A.M., Sandiford, S.L., Souza-Neto, J.A., Mulenga, M., Dimopoulos, G., 2011a. Natural microbe-mediated refractoriness to *Plasmodium* infection in *Anopheles gambiae*. Science 332, 855–858.

Cirimotich, C.M., Ramirez, J.L., Dimopoulos, G., 2011b. Native microbiota shape insect vector competence for human pathogens. Cell Host Microbe 10, 307–310.

Coon, K.L., Vogel, K.J., Brown, M.R., Strand, M.R., 2014. Mosquitoes rely on their gut microbiota for development. Mol. Ecol. 23, 2727–2739.

De Vooght, L., Caljon, G., Van Hees, J., Van Den Abbeele, J., 2015. Paternal transmission of a secondary symbiont during mating in the viviparous tsetse Fly. Mol. Biol. Evol. 32, 1977–1980.

Diaz Heijtz, R., Wang, S., Anuar, F., Qian, Y., Bjorkholm, B., Samuelsson, A., Hibberd, M.L., Forssberg, H., Pettersson, S., 2011. Normal gut microbiota modulates brain development and behavior. Proc. Natl. Acad. Sci. U.S.A. 108, 3047–3052.

Doudoumis, V., Alam, U., Aksoy, E., Abd-Alla, A.M., Tsiamis, G., Brelsfoard, C., Aksoy, S., Bourtzis, K., 2013. Tsetse-*Wolbachia* symbiosis: comes of age and has great potential for pest and disease control. J. Invertebr Pathol. (112 Suppl.), S94–S103.

Douglas, A.E., 2014. Symbiosis as a general principle in eukaryotic evolution. Cold Spring Harb. Perspect. Biol. 6.

Eleftherianos, I., Xu, M., Yadi, H., Ffrench-Constant, R.H., Reynolds, S.E., 2009. Plasmatocyte-spreading peptide (PSP) plays a central role in insect cellular immune defenses against bacterial infection. J. Exp. Biol. 212, 1840–1848.

Ferjoux, G., Auge, B., Boyer, K., Haenlin, M., Waltzer, L., 2007. A GATA/RUNX cis-regulatory module couples *Drosophila* blood cell commitment and differentiation into crystal cells. Dev. Biol. 305, 726–734.

Frentiu, F.D., Zakir, T., Walker, T., Popovici, J., Pyke, A.T., van den Hurk, A., McGraw, E.A., O'Neill, S.L., 2014. Limited dengue virus replication in field-collected *Aedes aegypti* mosquitoes infected with *Wolbachia*. PLoS Negl. Trop. Dis. 8, e2688.

Gee, J.D., 1975. Diuresis in the tsetse fly *Glossina austeni*. J. Exp. Biol. 63, 381–390.

Geiger, A., Fardeau, M.L., Njiokou, F., Ollivier, B., 2013. *Glossina* spp. gut bacterial flora and their putative role in fly-hosted trypanosome development. Front. Cell. Infect. Microbiol. 3, 34.

Haines, L.R., 2013. Examining the tsetse teneral phenomenon and permissiveness to trypanosome infection. Front. Cell. Infect. Microbiol. 3, 84.

Hartenstein, V., 2006. Blood cells and blood cell development in the animal kingdom. Annu. Rev. Cell Dev. Biol. 22, 677–712.

Hegedus, D., Erlandson, M., Gillott, C., Toprak, U., 2009. New insights into peritrophic matrix synthesis, architecture, and function. Annu. Rev. Entomol. 54, 285–302.

Honti, V., Csordas, G., Kurucz, E., Markus, R., Ando, I., 2014. The cell-mediated immunity of *Drosophila melanogaster*: hemocyte lineages, immune compartments, microanatomy and regulation. Dev. Comp. Immunol. 42, 47–56.

Hu, C., Aksoy, S., 2006. Innate immune responses regulate trypanosome parasite infection of the tsetse fly *Glossina morsitans morsitans*. Mol. Microbiol. 60, 1194–1204.

Johnson, K.N., 2015. The impact of *Wolbachia* on virus infection in mosquitoes. Viruses 7, 5705–5717.

Jupatanakul, N., Sim, S., Dimopoulos, G., 2014. The insect microbiome modulates vector competence for arboviruses. Viruses 6, 4294–4313.

Kambris, Z., Cook, P.E., Phuc, H.K., Sinkins, S.P., 2009. Immune activation by life-shortening *Wolbachia* and reduced filarial competence in mosquitoes. Science 326, 134–136.

Lamiable, O., Arnold, J., de Faria, I.J., Olmo, R.P., Bergami, F., Meignin, C., Hoffmann, J.A., Marques, J.T., Imler, J.L., 2016. Analysis of the contribution of hemocytes and autophagy to *Drosophila* antiviral immunity. J. Virol. 90, 5415–5426.

Lebestky, T., Chang, T., Hartenstein, V., Banerjee, U., 2000. Specification of Drosophila hematopoietic lineage by conserved transcription factors. Science 288, 146–149.

Lehane, M.J., 1997. Peritrophic matrix structure and function. Annu. Rev. Entomol. 42, 525–550.

Lemaitre, B., Hoffmann, J., 2007. The host defense of *Drosophila melanogaster*. Annu. Rev. Immunol. 25, 697–743.

Lindh, J.M., Lehane, M.J., 2011. The tsetse fly *Glossina fuscipes fuscipes* (Diptera: *Glossina*) harbours a surprising diversity of bacteria other than symbionts. Antonie Van Leeuwenhoek 99, 711–720.

Maltz, M.A., Weiss, B.L., O'Neill, M., Wu, Y., Aksoy, S., 2012. OmpA-mediated biofilm formation is essential for the commensal bacterium *Sodalis glossinidius* to colonize the tsetse fly gut. Appl. Environ. Microbiol. 78, 7760–7768.

Margulis, L., 1993. Origins of species: acquired genomes and individuality. Biosystems 31, 121–125.

Mbewe, N.J., Mweempwa, C., Guya, S., Wamwiri, F.N., 2015. Microbiome frequency and their association with trypanosome infection in male *Glossina morsitans centralis* of Western Zambia. Vet. Parasitol. 211, 93–98.

McMeniman, C.J., Lane, R.V., Cass, B.N., Fong, A.W., Sidhu, M., Wang, Y.F., O'Neill, S.L., 2009. Stable introduction of a life-shortening *Wolbachia* infection into the mosquito *Aedes aegypti*. Science 323, 141–144.

Michalkova, V., Benoit, J.B., Weiss, B.L., Attardo, G.M., Aksoy, S., 2014. Vitamin B6 generated by obligate symbionts is critical for maintaining proline homeostasis and fecundity in tsetse flies. Appl. Environ. Microbiol. 80, 5844–5853.

Moran, N.A., McCutcheon, J.P., Nakabachi, A., 2008. Genomics and evolution of heritable bacterial symbionts. Annu. Rev. Genet. 42, 165–190.

Moran, N.A., 2006. Symbiosis. Curr. Biol. 16, R866–R871.

Moran, N.A., 2007. Symbiosis as an adaptive process and source of phenotypic complexity. Proc. Natl. Acad. Sci. U.S.A. (104 Suppl. 1), 8627–8633.

Narasimhan, S., Fikrig, E., 2015. Tick microbiome: the force within. Trends Parasitol. 31, 315–323.

Narasimhan, S., Rajeevan, N., Liu, L., Zhao, Y.O., Heisig, J., Pan, J., Eppler-Epstein, R., Deponte, K., Fish, D., Fikrig, E., 2014. Gut microbiota of the tick vector *Ixodes scapularis* modulate colonization of the Lyme disease spirochete. Cell Host Microbe 15, 58–71.

Nicholson, J.K., Holmes, E., Kinross, J., Burcelin, R., Gibson, G., Jia, W., Pettersson, S., 2012. Host-gut microbiota metabolic interactions. Science 336, 1262–1267.

Pais, R., Lohs, C., Wu, Y., Wang, J., Aksoy, S., 2008. The obligate mutualist *Wigglesworthia glossinidia* influences reproduction, digestion, and immunity processes of its host, the tsetse fly. Appl. Environ. Microbiol. 74, 5965–5974.

Pan, X., Zhou, G., Wu, J., Bian, G., Lu, P., Raikhel, A.S., Xi, Z., 2012. *Wolbachia* induces reactive oxygen species (ROS)-dependent activation of the Toll pathway to control dengue virus in the mosquito *Aedes aegypti*. Proc. Natl. Acad. Sci. U.S.A. 109, E23–E31.

Parsons, B., Foley, E., 2016. Cellular immune defenses of *Drosophila melanogaster*. Dev. Comp. Immunol. 58, 95–101.

Perotti, M.A., Allen, J.M., Reed, D.L., Braig, H.R., 2007. Host-symbiont interactions of the primary endosymbiont of human head and body lice. FASEB J. 21, 1058–1066.

Ramirez, J.L., Short, S.M., Bahia, A.C., Saraiva, R.G., Dong, Y., Kang, S., Tripathi, A., Mlambo, G., Dimopoulos, G., 2014. *Chromobacterium* Csp_P reduces malaria and dengue infection in vector mosquitoes and has entomopathogenic and in vitro anti-pathogen activities. PLoS Pathog. 10, e1004398.

Rances, E., Ye, Y.H., Woolfit, M., McGraw, E.A., O'Neill, S.L., 2012. The relative importance of innate immune priming in *Wolbachia*-mediated dengue interference. PLoS Pathog. 8, e1002548.

Rio, R.V.M., Symula, R.E., Wang, J., Lohs, C., Wu, Y., Snyder, A.K., Bjornson, R.D., Oshima, K., Biehl, B.S., Perna, N.T., Hattori, M., Aksoy, S., 2012. Insight into the transmission biology and species-specific functional capabilities of tsetse (Diptera: Glossinidae) obligate symbiont *Wigglesworthia*. mBio. 3, e00240–11.

Rio, R.V., Attardo, G.M., Weiss, B.L., 2016. Grandeur alliances: symbiont metabolic integration and obligate arthropod hematophagy. Trends Parasitol. http://dx.doi.org/10.1016/j.pt.2016.05.002.

Rose, C., Belmonte, R., Armstrong, S.D., Molyneux, G., Haines, L.R., Lehane, M.J., Wastling, J., Acosta-Serrano, A., 2014. An investigation into the protein composition of the teneral *Glossina morsitans morsitans* peritrophic matrix. PLoS Negl. Trop. Dis. 8, e2691.

Rotureau, B., Van Den Abbeele, J., 2013. Through the dark continent: African trypanosome development in the tsetse fly. Front. Cell. Infect. Microbiol. 3, 53.

Rudenko, G., 2011. African trypanosomes: the genome and adaptations for immune evasion. Essays Biochem. 51, 47–62.

Sassera, D., Beninati, T., Bandi, C., Bouman, E.A., Sacchi, L., Fabbi, M., Lo, N., 2006. '*Candidatus* Midichloria mitochondrii', an endosymbiont of the tick *Ixodes ricinus* with a unique intramitochondrial lifestyle. Int. J. Syst. Evol. Microbiol. 56, 2535–2540.

Segata, N., Baldini, F., Pompon, J., Garrett, W.S., Truong, D.T., Dabire, R.K., Diabate, A., Levashina, E.A., Catteruccia, F., 2016. The reproductive tracts of two malaria vectors are populated by a core microbiome and by gender- and swarm-enriched microbial biomarkers. Sci. Rep. 6, 24207.

Shaw, W.R., Attardo, G.M., Aksoy, S., Catteruccia, F., 2015. A comparative analysis of reproductive biology of insect vectors of human disease. Curr. Opin. Insect Sci. 10, 142–148.

Shaw, W.R., Marcenac, P., Childs, L.M., Buckee, C.O., Baldini, F., Sawadogo, S.P., Dabire, R.K., Diabate, A., Catteruccia, F., 2016. *Wolbachia* infections in natural *Anopheles* populations affect egg laying and negatively correlate with *Plasmodium* development. Nat. Commun. 7, 11772.

Smith, R.C., Barillas-Mury, C., Jacobs-Lorena, M., 2015a. Hemocyte differentiation mediates the mosquito late-phase immune response against *Plasmodium* in *Anopheles gambiae*. Proc. Natl. Acad. Sci. U.S.A. 112, E3412–E3420.

Smith, T.A., Driscoll, T., Gillespie, J.J., Raghavan, R., 2015b. A *Coxiella*-like endosymbiont is a potential vitamin source for the Lone Star tick. Genome Biol. Evol. 7, 831–838.

Snyder, A.K., Rio, R.V., 2015. "*Wigglesworthia morsitans*" folate (Vitamin B9) biosynthesis contributes to tsetse host fitness. Appl. Environ. Microbiol. 81, 5375–5386.

Snyder, A.K., Deberry, J.W., Runyen-Janecky, L., Rio, R.V., 2010. Nutrient provisioning facilitates homeostasis between tsetse fly (Diptera: Glossinidae) symbionts. Proc. Biol. Sci. 277, 2389–2397.

Toh, H., Weiss, B.L., Perkin, S.A., Yamashita, A., Oshima, K., Hattori, M., Aksoy, S., 2006. Massive genome erosion and functional adaptations provide insights into the symbiotic lifestyle of *Sodalis glossinidius* in the tsetse host. Genome Res. 16, 149–156.

van den Hurk, A.F., Hall-Mendelin, S., Pyke, A.T., Frentiu, F.D., McElroy, K., Day, A., Higgs, S., O'Neill, S.L., 2012. Impact of *Wolbachia* on infection with chikungunya and yellow fever viruses in the mosquito vector *Aedes aegypti*. PLoS Negl. Trop. Dis. 6, e1892.

Walshe, D.P., Lehane, M.J., Haines, L.R., 2011. Post eclosion age predicts the prevalence of midgut trypanosome infections in *Glossina*. PLoS One 6, e26984.

Wamwiri, F.N., Alam, U., Thande, P.C., Aksoy, E., Ngure, R.M., Aksoy, S., Ouma, J.O., Murilla, G.A., 2013. *Wolbachia, Sodalis* and trypanosome co-infections in natural populations of *Glossina austeni* and *Glossina pallidipes*. Parasit. Vectors 6, 232.

Wang, J., Wu, Y., Yang, G., Aksoy, S., 2009. Interactions between mutualist *Wigglesworthia* and tsetse peptidoglycan recognition protein (PGRP-LB) influence trypanosome transmission. Proc. Natl. Acad. Sci. U.S.A. 106, 12133–12138.

Wang, J., Weiss, B.L., Aksoy, S., 2013. Tsetse fly microbiota: form and function. Front. Cell. Infect. Microbiol. 3, 69.

Weiss, B., Aksoy, S., 2011. Microbiome influences on insect host vector competence. Trends Parasitol. 27, 514–522.

Weiss, B.L., Wang, J., Aksoy, S., 2011. Tsetse immune system maturation requires the presence of obligate symbionts in larvae. PLoS Biol. 9, e1000619.

Weiss, B.L., Maltz, M., Aksoy, S., 2012. Obligate symbionts activate immune system development in the tsetse fly. J. Immunol. 188, 3395–3403.

Weiss, B.L., Wang, J., Maltz, M.A., Wu, Y., Aksoy, S., 2013. Trypanosome infection establishment in the tsetse fly gut is influenced by microbiome-regulated host immune barriers. PLoS Pathog. 9, e1003318.

Weiss, B.L., Savage, A.F., Griffith, B.C., Wu, Y., Aksoy, S., 2014. The peritrophic matrix mediates differential infection outcomes in the tsetse fly gut following challenge with commensal, pathogenic, and parasitic microbes. J. Immunol. 193, 773–782.

Welburn, S.C., Maudlin, I., 1992. The nature of the teneral state in *Glossina* and its role in the acquisition of trypanosome infection in tsetse. Ann. Trop. Med. Parasitol. 86, 529–536.

Wernegreen, J.J., 2002. Genome evolution in bacterial endosymbionts of insects. Nat. Rev. Genet. 3, 850–861.

Werren, J.H., Baldo, L., Clark, M.E., 2008. *Wolbachia*: master manipulators of invertebrate biology. Nat. Rev. Microbiol. 6, 741–751.

World Health Organization Fact Sheet, 2016. .

Zhong, J., Jasinskas, A., Barbour, A.G., 2007. Antibiotic treatment of the tick vector *Amblyomma americanum* reduced reproductive fitness. PLoS One 2, e405.

CHAPTER

10

Host–Microbe Interactions: A Case for *Wolbachia* Dialogue

Sassan Asgari

The University of Queensland, Brisbane, QLD, Australia

INTRODUCTION

Wolbachia pipiens is an endosymbiotic bacterium found not only in many insect species, but also in some other arthropods and nematodes. It is believed that about 40%–65% of insect species carry *Wolbachia* (Hilgenboecker et al., 2008; Zug and Hammerstein, 2012). Although they are mainly transmitted maternally, horizontal transfer of *Wolbachia* has also been known (e.g., Brown and Lloyd, 2015). The main effects of *Wolbachia* on their hosts are reproductive manipulations such as cytoplasmic incompatibility (CI), male killing, and feminization (O'Neill et al., 1997). CI gives an advantage to infected females and is the main mechanism for rapid spread and maintenance of *Wolbachia* in host populations (Turelli and Hoffmann, 1995). It occurs when *Wolbachia*-infected males mate with uninfected females, producing no viable progeny. This effect can potentially be used for vector population control (Atyame et al., 2015; Bourtzis, 2008).

In addition to reproductive effects, it was discovered first in *Drosophila melanogaster* that *Wolbachia* blocks replication of a variety of RNA viruses (Hedges et al., 2008; Teixeira et al., 2008). Consequently, attempts have been made to transinfect mosquitoes with the endosymbiont to inhibit transmission of vector-borne pathogens because most mosquito species are not naturally infected with *Wolbachia* (Aliota et al., 2016; Bian et al., 2010; Hughes et al., 2011a,b; Moreira et al., 2009; van den Hurk et al., 2012; reviewed in Hughes and Rasgon, 2014). The outcomes to suppress dengue virus have been quite promising under field conditions (Frentiu et al., 2014) with the possibility of suppressing other viruses such as Chikungunya (Aliota et al., 2016) and Zika (Dutra et al., 2016) transmitted by *Aedes aegypti*. However, the mechanism of pathogen blocking and reproductive manipulations by *Wolbachia* still remain largely unknown.

Apart from scientific curiosity, optimal utilization and sustainability of *Wolbachia* for vector control or blocking transmission of pathogens require a good understanding of the symbiotic association. Along this line, identification of *Wolbachia* and host factors, which play a role in the establishment and maintenance of the endosymbiosis

Arthropod Vector: Controller of Disease Transmission, Volume 1
http://dx.doi.org/10.1016/B978-0-12-805350-8.00010-6

BOX 10.1

microRNA BIOGENESIS AND INTERACTION WITH TARGETS

microRNAs (miRNAs) are small noncoding RNAs of ~22 nucleotides (nt) that play important roles in gene regulation. miRNA genes are expressed as primary miRNA (pri-miRNA) transcripts by RNA polymerase II, similar to mRNAs. The pri-miRNA contains one or more stem loops that are cleaved from the stem by the ribonuclease Drosha in the nucleus producing a ~70 nt hairpin structure known as the precursor miRNA (pre-miRNA). Subsequently, pre-miRNA is transported into the cytoplasm facilitated by Exportin-5, where its hairpin head is cleaved by another ribonuclease, Dicer I, generating a miRNA duplex. The duplex triggers the formation of the RNA-induced silencing complex (RISC), in which one of the strands normally gets degraded. The mature miRNA guides the complex to target sequences (mRNAs or genomic sequences in the nucleus) by sequence complementarity, which is often not 100% in animals, but contains mismatches and bulges. The complementarity in the seed region (nucleotides 2–8 from the 5′ end of miRNA) is believed to be crucial for target recognition and function. Target sequences could be in the 3′UTR, open reading frame or the 5′UTR of mRNAs, or in the promoter region of genes on the genomic DNA. Depending on the context and level of complementarity, binding of miR-RISC complex to target sequences may have different outcomes, such as mRNA degradation, repression of translation, increase in mRNA stability, and activation of translation/transcription.

and cause the phenotypic effects on the host, are of paramount importance. This includes consideration of reciprocal gene regulation via molecular cross talk between *Wolbachia* and the host. Small noncoding RNAs (sncRNAs) have emerged as important regulators of gene expression and therefore may play a role in the interspecies interaction/communication. While this area of research is in its infancy, here the latest on the role of sncRNAs in *Wolbachia*–host interaction, with an emphasis on mosquitoes has been discussed. For more information in regard to the role of miRNAs in vector–pathogen interactions, readers may refer to reviews published on this topic (Asgari, 2014b; Hussain et al., 2016) (Box 10.1).

IMPACT OF *WOLBACHIA* ON MOSQUITO SMALL RNAs

It has been shown in various host–pathogen interactions, including in insects, that upon infection the small RNA profile of the host changes (Asgari, 2014b; Hussain and Asgari, 2014; Hussain et al., 2016). An initial microarray analysis of female *A. aegypti* mosquitoes stably transinfected with *w*MelPop-CLA strain of *Wolbachia* showed that the abundance of 13 miRNAs was significantly altered in *Wolbachia*-infected mosquitoes (Fig. 10.1). Follow-up northern blot analyses validated changes in 9 out of the 13 miRNAs because of *Wolbachia* infection. In comparison with noninfected mosquitoes (treated with tetracycline), two of these miRNAs, aae-miR-2940 and aae-miR-309a-2, were highly induced at 4d and 12d after mosquito emergence, followed by moderate induction of aae-miR-2943-1 and aae-miR-970 at 4d and 12d after emergence, and aae-miR-308* (* indicates the passenger strand) and aae-miR-2941–2 at 4d after emergence. The abundance of three miRNAs aae-miR-989, aae-miR-210, and aae-miR-988 declined in the presence of *Wolbachia* at 12d postemergence (Hussain et al., 2011).

A deep sequencing analysis of small RNAs in the cytoplasmic and nuclear fraction of

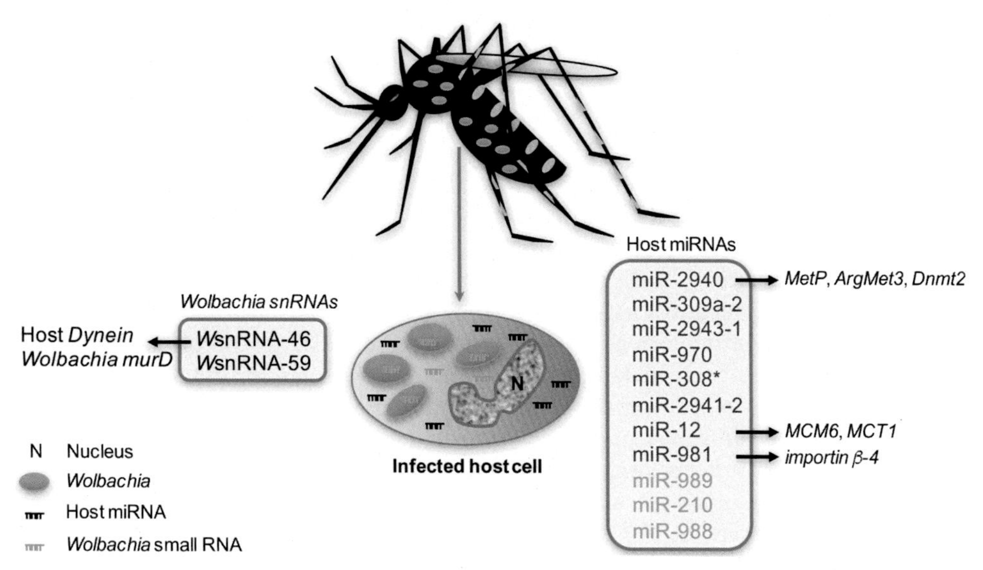

FIGURE 10.1 Host miRNAs and *Wolbachia* small noncoding RNAs (WsnRNA) regulate host and/or *Wolbachia* genes. In *Wolbachia*-infected *Aedes aegypti* mosquitoes, a number of host miRNAs are upregulated (red) or downregulated (green), the target gene(s) of some of them have been experimentally determined. These differentially expressed miRNAs contribute toward *Wolbachia* maintenance and may affect interactions of the cells with viruses that are transmitted by mosquitoes. WsnRNAs produced by *Wolbachia* regulate its own transcription but also appear to be transported into the host cell cytoplasm where they regulate host genes.

A. aegypti Aag2 cells with or without infection with *w*MelPop-CLA revealed that more miRNA types (81 miRNAs) were expressed in *Wolbachia*-infected cells in comparison to noninfected cells (71 miRNAs), suggesting induction of a number of miRNAs due to *Wolbachia* infection (Mayoral, Etebari, et al., 2014). Furthermore, there was an overall increase in the number of small RNAs between 18 and 30 nucleotides in both cellular fractions of *Wolbachia*-infected cells in comparison to those of noninfected Aag2 cells (Mayoral et al., 2014a,b). This increase coincided with elevated expression levels of Argonaute (Ago) proteins Ago1, Ago2, and Ago3 in *Wolbachia*-infected cells. In insects, Ago1 and, to some extent, Ago2 are the main components of the RISC involved in miRNA biogenesis, whereas Ago3 participates in PIWI-interacting RNA (piRNA) biogenesis (Hussain et al., 2016; Lucas et al., 2013). Data analyses also demonstrated changes in the distribution of a number of miRNAs between the cytoplasm and the nucleus (Mayoral et al., 2014a,b). For example, aae-miR-276-3p was found at 10 times higher levels in the nucleus as compared with the cytoplasm, or aae-miR-1 was only found in the cytoplasm of infected cells but not in the nucleus of noninfected cells. Another study showed that *Wolbachia* infection could reduce the relocation of Ago1 from the cytoplasm to the nucleus in cells from *A. aegypti* mosquitoes and *D. melanogaster* flies (Hussain et al., 2013a,b). This reduction may have implications for translocation of miRNAs normally associated with Ago1 and hence contribute to the differential association of miRNAs with the two cellular compartments. Reduction of Ago1 translocation to the nucleus was explained by a reduction in importin β-4, which is involved in Ago1 translocation to the nucleus (Hussain et al., 2013a,b). The reduction in importin β-4 is due to upregulation of aae-miR-981 in *Wolbachia*-infected cells, which negatively targets the protein.

In a recently published work, it was shown that the miRNA profile of *D. melanogaster* JW18 cells infected with *w*Mel strain of *Wolbachia* was not significantly changed when compared to JW18 cells without *Wolbachia* (treated with tetracycline) (Rainey et al., 2016). This outcome is not totally unexpected because studies indicate that the effects of *Wolbachia* on host could be *Wolbachia* strain- or host specific, both in terms of reproductive manipulations and virus blocking (Bourtzis et al., 2014; Martinez et al., 2012). In addition, hosts that have been in long-term association with *Wolbachia* as compared with those that have been transinfected appear to display less transcriptional effects (e.g., genes involved in immunity) (Bourtzis et al., 2000; Hughes et al., 2011a,b; Rancès et al., 2012; Xi et al., 2008). Furthermore, we have found that differentially expressed miRNAs in whole mosquitoes infected with *Wolbachia* may not always be significantly affected when tested in a cell line in the presence of the same *Wolbachia* strain (unpublished data). Along this line, temporal and tissue-specific expression of miRNAs has been shown in *D. melanogaster* and mosquitoes (Chen et al., 2014; Gu et al., 2013; Lucas et al., 2015; Ninova et al., 2014). Therefore, it is worth investigating *Wolbachia*'s effect on miRNAs in whole organisms.

MANIPULATION OF HOST miRNAs AS REGULATORS OF GENES INVOLVED IN *WOLBACHIA* MAINTENANCE

As mentioned earlier, because of *Wolbachia* infection the miRNA profile of the host mosquito is altered. Among those, aae-miR-2940-5p, which appears to be a mosquito-specific miRNA (based on miRBase), was highly induced in the presence of *w*MelPop-CLA (Hussain et al., 2011). Inhibition of the miRNA in mosquitoes by injection of the synthetic inhibitor (antagomir) of the miRNA led to significant reductions in the density of *Wolbachia* in mosquitoes suggesting the importance of this miRNA in *Wolbachia* maintenance in the host. Further investigations led to the identification of at least three targets of this miRNA (Fig. 10.1). *Metalloprotease m41 ftsH* (*MetP*) was the first target identified, which appears to be positively regulated by the miRNA. Interestingly, when *MetP* was silenced in mosquitoes via injection of double-stranded RNA to *MetP*, *Wolbachia* density significantly dropped similar to the observation when aae-miR-2940-5p was inhibited by injection of the synthetic antagomir. These outcomes are consistent with the positive regulation of MetP by aae-miR-2940-5p (Hussain et al., 2011). Induction of a homolog of *MetP* in the beetle *Callosobruchus chinensis* when infected with *Wolbachia* (Nikoh et al., 2008) shows that the protein's function in *Wolbachia*–host interaction may be conserved. However, it is not yet known how the metalloprotease could facilitate *Wolbachia* maintenance. MetP belongs to a group of conserved ATP-dependent metalloproteases that are membrane proteins localized to the inner mitochondrial membranes responsible for the quality control of membrane protein folding. The functional role of this protein has not been investigated in insects, but studies in HEK293 embryonic human cells derived from kidney revealed that silencing the gene in the cells led to accumulation of aberrant (oxidized or nonassembled) proteins in the inner mitochondrial membrane, abnormal morphology of cristae, cell proliferation, and high susceptibility to apoptotic stimuli (Stiburek et al., 2012). If MetP plays a similar role in insect cells, its function would be crucial for the well-being of the host cells and may also directly or indirectly affect *Wolbachia*.

Another target of aae-miR-2940-5p is *DNA methyltransferase* (*AaDnmt2*), which is negatively regulated by the miRNA (Zhang et al., 2013). As a consequence, in *w*MelPop-CLA infected mosquitoes, in which the miRNA is highly induced, significantly less *AaDnmt2* transcripts were found in the ovary and salivary gland tissues. There is only one DNA methyltransferase gene annotated in the *A. aegypti* genome, which is *AaDnmt2*, but the resulting protein has been shown to be rather an RNA methyltransferase (Schaefer et al., 2010; Tuorto et al., 2012). While limited genome methylation occurs in the mosquito (Ye et al., 2013), it is not clear whether AaDnmt2 or another unknown methyltransferase is involved in the methylation. Interestingly, less genome methylation was detected in *w*MelPop-CLA infected mosquitoes as compared to noninfected (tetracycline treated) mosquitoes (Ye et al., 2013), which coincides with downregulation of *AeDnmt2* in *Wolbachia*-infected mosquitoes (Zhang et al., 2013). When the gene was overexpressed in *w*MelPop-CLA–infected Aag2 cells, *Wolbachia* density significantly declined suggesting that this downregulation is important for *Wolbachia* maintenance. The negative effect of Dnmt2 overexpression on *Wolbachia* density was also observed in *D. melanogaster* (LePage et al., 2014). Provided that the functional role of AeDnmt2 in mosquitoes is unclear, it remains unknown how the reduction in the expression of the gene in the presence of *Wolbachia* benefits the endosymbiont. Methylation is an important epigenetic modification of the genomic DNA allowing flexibility in regulation of gene expression. In host–symbiont interactions, alterations of the host genome methylation may contribute to the differential expression of genes often identified in infected hosts. In several host–symbiont interactions, changes in genome methylation have been documented facilitating the establishment of symbiosis and playing an important role in conferring new traits to hosts, such as resistance to stress and microbial infections (reviewed in Asgari, 2014a). Similarly, alterations of genome methylation may contribute to several phenotypic effects identified in *Wolbachia*-infected mosquitoes, such as reproductive manipulations or pathogen blocking.

The third target of aae-miR-2940-5p is *arginine methyltransferase 3* (*AaArgM3*), which similar to MetP, is also positively regulated by the miRNA. Therefore, *AaArgM3* is expressed at a higher level in *Wolbachia*-infected mosquitoes as compared to noninfected mosquitoes (Zhang et al., 2014). Knockdown of the gene in *w*MelPop-CLA infected Aag2 cells resulted in reductions in *Wolbachia* density indicating the significance of the gene regulated by the miRNA in *Wolbachia*–host interaction. In addition to aae-miR-2940-5p, aae-miR-12 was also shown to be important in *Wolbachia* persistence because inhibition of the miRNA by transfection of the synthetic inhibitor led to reduced *Wolbachia* density (Osei-Amo et al., 2012). Two potential targets of aae-miR-12 were determined as *DNA replication licensing* (*MCM6*) and *monocarboxylate transporter* (*MCT1*), but how they facilitate *Wolbachia* replication/persistence remains unknown. Overall, these studies suggest that host miRNAs play important roles as regulators of gene expression in *Wolbachia*–mosquito interaction, and their alterations in the presence of the endosymbiont facilitate *Wolbachia*'s colonization/persistence in the host.

EFFECT OF ALTERATIONS OF HOST miRNAs BY *WOLBACHIA* ON HOST–VIRUS INTERACTIONS

Virus blocking by *Wolbachia* has been rather a recent discovery, and therefore its mechanism still remains largely unknown. What seems to be evident is that this property is not systemic in infected insects but depends on the presence of *Wolbachia* in the cell (Frentiu et al., 2010; Moreira et al., 2009), and *Wolbachia* density plays an important role in the extent of virus inhibition (Lu et al., 2012; Martinez et al., 2014; Osborne et al., 2012). However, research also shows that the presence of *Wolbachia* per se may not necessarily confer virus protection (Skelton et al., 2016). While the mechanism of virus blocking is unclear, potential contributors to this property, such as immune priming (mainly in mosquitoes transinfected with *Wolbachia*)

(Kambris et al., 2009; Pan et al., 2012; Rancès et al., 2012) and depletion of or competition for resources such as cholesterol and iron, have been identified (Caragata et al., 2013; Kremer et al., 2009; Molloy et al., 2016). Contribution of the short-interfering RNA (siRNA) pathway mediated by Dicer-2 has been ruled out in *Drosophila* (Hedges et al., 2012; Rainey et al., 2016).

Based on a study mentioned earlier, using JW18 cells and Semliki Forest virus (SFV), the investigators showed that the *Wolbachia* *w*Mel strain is likely to exert its virus blocking by inhibiting replication and/or translation of viral RNA without an effect on the host or *Wolbachia* transcriptome (Rainey et al., 2016). Notably, this effect is displayed at the very early stage of infection (less than 7h), and experimental evidence suggests that it is most likely present prior to virus entry/infection. However, it remains to be confirmed if this is a general mechanism exhibited by all *Wolbachia* strains that display virus blocking, and whether it similarly impacts different RNA viruses. The instance in *D. melanogaster* in which the *Wolbachia* *w*Mel strain (the same as in JW18) conferred protection against Flock House virus without affecting virus titer (Teixeira et al., 2008) does not seem to support the assumption. Nevertheless, lack of changes in miRNAs in the presence of *Wolbachia* with or without SFV infection (Rainey et al., 2016) indicates that miRNAs do not seem to play a role in virus inhibition in JW18–SFV interaction.

Unlike *w*Mel-infected JW18 cells, a number of altered miRNAs in *w*MelPop-CLA infected *A. aegypti* mosquitoes have been shown to affect the interaction of mosquito cells with viruses. For instance, *MetP*, as one of the targets of aae-miR-2940-5p that is positively regulated by the miRNA (Hussain et al., 2011), facilitates West Nile virus (WNV) replication; however, upon infection the virus specifically suppresses the miRNA as an antiviral response to reduce WNV replication (Slonchak et al., 2014). In *w*MelPop-CLA infected mosquitoes, the abundance of aae-miR-2940-5p is highly increased (Hussain et al., 2011), which should consequently benefit WNV because of the upregulation of MetP in the *Wolbachia*-infected cells. This benefit to WNV was in fact observed in *w*MelPop-CLA infected Aag2 cells but only at the genomic RNA and viral protein (envelope protein tested) levels and not secreted mature virion production (Hussain et al., 2013a,b). It was speculated that *Wolbachia* may interfere with the assembly or secretion of virions. In *Culex tarsalis* mosquitoes transiently transinfected with *w*AlbB strain of *Wolbachia*, WNV replication was also enhanced (Dodson et al., 2014); however, the mechanism of WNV enhancement in this instance is not known.

In another example, *AaDnmt2* downregulation by aae-miR-2940-5p in *Wolbachia*-infected mosquito cells was shown to reduce dengue virus serotype 2 (DENV-2) replication in the cells (Zhang et al., 2013). It seems AaDnmt2 facilitates DENV-2 replication because when it was overexpressed, DENV-2 replication increased. Therefore, suppression of AaDnmt2 levels in *Wolbachia*-infected cells may contribute toward virus blocking observed in these cells. Since overexpression of Dnmt2 leads to reductions in *Wolbachia* density in *A. aegypti* and *D. melanogaster* (LePage et al., 2014; Zhang et al., 2013), it is to the benefit of *Wolbachia* and the host insect to keep the expression of the gene at low levels to maintain higher *Wolbachia* density to inhibit virus replication. This is because infection of *A. aegypti* with DENV has been shown to impose fitness costs on the mosquito (Sylvestre et al., 2013). However, unlike *A. aegypti*, in a number of *Drosophila* species infected with *Wolbachia*, no effect on expression levels of Dnmt2 was found when *Wolbachia*-infected flies were compared with uninfected ones (Martinez et al., 2014). Furthermore, in this same study, no correlation between the expression levels of *Dnmt2* and protection against Drosophila C virus (DCV) was found. Therefore, regulation of Dnmt2 upon *Wolbachia* infection does not seem to be ubiquitous, or be the key mechanism underlying virus blocking noted in different insects. The difference could also stem from the nature of *Wolbachia* infection, i.e., natural versus transinfection. Several studies have shown that in insects devoid of *Wolbachia*, its artificial introduction by transinfection has a larger effect on the host immune system, physiology, and transcriptome as compared with insects naturally infected with *Wolbachia* with old association (reviewed in Hughes and Rasgon, 2014).

SMALL RNAs AS MEDIATORS OF DIALOGUE BETWEEN HOST AND *WOLBACHIA*

In the past few years, evidence has accumulated suggesting interspecies exchange of small noncoding RNAs, in particular between microorganisms and their hosts, with regulatory functions (reviewed in Knip et al., 2014). While prokaryotes commonly generate small noncoding RNAs that are mostly 50–250 nucleotide long (Miyakoshi et al., 2015), it appears that some bacteria could potentially produce miRNA-like small RNAs of less than 50 nt from stem-loop structures. For example, *Mycobacterium marinum*, an intracellular bacterium, was shown to express a 23 nt small RNA with characteristics of a miRNA (Furuse et al., 2014). This small RNA was found in association with the RISC complex of the host mouse cells suggesting that it could potentially regulate host genes, although it was found in low quantities.

A search for miRNA-like small RNAs in *w*MelPop-CLA infected *A. aegypti* mosquitoes resulted in the identification and characterization of two *Wolbachia* sncRNAs of ~30 nt, WsnRNA-46 and WsnRNA-49 (Mayoral et al., 2014a,b) (Fig. 10.1). Only in infected mosquitoes, both small RNAs are expressed in the ovaries, midgut, fat body, and salivary glands—tissues that have been shown to be infected with *Wolbachia* (Moreira et al., 2009). Interestingly, the WsnRNAs are also expressed in *D. melanogaster* flies infected with *w*MelPop and *Drosophila simulans* infected with *w*Au strain, suggesting that they are probably conserved small RNAs. Bioinformatics analysis indicated that the precursor sequences of both WsnRNAs are conserved in most strains belonging to the supergroup A *Wolbachia*.

The *Wolbachia* small RNAs were also found in the cytoplasm of the host cells suggesting their export into the host cells (Mayoral et al., 2014a,b). A bioinformatically determined target gene of WsnRNA-46 in *A. aegypti*, *Dynein*, was significantly induced in *w*MelPop-infected mosquitoes. Further analysis revealed that the target gene is positively regulated by WsnRNA-46 because injection of the synthetic mimic of the small RNA in mosquitoes led to increases in the transcript levels of *Dynein*. Also, transfection of WsnRNA-46 mimic into mosquito cells with a GFP reporter construct with target sequences of *Dynein* downstream of the *GFP* gene resulted in significant increases in the *GFP* transcript levels as compared with the control mimic or mutant mimics with point mutations in their seed region (nucleotides 2–8 from the 5′ end of miRNA). This suggests that the small RNAs produced from *Wolbachia* could be involved in host manipulation. Utilization of host microtubules and Dyncin by *Wolbachia* for anterior localization in the *D. melanogaster* oocytes has been established (Ferree et al., 2005). If microtubules and the motor protein Dynein are essential for *Wolbachia* maintenance and transport of membrane-bound vesicles in the host cells, modulation of *Dynein* could be beneficial for the endosymbiont.

In addition to regulation of host genes, WsnRNAs may also regulate *Wolbachia*'s own target genes. A target gene of WsnRNA-46 in *Wolbachia*, *UDP-N-acetylmuranomoyalanine-D-glutamate ligase* (*murD*), was downregulated when the inhibitor of the small RNA was transfected into *w*MelPop-infected Aag2 cells. In the follow-up experiments, in which the target sequences of WsnRNA-46 in *murD* were cloned downstream of the *GFP* reporter gene and transfected into mosquito cells, application of the mimic and the inhibitor of the small RNA led to reduction and increase in the GFP levels, respectively, validating the interaction (Mayoral et al., 2014a,b).

CONCLUSIONS

Wolbachia as a widespread endosymbiont of insects has attracted considerable attention because of its potential in insect control and suppression of vector-borne diseases. While the mechanisms underlying these host manipulations are largely unknown, efforts in this area have intensified after the discovery of *Wolbachia*'s virus-blocking property. It is apparent from these investigations that there are variations in the mechanisms depending on host species, *Wolbachia* strain, and the duration of association (natural vs. transinfection). Small noncoding RNAs have recently been recognized as a means for interspecies cross talk involved in regulating gene expression in hosts or parasites. Similarly, sncRNAs produced from *Wolbachia* or host miRNAs appear to play a role in the maintenance of *Wolbachia* and perhaps contribute to its pathogen-blocking effect. However, more research is needed in this area, and better understanding of their role in the interactions may provide mechanisms for their utilization in enhancing *Wolbachia*'s maintenance/density and sustainability, which are important for their applications.

Regardless of the lack of complete understanding of *Wolbachia*'s effects on the host, in particular virus blocking, attempts to utilize *Wolbachia* to block transmission of viruses, such as dengue virus, have been proven promising and may be extended to other viruses or vector-borne pathogens. Nevertheless, long-term successful utilization of *Wolbachia* requires better appreciation of the mechanisms behind the effects.

Acknowledgments

The author wishes to thank Dr. Grant Hughes from University of Texas Medical Branch for helpful comments on the manuscript. Work in the author's laboratory has been supported by the Australian Research Council (DP110102112) and National Health and Medical Research (APP1062983, APP1027110).

References

Aliota, M.T., Walker, E.C., Uribe Yepes, A., Dario Velez, I., Christensen, B.M., Osorio, J.E., 2016. The wMel strain of *Wolbachia* reduces transmission of Chikungunya virus in *Aedes aegypti*. PLoS Negl. Trop. Dis. 10, e0004677.

Asgari, S., 2014a. Epigenetic modifications underlying symbiotic-host interactions. Adv. Genet. 86, 253–276.

Asgari, S., 2014b. Role of microRNAs in arbovirus/vector interactions. Viruses 6, 3514–3534.

Atyame, C.M., Cattel, J., Lebon, C., Flores, O., Dehecq, J.S., Weill, M., Gouagna, L.C., Tortosa, P., 2015. *Wolbachia*-based population control strategy targeting *Culex quinquefasciatus* mosquitoes proves efficient under semi-field conditions. PLoS One 10, e0119288.

Bian, G., Xu, Y., Lu, P., Xie, Y., Xi, Z., 2010. The endosymbiotic bacterium *Wolbachia* induces resistance to Dengue virus in *Aedes aegypti*. PLoS Pathog. 6, e1000833.

Bourtzis, K., 2008. *Wolbachia*-based technologies for insect pest population control. Adv. Exp. Med. Biol. 627, 104–113.

Bourtzis, K., Dobson, S.L., Xi, Z., Rasgon, J.L., Calvitti, M., Moreira, L.A., Bossin, H.C., Moretti, R., Baton, L.A., Hughes, G.L., Mavingui, P., Gilles, J., 2014. Harnessing mosquito-*Wolbachia* symbiosis for vector and disease control. Acta Trop. 132, S150–S163.

Bourtzis, K., Pettigrew, M., O'Neill, S., 2000. *Wolbachia* neither induces nor suppresses transcripts encoding antimicrobial peptides. Insect Mol. Biol. 9, 635–639.

Brown, A.N., Lloyd, V.K., 2015. Evidence for horizontal transfer of *Wolbachia* by a *Drosophila* mite. Exp. Appl. Acarol. 66, 301–311.

Caragata, E.P., Rancès, E., Hedges, L.M., Gofton, A.W., Johnson, K.N., O'Neill, S.L., McGraw, E.A., 2013. Dietary cholesterol modulates pathogen blocking by *Wolbachia*. PLoS Pathog. 9, 1003459.

Chen, Y.W., Song, S., Weng, R., Verma, P., Kugler, J.M., Buescher, M., Rouam, S., Cohen, S.M., 2014. Systematic study of *Drosophila* microRNA functions using a collection of targeted knockout mutations. Dev. Cell 31, 784–800.

Dodson, B., Hughes, G., Paul, O., Matacchiero, A., Kramer, L., Rasgon, J., 2014. *Wolbachia* enhances West Nile virus (WNV) infection in the mosquito *Culex tarsalis*. PLoS Negl. Trop. Dis. 8, e2965.

Dutra, H.L., Rocha, M.N., Dias, F.B., Mansur, S.B., Caragata, E.P., Moreira, L.A., 2016. *Wolbachia* blocks currently circulating Zika virus isolates in Brazilian *Aedes aegypti* mosquitoes. Cell Host Microbe 19, 771–774.

Ferree, P., Frydman, H., Li, J., Cao, J., Wieschaus, E., Sullivan, W., 2005. *Wolbachia* utilizes host microtubules and dynein for anterior localization in the *Drosophila* oocyte. PLoS Pathog. 1, e14.

Frentiu, F.D., Robinson, J., Young, P.R., McGraw, E.A., O'Neill, S.A., 2010. *Wolbachia*-mediated resistance to Dengue virus infection and death at the cellular level. PLoS One 5, e13398.

Frentiu, F.D., Zakir, T., Walker, T., Popovici, J., Pyke, A.T., van den Hurk, A., McGraw, E.A., O'Neill, S.L., 2014. Limited dengue virus replication in field-collected *Aedes aegypti* mosquitoes infected with *Wolbachia*. PLoS Negl. Trop. Dis. 8, e2688.

Furuse, Y., Finethy, R., Saka, H., Xet-Mull, A., Sisk, D., Smith, K., Lee, S., Coers, J., Valdivia, R., Tobin, D., Cullen, B., 2014. Search for microRNAs expressed by intracellular bacterial pathogens in infected mammalian cells. PLoS One 9, e106434.

Gu, J., Hu, W., Wu, J., Zheng, P., Chen, M., James, A., Chen, X., Tu, Z., 2013. miRNA genes of an invasive vector mosquito, *Aedes albopictus*. PLoS One 8, e67638.

Hedges, L.M., Brownlie, J.C., O'Neill, S.L., Johnson, K.N., 2008. *Wolbachia* and virus protection in insects. Science 322, 702.

Hedges, L.M., Yamada, R., O'Neill, S.L., Johnson, K.N., 2012. The small interfering RNA pathway is not essential for *Wolbachia*-mediated antiviral protection in *Drosophila melanogaster*. Appl. Environ. Microbiol. 78, 6773–6776.

Hilgenboecker, K., Hammerstein, P., Schlattmann, P., Telschow, A., Werren, J.H., 2008. How many species are infected with *Wolbachia*?—A statistical analysis of current data. FEMS Microbiol. Lett. 218, 215–220.

Hughes, G.L., Koga, R., Xue, P., Fukatsu, T., Rasgon, J.L., 2011a. *Wolbachia* infections are virulent and inhibit the human malaria parasite *Plasmodium falciparum* in *Anopheles gambiae*. PLoS Pathog. 7, e1002043.

Hughes, G.L., Rasgon, J.L., 2014. Transinfection: a method to investigate *Wolbachia*-host interactions and control arthropod-borne disease. Insect Mol. Biol. 23, 141–151.

Hughes, G.L., Ren, X., Ramirez, J.L., Sakamoto, J.M., Bailey, J.A., Jedlicka, A.E., Rasgon, J.L., 2011b. *Wolbachia* infections in *Anopheles gambiae* cells: transcriptomic characterization of a novel host-symbiont interaction. PLoS Pathog. 7, e1001296.

Hussain, M., Asgari, S., 2014. MicroRNAs as mediators of insect host–pathogen interactions and immunity. J. Insect Physiol. 70, 151–158.

Hussain, M., Etebari, K., Asgari, S., 2016. Functions of small RNAs in mosquitoes. Adv. Insect Physiol. 51, 189–222.

Hussain, M., Frentiu, F.D., Moreira, L.A., O'Neill, S.L., Asgari, S., 2011. *Wolbachia* utilizes host microRNAs to manipulate host gene expression and facilitate colonization of the dengue vector *Aedes aegypti*. Proc. Natl. Acad. Sci. USA 108, 9250–9255.

Hussain, M., Lu, G., Torres, S., Edmonds, J.H., Kay, B.H., Khromykh, A.A., Asgari, S., 2013a. Effect of *Wolbachia* on replication of West Nile Virus in mosquito cell line and adult mosquitoes. J. Virol. 87, 851–858.

Hussain, M., O'Neill, S.L., Asgari, S., 2013b. *Wolbachia* interferes with the intracellular distribution of Argonaute 1 in the dengue vector *Aedes aegypti* by manipulating the host microRNAs. RNA Biol. 10, 1868–1875.

Kambris, Z., Cook, P.E., Phuc, H.K., Sinkins, S.P., 2009. Immune activation by life-shortening *Wolbachia* and reduced filarial competence in mosquitoes. Science 326, 134–136.

Knip, M., Constantin, M.E., Thordal-Christensen, H., 2014. Trans-kingdom cross-talk: small RNAs on the move. PLoS Genet. 10, e1004602.

Kremer, N., Voronin, D., Charif, D., Mavingui, P., Mollereau, B., Vavre, F., 2009. *Wolbachia* interferes with ferritin expression and iron metabolism in insects. PLoS Pathog. 5, 12.

LePage, D., Jernigan, K., Bordenstein, S., 2014. The relative importance of DNA methylation and Dnmt2-mediated epigenetic regulation on *Wolbachia* densities and cytoplasmic incompatibility. Peer J. 2, e678.

Lu, P., Bian, G., Pan, X., Xi, Z., 2012. *Wolbachia* induces density-dependent inhibition to dengue virus in mosquito cells. PLoS Negl. Trop. Dis. 6, e1754.

Lucas, K., Myles, K., Raikhel, A., 2013. Small RNAs: a new frontier in mosquito biology. Trends Parasitol. 29, 295–303.

Lucas, K., Roy, S., Ha, J., Gervaise, A., Kokoza, V., Raikhel, A., 2015. MicroRNA-8 targets the Wingless signaling pathway in the female mosquito fat body to regulate reproductive processes. Proc. Natl. Acad. Sci. U S A 112, 1440–1445.

Martinez, J., Duplouy, A., Woolfit, M., Vavre, F., O'Neill, S.L., Varaldi, J., 2012. Influence of the virus LbFV and of *Wolbachia* in a host-parasitoid interaction. PLoS One 7, e35081.

Martinez, J., Longdon, B., Bauer, S., Chan, Y.S., Miller, W.J., Bourtzis, K., Teixeira, L., Jiggins, F.M., 2014. Symbionts commonly provide broad spectrum resistance to viruses in insects: a comparative analysis of *Wolbachia* strains. PLoS Pathog. 10, e1004369.

Mayoral, J., Etebari, K., Hussain, M., Khromykh, A., Asgari, S., 2014a. *Wolbachia* infection modifies the profile, shuttling and structure of microRNAs in a mosquito cell line. PLoS One 9, e96107.

Mayoral, J.G., Hussain, M., Joubert, D.A., Iturbe-Ormaetxe, I., O'Neill, S.L., Asgari, S., 2014b. *Wolbachia* small non-coding RNAs and their role in cross-kingdom communications. Proc. Natl. Acad. Sci. U S A 111, 18721–18726.

Miyakoshi, M., Chao, Y., Vogel, J., 2015. Regulatory small RNAs from the 3′ regions of bacterial mRNAs. Curr. Opin. Microbiol. 24, 132–139.

Molloy, J.C., Sommer, U., Viant, M.R., Sinkins, S.P., 2016. *Wolbachia* modulates lipid metabolism in *Aedes albopictus* mosquito cells. Appl. Environ. Microbiol. 82, 3109–3120.

Moreira, L.A., Iturbe-Ormaetxe, I., Jeffery, J.A., Lu, G.J., Pyke, A.T., Hedges, L.M., Rocha, B.C., Hall-Mendelin, S., Day, A., Riegler, M., Hugo, L.E., Johnson, K.N., Kay, B.H., McGraw, E.A., van den Hurk, A.F., Ryan, P.A., O'Neill, S.L., 2009. A *Wolbachia* symbiont in *Aedes aegypti* limits infection with dengue, chikungunya, and *Plasmodium*. Cell 139, 1268–1278.

Nikoh, N., Tanaka, K., Shibata, F., Kondo, N., Hizume, M., Shimada, M., Fukatsu, T., 2008. *Wolbachia* genome integrated in an insect chromosome: evolution and fate of laterally transferred endosymbiont genes. Genome Res. 18, 272–280.

Ninova, M., Ronshaugen, M., Griffiths-Jones, S., 2014. Conserved temporal patterns of microRNA expression in *Drosophila* support a developmental hourglass model. Genome Biol. Evol. 6, 2459–2467.

O'Neill, S., Hoffmann, A.A., Werren, J.H., 1997. Influential Passengers: Inherited Microorganisms and Arthropod Reproduction. Oxford University Press, Oxford.

Osborne, S.E., Iturbe-Ormaetxe, I., Brownlie, J.C., O'Neill, S.L., Johnson, K.N., 2012. Antiviral protection and the importance of *Wolbachia* density and tissue tropism in *Drosophila*. Appl. Environ. Microbiol. 78, 6922–6929.

Osei-Amo, S., Hussain, M., O'Neill, S.L., Asgari, S., 2012. *Wolbachia*-induced aae-miR-12 miRNA negatively regulates the expression of MCT1 and MCM6 genes in *Wolbachia*-infected mosquito cell line. PLoS One 7, e50049.

Pan, X., Zhou, G., Wu, J., Bian, G., Lu, P., Raikhel, A.S., Xi, Z., 2012. *Wolbachia* induces reactive oxygen species (ROS)-dependent activation of the Toll pathway to control dengue virus in the mosquito *Aedes aegypti*. Proc. Natl. Acad. Sci. U S A 109, E23–E31.

Rainey, S.M., Martinez, J., McFarlane, M., Juneja, P., Sarkies, P., Lulla, A., Schnettler, E., Varjak, M., Merits, A., Miska, E.A., Jiggins, F.M., Kohl, A., 2016. *Wolbachia* blocks viral genome replication early in infection without a transcriptional response by the endosymbiont or host small RNA pathways. PLoS Pathog. 12, e1005536.

Rancès, E., Ye, Y.H., Woolfit, M., McGraw, E.A., O'Neill, S.L., 2012. The relative importance of innate immune priming in *Wolbachia*-mediated dengue interference. PLoS Pathog. 8, e1002548.

Schaefer, M., Pollex, T., Hanna, K., Tuorto, F., Meusburger, M., Helm, M., Lyko, F., 2010. RNA methylation by Dnmt2 protects transfer RNAs against stress induced cleavage. Genes Dev. 24, 1590–1595.

Skelton, E., Rancès, E., Frentiu, F., Kusmintarsih, E., Iturbe-Ormaetxe, I., Caragata, E., Woolfit, M., O'Neill, S., 2016. A native *Wolbachia* endosymbiont does not limit dengue virus infection in the mosquito *Aedes notoscriptus* (Diptera: Culicidae). J. Med. Entomol. 53, 401–408.

Slonchak, A., Hussain, M., Torres Morales, S., Asgari, S., Khromykh, A.A., 2014. Expression of mosquito microRNA aae-miR-2940-5p is down-regulated in response to West Nile virus infection to restrict viral replication. J. Virol. 88, 8457–8467.

Stiburek, L., Cesnekova, J., Kostkova, O., Fornuskova, D., Vinsova, K., Wenchich, L., Houstek, J., Zeman, J., 2012. YME1L controls the accumulation of respiratory chain subunits and is required for apoptotic resistance, cristae morphogenesis, and cell proliferation. Mol. Biol. Cell 23, 1010–1023.

Sylvestre, G., Gandini, M., Maciel-de-Freitas, R., 2013. Age-dependent effects of oral infection with dengue virus on *Aedes aegypti* (Diptera: Culicidae) feeding behavior, survival, oviposition success and fecundity. PLoS One 8, e59933.

Teixeira, L., Ferreira, A., Ashburner, M., 2008. The bacterial symbiont *Wolbachia* induces resistance to RNA viral infections in *Drosophila melanogaster*. PLoS Biol. 6, 2753–2763.

Tuorto, F., Liebers, R., Musch, T., Schaefer, M., Hofmann, S., Kellner, S., Frye, M., Helm, M., Stoecklin, G., Lyko, F., 2012. RNA cytosine methylation by Dnmt2 and NSun2 promotes tRNA stability and protein synthesis. Nat. Stru Mol. Biol. 19, 900–905.

Turelli, M., Hoffmann, A.A., 1995. Cytoplasmic incompatibility in *Drosophila simulans*: dynamics and parameter estimates from natural populations. Genetics 140, 1319–1338.

van den Hurk, A.F., ll-Mendelin, S., Pyke, A.T., Frentiu, F.D., McElroy, K., Day, A., Higgs, S., O'Neill, S.L., 2012. Impact of *Wolbachia* on infection with chikungunya and yellow fever viruses in the mosquito vector *Aedes aegypti*. PLoS Negl. Trop. Dis. 6, e1892.

Xi, Z., Gavotte, L., Xie, Y., Dobson, S.L., 2008. Genome-wide analysis of the interaction between the endosymbiotic bacterium *Wolbachia* and its *Drosophila* host. BMC Genomics 9, 1.

Ye, Y.H., Woolfit, M., Huttley, G.A., Rancès, E., Caragata, E.P., Popovici, J., O'Neill, S.L., McGraw, E.A., 2013. Infection with a virulent strain of *Wolbachia* disrupts genome wide-patterns of cytosine methylation in the mosquito *Aedes aegypti*. PLoS One 8, e66482.

Zhang, G., Hussain, M., Asgari, S., 2014. Regulation of arginine methyltransferase 3 by a *Wolbachia*-induced microRNA in *Aedes aegypti* and its effect on *Wolbachia* and dengue virus replication. Insect. Biochem. Mol. Biol. 53, 81–88.

Zhang, G., Hussain, M., O'Neill, S.L., Asgari, S., 2013. *Wolbachia* uses a host microRNA to regulate transcripts of a methyltransferase contributing to dengue virus inhibition in *Aedes aegypti*. Proc. Natl. Acad. Sci. U.S.A. 110, 10276–10281.

Zug, R., Hammerstein, P., 2012. Still a host of hosts for *Wolbachia*: analysis of recent data suggests that 40% of terrestrial arthropod species are infected. PLoS One 7, e38544.

CHAPTER

11

The Gut Microbiota of Mosquitoes: Diversity and Function

Michael R. Strand

University of Georgia, Athens, GA, United States

INTRODUCTION

The digestive system of animals serves as the primary site for nutrient absorption, water balance, and waste excretion. Microorganisms also reside in the digestive tract of most animals where they are collectively referred to as the gut microbiota. The gut microbiota of most vertebrates and invertebrates consists primarily of bacteria, which have also been implicated in functions ranging from diet sensing and metabolism to immunity and behavior (Nicholson et al., 2012; Engel and Moran, 2013; Lee and Brey, 2013; Goodrich et al., 2016). It has long been known that the digestive tract of mosquitoes contains microbes (Hinman, 1930), but most studies on community composition and function have been conducted in the last 10 years (Minard et al., 2013). An estimated 3500 species of mosquitoes exist worldwide. All reside in the family *Culicidae* which is subdivided into two subfamilies (*Anophelinae* and *Culicinae*) and ~54 genera (Reidenbach et al., 2009; Wilkerson et al., 2015). Like most of the mosquito literature, however, the majority of studies on the gut microbiota focus on human disease vectors such as the yellow fever mosquito, *Aedes aegypti,* and African malaria mosquito, *Anopheles gambiae.* In this chapter, I summarize current knowledge of mosquito gut microbiota diversity and function. Other disease vectors such as ticks and tsetse flies also have a gut microbiota or other symbionts. However, I largely restrict this discussion to mosquitoes because chapters elsewhere in this volume address symbiont diversity and function in these and other arthropods.

ACQUISITION AND COMMUNITY DIVERSITY OF THE MOSQUITO GUT MICROBIOTA

Many factors related to habitat, diet, and life stage can affect the community of microbes that colonize the digestive tract of insects. In the case of mosquitoes, all species are aquatic during their immature stages but also exhibit habitat preferences in terms of oviposition site and where offspring forage (Clements, 1992). Larval

Arthropod Vector: Controller of Disease Transmission, Volume 1
http://dx.doi.org/10.1016/B978-0-12-805350-8.00011-8

feeding behavior also varies although diet analysis indicates most species consume organic detritus, microorganisms, and small invertebrates (Merritt et al., 1992). Mosquitoes pupate in the larval habitat, but adults are terrestrial with both sexes imbibing water and feeding on various nectar sources (Foster, 1995). Most ,but not all, species are also anautogenous, which means that adult females must consume at least one blood meal from a vertebrate host for each clutch of eggs they lay (Briegel, 2003).

The first studies of mosquito gut microbes used thioglycolate broth and other culture methods to show that bacteria were present in larvae (Hinman, 1930; Rozeboom, 1935; Chao and Wistreich, 1959; Jones and DeLong, 1961; Ferguson and Micks, 1961). Flourescent dyes such as acridine orange or 4′6-diamidino-2-phenylindole (DAPI) have also been used together with morphology and culture assays to estimate the abundance of bacteria in the gut during different life stages and to identify other organisms including protists, algae, fungi, and rotifers (Hinman, 1930; Walker et al., 1988; Merritt et al., 1990; DeMaio et al., 1996).

Culture-dependent methods combined with sequencing have since been used to isolate and identify a number of bacteria from larvae and adults (DeMaio et al., 1996; Lindh et al., 2005; Rani et al., 2009; Dong et al., 2009; Chouaia et al., 2010; Gusmao et al., 2010; Cirimotich et al., 2011; Djadid et al., 2011; Terenius et al., 2012; Chavshin et al., 2012; Coon et al., 2014). These studies have been most important from the perspective of identifying gut community members for functional assays, but do not provide a comprehensive picture of community diversity because operational taxonomic units (OTUs) or species that are not amenable to the culture methods used will not be identified. Thus, deep sequencing of primarily 16S rRNA gene domains has also been used to assess bacterial diversity in a culture-independent manner (Wang et al., 2011; Zouache et al., 2011; Boissiere et al., 2012; Chavshin et al., 2012; Osei-Poku et al., 2012; Coon et al., 2014; Gimonneau et al., 2014; Minard et al., 2014; Duguma et al., 2015; Yadav et al., 2015; Muturi et al., 2016a; Buck et al., 2016). A majority of these studies focus only on adults, which has primarily provided insights about how the gut microbiota varies between species or individuals as a function of collection site. However, some studies have examined bacterial diversity across life stages, which provides insights into how mosquitoes acquire their gut microbiota and how community composition changes over the course of development (Wang et al., 2011; Gimonneau et al., 2014; Coon et al., 2014; Duguma et al., 2015).

Bacterial Diversity in the Gut Is Low

The above literature overall indicates that bacterial diversity in the mosquito gut is much lower (<200 OTUs) than in vertebrates but is broadly similar to holometabolous insects such as *Drosophila melanogaster* and the honeybee *Apis melifera* (Shin et al., 2011; Wong et al., 2011; Martinson et al., 2011; Engel et al., 2012). Results also indicate that diversity is usually lower in laboratory cultures than field collected mosquitoes (Wang et al., 2011; Boissiere et al., 2012; Coon et al., 2014, 2016a). Most identified bacteria are Gram-negative aerobes or facultative anaerobes that belong to four phyla: Proteobacteria, Firmicutes, Bacteroidetes, and Actinobacteria. Bacteria in these phyla are also commonly reported in the digestive tracts of other insects (Yun et al., 2014). While recurrent detection of bacteria in particular phyla has been suggested as evidence that mosquitoes have a core gut microbiota (Zouache et al., 2012; Minard et al., 2013; Dennison et al., 2014), this level of taxonomic resolution is too coarse to support such a conclusion. Individual studies have also noted that certain genera of bacteria including *Actinobacteria, Chryseobacterium, Elizabethkingia, Enterobacter, Pseudomonas,* and *Thorselia* are more abundant than others (Wang et al., 2011; Boissiere et al., 2012; Chavshin et al.,

2012; Coon et al., 2014; Gimonneau et al., 2014; Minard et al., 2014; Duguma et al., 2015; Muturi et al., 2016a). However, comparing results across studies does not identify particular genera or species of bacteria that are consistently the most abundant community members in a particular species or life stage. In addition, while several viruses and yeast have been identified in mosquitoes (Gusmao et al., 2010; Ricci et al., 2011; Hegde et al., 2015), recently other microbes such as fungi have been characterized as constituents of the mosquito gut microbiota (Muturi et al., 2016b).

Most Gut Bacteria Are Acquired From the Environment

Some insects including termites, several social Hymenoptera, certain Heteroptera, and cockroaches directly acquire gut microbes from parents or other individuals by vertical transmission, trophallaxis, coprophagy, or consumption of microbes on the surface of eggs (Beard et al., 2002; Nikoh et al., 2011; Anderson et al., 2012; Martinson et al., 2011). Maternally transmitted beneficial symbionts such as *Sodalis glossinidius* in the tsetse fly also exist both intracellularly and in the gut lumen (Attardo et al., 2008). In contrast, other insects such as drosophilids and Lepidoptera predominantly acquire their gut microbiota from the environment (Broderick and Lemaitre, 2012; Engel and Moran, 2013).

Two lines of evidence indicate that mosquitoes also predominantly acquire their gut microbiota from the environment. First, field surveys of adults reveal high within and between species variability in gut community composition (Zouache et al., 2011; Bossiere et al., 2012; Osei-Poku et al., 2012; Minard et al., 2013; Gimonneau et al., 2014; Muturi et al., 2016a; Buck et al., 2016). These patterns would not be expected if mosquitoes predominantly acquire their gut microbiota directly from parents or other individuals. Second, studies of mosquitoes across life stages show that aquatic habitats contain more diverse bacterial communities than detected in larvae (Ponnusamy et al., 2008; Wang et al., 2011; Dada et al., 2014; Coon et al., 2014). However, the OTUs in larvae also near fully overlap with the aquatic environment although relative abundance differs (Gimonneau et al., 2014; Coon et al., 2014). This pattern strongly suggests that a subset of the bacteria in the water where larvae develop colonize the digestive tract, which is followed by certain OTUs increasing in abundance while others decline or fail to persist (Fig. 11.1).

Moll et al. (2001) noted that mosquito larvae avoid bacteria in the meconium they expel at metamorphosis, and used culture-based methods to conclude that *Aedes*, *Culex*, and *Anopheles* adults emerge with few or no gut bacteria. This further led to the suggestion that adults do not transstadially acquire gut bacteria from larvae but may reacquire bacteria detected in larvae by imbibing water after emerging from the pupal stage (Moll et al., 2001; Lindh et al., 2008; Terenius et al., 2012). Coon et al. (2014) used deep sequencing of 16S rDNAs together with controlled laboratory experiments to formally test whether adult *A. aegypti* emerge from the pupal stage with a gut microbiota or reacquire gut bacteria by other means. Results showed that the bacterial community in the adult gut was less diverse, and relative abundance of community members differs from larvae. However, the OTUs present fully overlapped with those in the larval community. Combined with other controls, these data supported that adults transstadially acquire a portion of the gut microbiota in larvae (Fig. 11.1). Other data sets also show that bacterial diversity is lower in adults than larvae (Wang et al., 2011; Boissiere et al., 2012; Gimonneau et al., 2014), while two studies similarly conclude that adult *Anopheles stephensi* and *Culex tarsalis* acquire gut bacteria transstadially (Duguma et al., 2015; Chavshin et al., 2015). Feeding on extrafloral nectaries or imbibing water could modify the microbiota in adults (Lindh et al., 2008; Crotti et al., 2010;

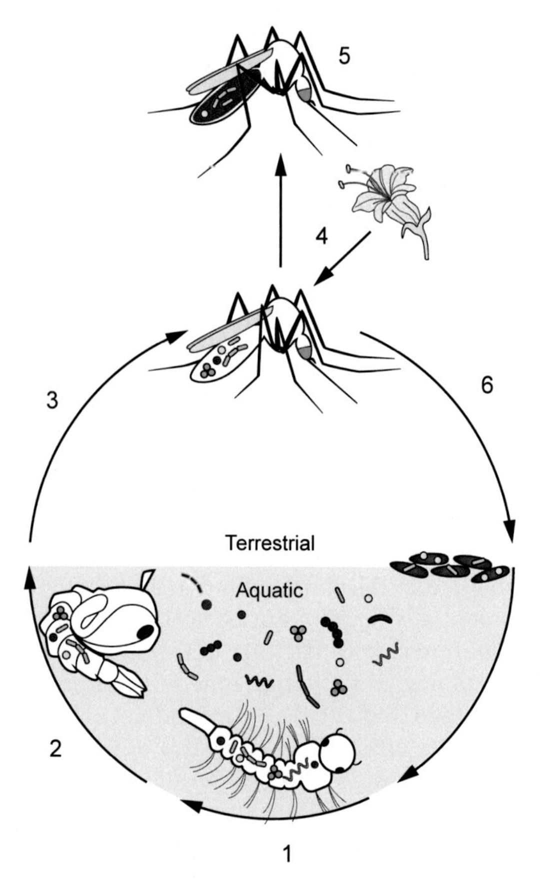

FIGURE 11.1 Schematic illustrating the acquisition and transmission of the mosquito gut microbiota. Mosquito larvae ingest bacteria and other microbes present in their aquatic habitat. Some of these community members persist and increase in abundance in the gut while others decline or fail to establish (1). Some members of the larval gut microbiota persist in the pupal stage (2) and are transstadially transmitted to adults (3). Relative abundance of some community members in adults differs from the larval stage while diversity is lower. Adults potentially acquire other bacteria from the environment through feeding on extrafloral nectaries (4). Bloodfeeding by females results in some gut community members increasing in abundance while other community members fail to persist (5). Eggs females lay have some bacteria on their surface that are gut community members (6).

Wang et al., 2011), although this has not been experimentally examined. In contrast, several studies document that consumption of a blood meal by adult females stimulates proliferation of some gut community members but reduces overall diversity (Gaio et al., 2011; Oliveira et al., 2011; Wang et al., 2011; Zouache et al., 2012; Coon et al., 2014) (Fig. 11.1).

Some Gut Bacteria Can Be Acquired Directly

While mosquitoes acquire most gut community members from the environment, evidence supports that mosquitoes can directly acquire some bacteria present in the digestive tract (Fig. 11.1). One example is the Alphaproteobacteria in the genus *Asaia*, which have been identified in some populations of *A. gambiae*, *Anopheles stephensae*, and *A. aegypti* (Crotti et al., 2010; Damiani et al., 2010). Adult mosquitoes may acquire *Asaia* from extrafloral nectaries, but horizontal transmission through copulation or feeding on a shared resource can occur. *Asaia* have also been detected in other organs including the reproductive tract and on the surface of eggs, which can result in direct transmission to offspring after hatching (Damiani et al., 2010). Proteobacteria and Actinobacteria present in the aquatic habitats and guts of larvae have similarly been detected in the salivary glands and reproductive system of *A. gambiae* and *Anopheles coluzzii* (Gimonneau et al., 2014; Sharma et al., 2014; Tchioffo et al., 2016). Adult mosquitoes also deposit other genera of bacteria besides *Asaia* on the surface of eggs (Chao et al., 1963). These include *Chryseobacterium*, *Delftia*, *Actinobacter*, and *Stenotrophomonas* identified in the aquatic habitat and digestive tract of some *A. aegypti* (Coon et al., 2014). How *Asaia* and other bacteria become associated with other tissues besides the gut, as well as the precise origin (hindgut, ovaries, oviducts) of the bacteria deposited on the surface of eggs, is unclear.

On the other hand, several lines of evidence strongly support that no bacteria identified as gut community members are present in mosquito eggs (Coon et al., 2014). The only bacteria thus far detected in the eggs of multiple species of mosquitoes are *Wolbachia,* which are well-known intracellular Alphaproteobacteria that are maternally transmitted by many arthropods (Hegde et al., 2015). An intracellular *flavobacterium* in the genus *Blattabacterium* has also been reported in one population of *Aedes albopictus* but thus far not elsewhere (Zouache et al., 2012). Neither *Wolbachia* nor *Blattabacterium,* however, are members of the gut microbiota.

The Gut as a Habitat for Microbes

The literature at large distinguishes between microbes that establish residency in the digestive tract (autochthonous) and community members that are transient (allochthonous). In vertebrates, resident community members are thought to be more closely associated with the gut epithelium while transient community members are in the lumen together with consumed or digested food (Nava and Stappenbeck, 2011). In contrast, the digestive tract of mosquitoes and other holometabolous insects exhibit several features that differ from vertebrates and could potentially affect where resident versus transient community members reside. That insects grow and develop by molting also markedly differs from vertebrates and could potentially have a large effect on colonization and persistence of gut microbes.

Food ingested by mosquito larvae moves through the foregut to the midgut where digestion occurs under strongly alkaline conditions (pH 11) (Boudko et al., 2001). Nitrogenous waste collected by the Malphigian tubules together with food waste are then expelled from the hindgut, which also has important functions in osmotic regulation. The foregut and hindgut are lined with a thin cuticle while epithelial cells in proximity to the gastric caecae secrete caecal and peritrophic matrices (Volkman and Peters, 1989; Edwards and Jacobs-Lorena, 2000). Together, these barriers prevent many microbes from directly contacting gut epithelia. The fore and hindgut cuticle plus the peritrophic matrix of the midgut are also shed and replaced during molting, which could affect colonization dynamics. As previously noted, portions of the larval gut together with bacteria are expelled at pupation (Moll et al., 2001), which is followed by gut tissue histolysis and remodeling during the pupal stage. The resulting adult foregut consists of a cuticular pharyngeal pump in the head plus dorsal diverticula and a prominent crop in the thorax, which stores ingested water and nectar. A blood meal in contrast directly enters the midgut where it is digested by multiple proteases and concentrated through water removal within a peritrophic membrane (Isoe et al., 2009; Drake et al., 2015). The lumen of the adult midgut in female *A. aegypti, A. gambiae,* and select other species is alkaline but average pH (8.0–9.5) is lower than in larvae (Corena et al., 2005). The adult hindgut also has similar functions as in larvae.

Little information is currently available on the spatial distribution of the gut microbiota in relation to gut morphology. Observations using DAPI staining suggest the majority of bacteria in larvae reside in the midgut within the endoperitrophic space (Walker et al., 1988; Merritt et al., 1990). A *Pseudomonas* from *A. stephensi* transformed to express green fluorescent protein (GFP) was detected in the midgut and Malpighian tubules of larvae, pupae, and adults (Chavshin et al., 2015). Colonization of *A. stephensi* with GFP-expressing *Escherichia coli* was also detected in the midgut and Malpighian tubules with the latter being noted as a potential refuge for bacteria that are transstadially transmitted from larvae to adults because these organs are not remodeled during the pupal stage (Chavshin et al., 2015). GFP-tagged *Asaia* are predominantly detected in the midgut of adult *A. gambiae* but as previously noted are also detected in other tissues (Damiani et al., 2010). The effects of gut clearance during a molt on the distribution of these or other gut bacteria in contrast is unknown. Whether pH and digestive activity in the midgut or nitrogenous waste in the hindgut impact which bacteria in the aquatic environment colonize the mosquito gut is also unknown. However, studies in adult *A. aegypti* do show that digestion of hemoglobin after a blood meal reduces reactive oxygen species (ROS) produced by midgut epithelial cells, which correlates with proliferation of some bacteria (Oliveira et al., 2011).

Lastly, microbial traits that operate within and between species can affect gut community composition (Fig. 11.2). For example, studies of gnotobiotic *A. aegypti* larvae colonized by a single member of the gut community showed that abundant *Microbacterium* present in conventionally reared larvae could not colonize the larval gut (Coon et al., 2014). This finding indicates that colonization and/or persistence of this community member depends on the presence of other bacteria. *Asaia* has been linked to increased abundance of *Actinobacter* in adult *A. albopictus* (Minard et al., 2013), while *Serratia marcescens* has been linked to reduced abundance of *Sphingomonas* and *Burkholderiaceae* in *A. aegypti* (Terenius et al., 2012). An *Enterobacter* in *A. gambiae* also shows antibacterial activity against several bacteria, which could affect gut community composition (Bahia et al., 2014).

FUNCTIONS OF THE GUT MICROBIOTA IN MOSQUITOES

Understanding of gut microbiota function varies greatly across insects with little or no information for most taxa, while model species such as *D. melanogaster* have been studied from several perspectives including affects on immunity, metabolism, growth, and behavior (Buchon et al., 2013; Lee and Brey, 2013; Dobson et al., 2015; Thibert et al., 2016). Intermediate between these extremes are termites and certain Coleoptera that have been studied primarily from the perspective of digestion, and social Hymenoptera, such as honeybees, that have been studied in terms of functional diversity of gut bacteria and interactions with pathogens (Crotti et al., 2013; Engel and Moran, 2013). Functional studies in mosquitoes have primarily focused on the role of the gut microbiota in vector competence. Some insights have also accrued on *Wolbachia* transmission, nutrient acquisition, development, and oviposition behavior.

Vector Competence

Mosquitoes can potentially consume human or other vertebrate pathogens when they blood-feed. In most cases the acquired organism fails to persist. However, several species of *Plasmodium*, certain viruses, and some filarial nematodes have evolved to invade the midgut of some mosquito species, propagate, and transmit themselves when a female consumes another blood meal. Mosquitoes permissive to such infections are thus referred to as vector competent. The peritrophic matrix and midgut epithelium serve as barriers for entry of many microbes while several components of the mosquito immune system have been implicated in defense against *Plasmodium* and different arboviruses (Saraiva et al., 2016). The first evidence the gut microbiota can affect vector competence is derived from studies of anopheline mosquitoes treated with antibiotics or bacteria-binding antibodies. This reduced the abundance of bacteria in the digestive tract but increased infection rates by *Plasmodium* (Pumpini et al., 1993, 1996; Gonzalez-Ceron et al., 2003). Antibiotics in human blood meals have also been shown to affect the gut microbiota and *P. falciparum* infection in *A. gambiae* (Gendrin et al., 2015), while antibiotic treatment of *A. aegypti* increases titers of dengue virus serotype 2 (DENV-2) (Xi et al., 2008).

Experimental evidence supports two mechanisms for why reducing gut bacteria increases *Plasmodium* or DENV infection (Fig. 11.2A). The first is that the gut microbiota primes the

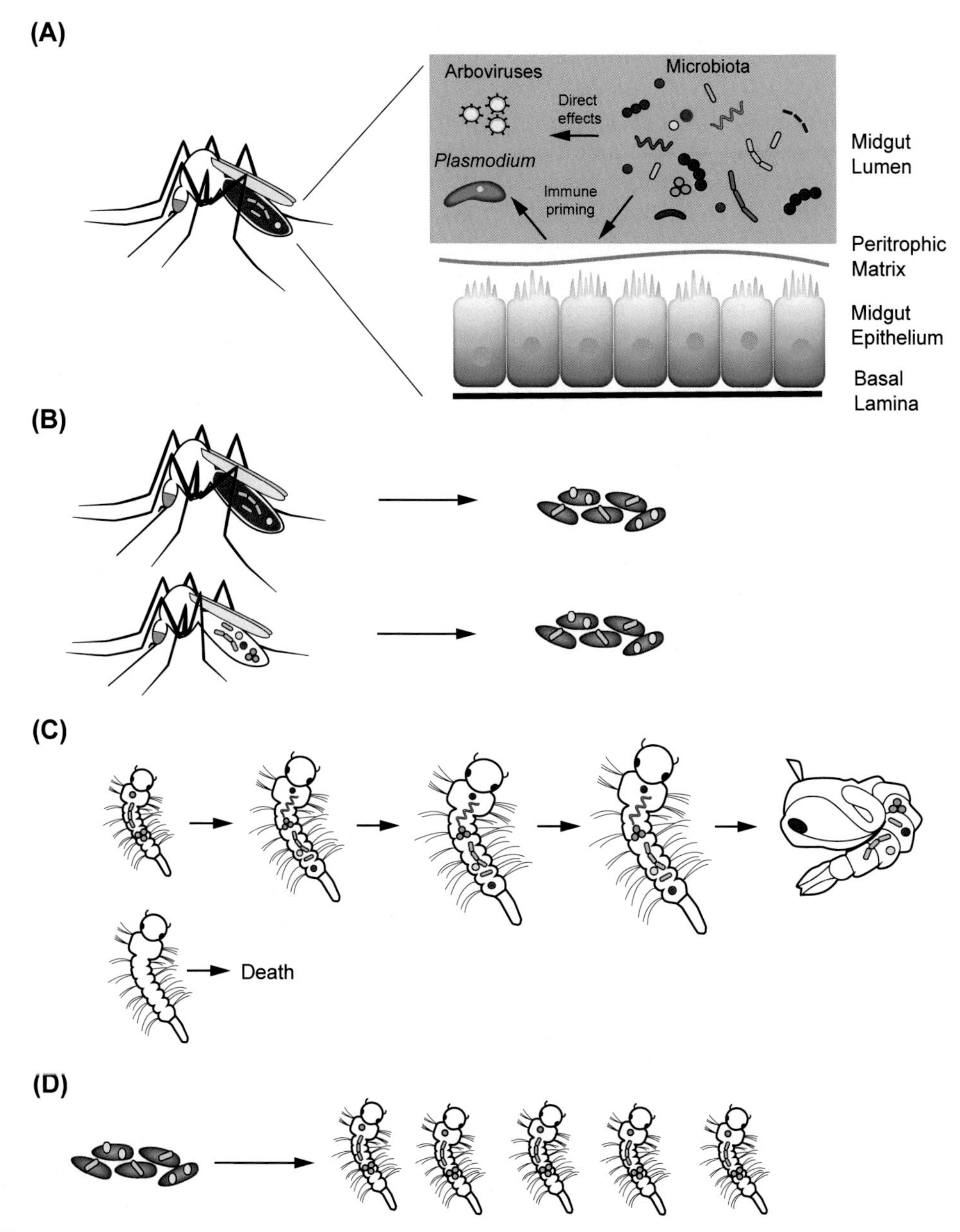

FIGURE 11.2 **Four main functional roles have been identified for the mosquito gut microbiota.** (A) Vector competency. To the left is an adult female after consuming a blood meal while to the right is a schematic of the midgut. The gut microbiota can negatively affect infection by some arboviruses and *Plasmodium* through direct production of factors or immune priming. The gut microbiota can also positively affect arbovirus infection. (B) Nutrient acquisition and egg production. The gut microbiota in anautogenous *Aedes aegypti* has been implicated in blood meal digestion while in the autogenous mosquito *Aedes atropalpus* the gut microbiota strongly affects egg production. (C) Larval development. Several mosquito species develop successfully with a gut microbiota, but axenic larvae fail to develop beyond the first instar. (D) Egg hatching. Bacteria on the surface affect egg hatching rates in *A. aegypti*.

mosquito immune system to produce factors that have anti-*Plasmodium* and antiviral activity. Studies initially conducted in *D. melanogaster* show that recognition molecules such as peptidoglycan recognition protein family members (PGRPs) are expressed in the digestive tract and play important surveillance roles that include distinguishing resident gut bacteria from potential pathogens (Buchon et al., 2013; Lee and Brey, 2013). These surveillance functions modulate immune pathways that regulate expression of antimicrobial peptides (AMPs), ROS, and other defense molecules. The absence of gut bacteria has also been shown to reduce AMP and ROS expression. Studies in mosquitoes have identified orthologous recognition molecules (Saraiva et al., 2016). In *A. aegypti*, experiments further show that antibiotic treatment reduces expression of AMPs and selects other immune genes which correlates with increases in *Plasmodium* or DENV-2 titers (Xi et al., 2008; Ramirez et al., 2014), while in *A. gambiae* gut bacteria activate the immune deficiency (Imd) and Toll pathways that have anti-*Plasmodium* functions (Dong et al., 2009; Eappen et al., 2013; Bahia et al., 2014; Ramirez et al., 2014) (Fig. 11.2).

The second mechanism is that some bacteria in the digestive tract produce factors that have antiviral and/or anti-*Plasmodium* activity (Fig. 11.2A). Indirect support for this derives from field studies that have identified correlations between the presence of certain bacteria in *A. gambiae* and *P. falciparum* infection (Bossiere et al., 2012). Direct support comes from isolating several gut community members including *Enterobacter, Chromobacterium, Proteus,* and *Paenibacillus* that in bioassays directly inhibit *Plasmodium* or arboviruses (Joyce et al., 2011; Cirimotich et al., 2011; Bahia et al., 2014; Ramirez et al., 2014). The identity of the bacterial factors responsible for these activities is unknown.

Infection by certain strains of *Wolbachia* also disrupts infection of *A. aegypti* by DENV and Zika virus (ZIKV) and *A. stephensi* by *Plasmodium* (Bian et al., 2013; Dutra et al., 2016). This too has been linked to immune priming (Pan et al., 2012). However, other studies implicate competition for metabolic resources and microRNA (miRNA) expression in the antiviral effects of *Wolbachia* (Hussain et al., 2011; Caragata et al., 2013). While resource competition or small RNAs could play a role in the antipathogen activity of the gut microbiota, no evidence currently supports this.

Counterbalancing the view that the mosquito gut microbiota can antagonize infection by some human pathogens is the result that gut microbes promote infection by others (Fig. 11.2A). This includes data that O'nyong nyong virus (ONNV) requires the gut microbiota for infection of *A. gambiae* (Carissimo et al., 2015) and that *Serratia* species increases replication of DENV-2 and chikungunya virus (CHIKV) in *A. aegypti* (Apte-Deshpande et al., 2012, 2014). Studies in *A. albopictus* further indicate CHIKV infection affects diversity of the gut microbiota with some community members (*Enterobacteriaceae*) increasing in abundance while certain Alphaproteobacteria and Gammaproteobacteria decline (Zouache et al., 2012). In summary, the role of the gut microbiota in vector competence is clearly complex with host, community composition, and pathogen all having an important bearing on interactions and outcomes.

Wolbachia Transmission

Closely related to the topic of vector competence is the effect of gut microbiota on potential pathogens and other organisms that can infect mosquitoes but are not transmitted to vertebrates. Studies of other insects provide several examples of the gut microbiota enhancing resistance to infection by pathogenic bacteria and protozoans (Engel and Moran, 2013). In mosquitoes, the only data currently in the literature focus on *Asaia* and *Wolbachia* where each can influence the establishment and persistence of the other (Hughes et al., 2014; Rossi et al., 2015). Although *Asaia* is present in the mosquito gut, its ability to disrupt *Wolbachia* transmission could occur through intracellular interactions in the ovaries and possibly other organs where both bacteria are detected (Rossi et al., 2015).

Nutrient Acquisition and Digestion

While gut microbiota play important roles in digestion and nutrient acquisition in several organisms, few studies have examined these issues in mosquitoes. Gaio et al. (2011) treated adult female *A. aegypti* with different antibiotics and observed that carbenicillin and tetracycline caused large reductions in the abundance of culturable gut bacteria, which correlated with some delays in blood meal digestion and a small but statistically significant reduction in the number of eggs females laid. *Enterobacter* and *Serratia* species isolated in the study also exhibited hemolytic activity, which suggested these community members produce factors with roles in blood meal digestion (Fig. 11.2B). These factors could also interact with trypsin-like proteases that *A. aegypti* midgut cells secrete, which are also involved in digestion (Isoe et al., 2009). A second study implicates the gut microbiota in nutrient acquisition by autogenous mosquitoes, which produce eggs without blood feeding (Coon et al., 2016b). *Aedes atropalpus* is a facultatively autogenous species that produces a similar number of eggs in its first gonadotropic cycle as blood-fed *A. aegypti* but uses only nutrient reserves acquired during the larval stage to do so (Gulia-Nuss et al., 2012). Under identical rearing conditions, similar communities of bacteria are present in the aquatic habitat of both *A. aegypti* and *A. atropalpus*, but select genera of Bacteroidetes and Actinobacteria dominate the gut community in larvae of the former while certain Betaproteobacteria dominate in the latter (Coon et al., 2014). Coon et al. (2016b) produced gnotobiotic *A. aegypti* and *A. atropalpus* that were singly colonized by these abundant community members. Several supported growth to adulthood and egg production after consumption of a blood meal by *A. aegypti* to levels that were equivalent to conventionally reared females with a normal gut microbiota. In contrast, only one Betaproteobacterium, *Comamonas*, supported growth to adulthood and egg production to equivalent levels as conventional controls in *A. atropalpus* (Coon et al., 2016b). Thus, while several gut community members equivalently supported nutrient acquisition for growth and egg production by *A. aegypti*, only certain community members did so in *A. atropalpus* (Fig. 11.2B).

Development and Survival

Several lines of evidence support roles for the gut microbiota in development of the digestive tract and immune system of *D. melanogaster* and model vertebrates such as mice (Buchon et al., 2013; Gensollen et al., 2016). In contrast, the gut microbiota is not essential for growth and survival because axenic progeny of both can be reared for generations without a gut microbiota if provided a nutritionally adequate diet. A few insects with highly specialized diets do rely on symbionts for survival but other insects with more general diets such as larval stage *D. melanogaster* also do not require gut bacteria for survival (Engel and Moran, 2013).

Early studies of mosquito larvae reported that reducing the abundance of microbes in the aquatic environment elevated mortality and/or delayed growth (Hinman, 1930; Rozeboom, 1935; Ferguson and Micks, 1961; Chao et al., 1963). Antibiotic treatment caused *A. stephensi* larvae to develop asynchronously, while addition of antibiotic-resistant *Asaia* restored growth rates to control levels (Chouaia et al., 2012). Coon et al. (2014) in contrast used the approach of surface sterilizing eggs to produce axenic *A. aegypti, A. atropalpus,* and *A. gambiae* larvae that polymerase chain reaction and culture-based assays confirmed had no gut microbiota. Strikingly, larvae of each species died at first instars without molting when provided a nutritionally adequate but germ-free diet (Fig. 11.2C). In contrast, inoculation of axenic larvae with normal gut microbiota or colonization of larvae with several individual members of the gut community fully rescued development into adults (Coon et al., 2014) (Fig. 11.2C). Collectively, these studies identify a role for the gut microbiota in mosquito development. However, the variable outcomes reported in regard to survival could indicate that mosquitoes differ in either their requirement for gut bacteria or methodology, such as using antibiotics versus egg sterilization. Another potentially important consideration is whether mosquitoes are infected by intracellular symbionts such as *Wolbachia* that were absent in the cultures studied by Coon et al. (2014) but can affect mosquito fitness both positively and negatively (Brownlie et al., 2009; Almeida et al., 2011; Ross et al., 2016). However, recent results by Coon et al. (2016a) definitively show that axenic larvae from several species that are maintained under sterile conditions do not develop, that the presence of *Wolbachia* does not influence this outcome, and that antibiotic treatment does not fully eliminate the gut microbiota of mosquitoes, which likely underlies why studies that have used antibiotics do not observe that larvae cease to grow. Going forward key needs are to understand mechanistically how gut bacteria interact with the mosquito host to mediate growth, molting, and survival.

Oviposition and Egg Hatching

In addition to colonizing the digestive tract, bacteria in aquatic environments have also been implicated in mediating oviposition site preferences and egg hatching. Experiments with *A. aegypti* suggest that bacteria produce carboxylic acids and methyl ester metabolites that stimulate oviposition by *A. aegypti* (Ponnusamy et al., 2008). The identity of the bacteria producing these oviposition stimulants remains uncharacterized, although it is possible that some include species that also colonize the larval gut. Many *Aedes* sp. also lay eggs that can remain dormant for long periods before hatching when favorable aquatic conditions are available for larvae. Several factors have been suggested to regulate egg hatching including bacteria (Byttebier et al., 2014). *A. aegypti* eggs with a greater abundance of bacteria on their surface have been reported to hatch sooner than eggs with fewer bacteria (Gillett et al., 1977), while experiments comparing the presence and absence of bacteria in the aquatic environment also show differences in egg hatching rates (Ponnusamy et al., 2011) (Fig. 11.2D). How bacteria stimulate hatching is unknown but bacteria on the surface of eggs that promote hatching are likely members of the adult gut microbiota.

CONCLUDING REMARKS

Recent results have greatly advanced understanding of within and between species diversity of the mosquito gut microbiota. Literature also supports that larvae acquire most gut community members from the environment, some members of the larval community are transstadially transmitted to adults, and blood-feeding can substantially alter gut community composition. Issues in need of further investigation include better understanding of how the gut environment shapes community composition, how host species or genotype influences these interactions, and how different microbes influence community assembly.

Gut community members intrinsically important in vector competence have been suggested as candidates for use in disease control (Dennison et al., 2014). The genetic modification of gut bacteria has also been proposed as a strategy for altering vector competency (paratransgenesis). Both *Asaia* and *Pantoea* have been examined in this context because they are amenable to genetic manipulation and can be directly transmitted (Minard et al., 2013). However, current evidence does not indicate any member of the mosquito gut microbiota is transmitted to offspring with high efficiency over multiple generations, which suggests using gut bacteria as control tools may require other strategies for delivery. Insights into gut microbiota functions in mosquitoes have also progressed in select areas besides vector competency. However, understanding of the mechanisms that underlie how gut bacteria positively or negatively influence vector competence remains relatively weak in several respects, while understanding of how gut bacteria influence nutrient acquisition and larval development is weaker still. Future studies are certain to shed new light on these and other study areas.

Acknowledgments

Research from the author cited here is supported by the National Institutes of Health (R01AI1033108 and R01AI106892) and Georgia Agricultural Experiment Station. We thank J.A. Johnson for her assistance with figure preparation.

References

Almeida, F.D., Moura, A.S., Cardoso, A.F., et al., 2011. Effects of *Wolbachia* on fitness of *Culex quinquefasciatus* (Diptera; Culicidae). Infect. Genetics Evol. 11, 2138–2143.

Anderson, K.E., Russell, J.A., Moreau, C.S., Kautz, S., Sullam, K.E., Hu, Y., Basinger, U., Mott, B.M., Buck, N., Wheeler, D.E., 2012. Highly similar microbial communities are shared among related and tropically similar ant species. Mol. Ecol. 21, 2282–2296.

Apte-Deshpande, A., Paingankar, M., Gokhale, M.D., Deobagkar, D.N., 2012. *Serratia odorifera* a midgut inhabitant of *Aedes aegypti* mosquito enhances its susceptibility to dengue-2 virus. PLoS One 7, e40401.

Apte-Deshpande, A.D., Paingankar, M.S., Gkhae, M.D., Deobagkar, D.N., 2014. *Serratia odorifera* mediated enhancement in susceptibility of *Aedes aegypti* for chikungunya virus. Indian J. Med. Res. 139, 762–768.

Attardo, G.M., Lohs, C., Heddi, A., Alam, U.H., Yildirim, S., Aksoy, S., 2008. Analysis of milk gland structure and function in *Glossina mirsitans*: milk production, symbiont populations and fecundity. J. Insect Physiol. 54, 1236–1242.

Bahia, A.C., Dong, Y., Blumberg, B.J., Miambo, G., Tripathi, A., Ben Marzouk-Hidalgo, O.J., Chandra, R., Dimopoulos, G., 2014. Exploring *Anopheles* gut bacteria for *Plasmodium* blocking activity. Environ. Microbiol. 1462–2920.

Beard, C.B., Cordon-Rosales, C., Durvasula, R.V., 2002. Bacterial symbionts of the Triatominae and their potential use in control of Chagas diease transmission. Ann. Rev. Entomol. 47, 123–141.

Bian, G., Joshi, D., Dong, Y., Lu, P., Zhou, G., Pan, X., Xu, Y., Dimopoulos, G., Xi, Z., 2013. *Wolbachia* invades *Anopheles stephensi* populations and induces refractoriness to *Plasmodium* infection. Science 340, 748–751.

Boissiere, A., Tchioffo, M.T., Bachar, D., et al., 2012. Midgut microbiota of the malaria mosquito vector *Anopheles gambiae* and interactions with *Plasmodium falciparum* infection. PLoS Pathog. 8, e1002742.

Boudko, D.Y., Moroz, L.L., Harvey, W.R., Linser, P.J., 2001. Alkalinization by chloride/bicarbonate pathway in larval mosquito midgut. Proc. Natl. Acad. Sci. U.S.A. 98, 15354–15359.

Briegel, H., 2003. Physiological bases of mosquito ecology. J. Vector Ecol. 28, 1–11.

Broderick, N.A., Lemaitre, B., 2012. Gut-associated microbes of *Drosophila melanogaster*. Gut Microbes 3, 307–321.

Brownlie, J.C., Cass, B.N., Riegler, M., 2009. Evidence for metabolic provisioning by a common invertebrate endosymbiont, *Wolbachia pipientis*, during periods of nutritional stress. PLoS Pathog. 5, e1000368.

Buchon, N., Broderick, N.A., Lemaitre, B., 2013. Gut homeostasis in a microbial world: insights from *Drosophila melanogaster*. Nat. Rev. Microbiol. 11, 615–626.

Buck, M., Nilsson, L.K., Brunius, C., Dabire, R.K., Hopkins, R., Terenius, O., 2016. Bacterial associations reveal spatial population dynamics in *Anopheles gambiae* mosquitoes. Sci. Rep. 10 (6), 22806.

Byttebier, B., De Mayo, M.S., Fischer, S., 2014. Hatching response of *Aedes aegypti* (Diptera: Culicidae) eggs at low temperatures: effects of hatching media and storage conditions. J. Med. Entomol. 51, 97–103.

Caragata, E.P., Rances, E., Hedges, L.M., Gofton, A.W., Johnson, K.N., O'Neill, S.L., McGraw, E.A., 2013. Dietary cholesterol modulates pathogen blocking by *Wolbachia*. PLoS Pathog. 9, e1003459.

Carissimo, G., Pondeville, E., McFarlane, M., Dietrich, I., Mitri, C., Bischoff, E., Antoniewski, C., Bourgouin, c, Failloux, A.B., Kohl, A., Vernick, K.D., 2015. Antiviral immunity of *Anopheles gambiae* is highly comparatmentalized, with distinct roles for RNA interference and gut bacteria. Proc. Natl. Acad. Sci. U.S.A. 112, E176–E185.

Chavshin, A.R., Oshaghi, M.A., Vatandoost, H., et al., 2012. Identification of bacterial microflora in the midgut of the larvae and adult of wild caught *Anopheles stephensi*: a step toward finding suitable paratransgenesis candidates. Acta Tropica 121, 129–134.

Chavshin, A.R., Oshaghi, M.A., Vatandoost, H., Yakhchali, B., Zarenejad, F., Terenius, O., 2015. Malphighian tubules are important determinants of *Pseudomans transstadial* transmission and longtime persistence in *Anopheles stephensi*. Parasites Vectors 21 (8), 36.

Chao, J., Wistreich, G.A., 1959. Microbial isolations from the mid-gut of *Culex tarsalis* Coquillet. J. Insect Pathology 1, 311–318.

Chao, J., Wistreich, G.A., Moore, J., 1963. Failure to isolate microorganisms from within mosquito eggs. Microbial isolations from the mid-gut of *Culex tarsalis* Coquillet. Ann. Entomological Society of America 56, 559–561.

Chouaia, B., Rossi, P., Montagna, M., et al., 2010. Molecular evidence for multiple infections as revealed by typing of *Asaia* bacterial symbionts of four mosquito species. App. Environ. Microbiol. 76, 7444–7450.

Chouaia, B., Rossi, P., Epis, S., et al., 2012. Delayed larval development in *Anopheles* mosquitoes deprived of *Asaia* bacterial symbionts. BMC Microbiol. 12, S2.

Cirimotich, C.M., Dong, Y.M., Clayton, A.M., et al., 2011. Natural microbe-mediated refractoriness to *Plasmodium* infection in *Anopheles gambiae*. Science 332, 855–858.

Clements, A.N., 1992. The Biology of Mosquitoes. Development, Nutrition, and Reproduction, vol. 1. Chapman & Hall, New York.

Coon, K.L., Vogel, K.J., Brown, M.R., Strand, M.R., 2014. Mosquitoes rely on their gut microbiota for development. Mol. Ecol. 23, 2727–2739.

Coon, K.L., Brown, M.R., Strand, M.R., 2016a. Mosquitoes host communities of bacteria that are essential for development but vary greatly between local habitats. Mol. Ecol. 22, 5806–5826.

Coon, K.L., Brown, M.R., Strand, M.R., 2016b. Gut bacteria differentially affect egg production in the anautogenous mosquito *Aedes aegypti* and autogenous mosquito *Aedes atropalpus* (Diptera: Culicidae). Parasites Vectors.9, 375.

Corena, M.D.P., Vanekeris, L., Salazar, M.I., Bowers, D., Fiedler, M.M., Sliverman, D., Tu, C., Linser, P.J., 2005. Carbonic anhydrase in the adult mosquito midgut. J. Exp. Biol. 208, 3263–3273.

Crotti, E., Rizzi, A., Chouaia, B., Ricci, I., Favia, G., Alma, A., Sacchi, L., Bourtzis, K., Mandrioli, M., Cherif, A., Bandi, C., Daffonchio, D., 2010. Acetic acid bacteria, newly emerging symbionts of insects. App. Environ. Microbiol. 76, 6963–6970.

Crotti, E., Sansonno, L., Prosdocimi, E.M., Vacchini, V., Hamdi, C., Cherif, A., Gonella, E., Marzorati, M., Balloi, A., 2013. Microbial symbionts of honybees: a promising tool to improve honeybee health. New Biotechnol. 30, 716–722.

Dada, N., Jumas-Bilak, E., Manguin, S., Seidu, R., Stenstrom, T.A., Overgaard, H.J., 2014. Comparative assessment of the bacterial communities associated with *Aedes aegypti* larvae and water from domestic water storage containers. Parasites Vectors 7, 391.

Damiani, C., Ricci, I., Crotti, E., et al., 2010. Mosquito-bacteria symbiosis: the case of *Anopheles gambiae* and *Asaia*. Microbial Ecol. 60, 644–654.

DeMaio, J., Pumpuni, C.B., Kent, M., Beier, J.C., 1996. The midgut bacterial flora of wild *Aedes triseriatus*, *Culex pipiens*, and *Psorophora columbiae* mosquitoes. Am. J. Trop. Med. Hyg. 54, 219–223.

Dennison, N.J., Jupatanakul, N., Dimopoulos, G., 2014. The mosquito microbiota influences vector competence for human pathogens. Curr. Opin. Insect Sci. 1, 6–13.

Djadid, N.D., Jazayeri, H., Raz, A., Favia, G., Ricci, I., Zakeri, S., 2011. Identification of the midgut microbiota of *An. stephensi* and *An. maculipennis* for their application as a paratransgenic tool against malaria. PLoS One 6, e28484.

Dobson, A.J., Chaston, J.M., Newell, P.D., Donahue, L., Hermann, S.L., Sannino, D.R., Westmiller, S., Wong, A.C., Clark, A.G., Lazzaro, B.P., Douglas, A.E., 2015. Host genetic determinants of microbiota-dependent nutrition revealed by genome-wide analysis of *Drosophila melanogaster*. Nat. Commun. 18 (6), 6312.

Dong, Y.M., Manfredini, F., Dimopoulos, G., 2009. Implication of the mosquito midgut microbiota in defense against malaria parasites. PLoS Pathog. 5, 1000423.

Drake, L.L., Rodriguez, S.D., Hansen, I.A., 2015. Functional characterization of aquaporins and aquaglyceroporins of the yellow fever mosquito, *Aedes aegypti*. Sci. Rep. 15 (5), 7795.

Duguma, D., Hall, M.W., Rugman-Jones, P., et al., 2015. Developmental succession of the microbiome of *Culex* mosquitoes. BMC Microbiol. 15, 140.

Dutra, H.L.C., Rocha, M.N., Dias, F.B.S., Mansur, S.B., Caragata, E.P., Moreira, L.A., 2016. *Wolbachia* blocks currently circulating zika virus isolates in Brazilian *Aedes aegypti* mosquitoes. Cell Host Microbe 19, 1–4.

Eappen, A.G., Smith, R.C., Jacobs-Lorena, M., 2013. Enterobacter-activated mosquito immune response in *Plasmodium* involve activation of SRPN6 in *Anopheles stephensi*. PLoS One 8, e62937.

Edwards, M.J., Jacobs-Lorena, M., 2000. Permeability and disruption of the peritrophic matrix and caecal membrane from *Aedes aegypti* and *Anopheles gambiae* mosquito larvae. J. Insect Physiol. 46, 1313–1320.

Engel, P., Moran, N.A., 2013. The gut microbiota of insects-diversity in structure and function. FEMS Microbiol. Rev. 37, 699–735.

Engel, P., Martinson, V.G., Moran, N.A., 2012. Functional diversity within the simple gut microbiota of the honey bee. Proc. Natl. Acad. Sci. U.S.A. 109, 11002–11007.

Ferguson, M.J., Micks, D.W., 1961. Microorganisms associated with mosquitoes: I. Bacteria isolated from the mid-gut of adult *Culex fatigans* Wiedemann. J. Insect Pathology 3, 112–119.

Foster, W.A., 1995. Mosquito sugar feeding and reproductive energetics. Ann. Rev. Entomol. 40, 443–474.

Gaio, A.D., Gusmão, D.S., Santos, A.V., Berbert-Molina, M.A., Pimenta, P.F.P., Lemos, F.J.A., 2011. Contribution of midgut bacteria to blood digestion and egg production in *Aedes aegypti* (Diptera: Culicidae) (L.). Parasites Vectors 4, 105.

Gendrin, M., Rodgers, F.H., Yerbanga, R.S., Ouedraogo, J.B., Basanez, M.G., Cohuet, A., Christophides, G.K., 2015. Antibiotics in ingested human blood affect the mosquito microbiota and capacity to transmit malaria. Nat. Commun. 6 (6), 5921.

Gensollen, T., Iyer, S.S., Kasper, D.L., Blumberg, R.S., 2016. How colonization by microbiota in early life shapes the immune system. Science 352, 539–544.

Gillett, J.D., Roman, E.A., Phillips, V., 1977. Erratic hatching in *Aedes* eggs: a new interpretation. Proc. R. Soc. B 196, 223–232.

Gimonneau, G., Tchioffo, M.T., Abate, L., et al., 2014. Composition of *Anopheles coluzzii* and *Anopheles gambiae* microbiota from larval to adult stages. Infect. Genetics Evol. 28, 715–724.

Gonzalez-Ceron, L., Santilian, F., Rodriguez, M.H., Mendez, D., Hernandez-Avila, J.E., 2003. Bacteria in midguts of field-collected *Anopheles albimanus* block *Plasmodium vivax* sporogonic development. J. Med. Entomol. 40, 371–374.

Goodrich, J.K., Davenport, E.R., Water, J.L., Clark, A.G., Ley, R.E., 2016. Cross-species comparisons of host genetic association with the microbiome. Science 352, 532–538.

Gulia-Nuss, M., Eum, J.H., Strand, M.R., Brown, M.R., 2012. Ovary ecdysteroidogenic hormone activates egg maturation in the mosquito *Georgecraigius atropalpus* after adult eclosion or a blood meal. J. Exp. Biol. 215, 3758–3767.

Gusmão, D.S., Santos, A.V., Marini, D.C., Bacci, M., Berbert-Molina, M.A., Lemos, F.J.A., 2010. Culture-dependent and culture-independent characterization of microorganisms associated with *Aedes aegypti* (Diptera: Culicidae) (L.) and dynamics of bacterial colonization in the midgut. Acta Tropica 115, 275–281.

Hegde, S., Rasgon, J.L., Hughes, G.L., 2015. The microbiome modulates arbovirus transmission in mosquitoes. Curr. Opin. in Virology 15, 97–102.

Hinman, E.H., 1930. A study of the food of mosquito larvae. Am. J. Hyg. 12, 238–270.

Hughes, G.L., Dodson, B.L., Johnson, R.M., Murdock, C.C., Tsujimoto, H., Suzuki, Y., Patt, A.A., Cui, L., Nossa, C.W., Barry, R.M., Sakamoto, J.M., Hornett, E.A., Rasgon, J.L., 2014. Native microbiome impedes vertical transmission of *Wolbachia* in *Anopheles* mosquitoes. Proc. Natl. Acad. Sci. U.S.A. 111, 12498–12503.

Hussain, M., Frentiu, F.D., Moreira, L.A., O'Neill, S.L., Asgari, S., 2011. *Wolbachia* uses host microRNAs to manipulate host gene expression and facilitate colonization of the dengut vector *Aedes aegypti*. Proc. Natl. Acad. Sci. U.S.A. 108, 9250–9255.

Isoe, J., Rascon Jr., A.A., Kunz, S., Miesfeld, R.L., 2009. Molecular genetic analysis of midgut serine proteases in *Aedes aegypti*. Insect Biochemistry and Mol. Biol. 39, 903–912.

Jones, W.L., DeLong, D.M., 1961. A simplified technique for sterilizing the surface of *Aedes aegypti* eggs. J. Economic Entomol. 54, 813–814.

Joyce, J.D., Nogueira, J.R., Bales, A.A., Pittman, K.E., Anderson, J.R., 2011. Interactions between La Crosse virus and bacteria isolated from the digestive tract of *Aedes albopictus* (Diptera: Culicidae). J. Med. Entomol. 48, 389–394.

Lee, W.-J., Brey, P.T., 2013. How microbiomes influence metazoan development: insights form history and *Drosophila* modeling of gut-microbe interactions. Ann. Rev. Cell Biol. 29, 571–592.

Lindh, J.M., Terenius, O., Faye, I., 2005. 16S rRNA gene-based identification of midgut bacteria from field-caught *Anopheles gambiae* sensu lato and *a. funestus* mosquitoes reveals new species related to known insect symbionts. App. Environ. Microbiol. 71, 7217–7223.

Lindh, J.M., Borg-Karlson, A.-K., Faye, I., 2008. Transstadial and horizontal transfer of bacteria within a colony of *Anopheles gambiae* (Diptera: Culicidae) and oviposition response to bacteria-containing water. Acta Tropica 107, 242–250.

Martinson, V.G., Danforth, B.N., Minckley, R.L., Rueppell, O., Tingek, S., Moran, N.A., 2011. A simple and distinctive microbiota associated with honey bees and bumble bees. Mol. Ecol. 20, 619–628.

Merritt, R.W., Olds, E.J., Walker, E.D., 1990. Natural food and feeding ecology of larval *Coquillettida perturbans*. J. Am. Mosq. Control Assoc. 6, 35–42.

Merritt, R.W., Dadd, R.H., Walker, E.D., 1992. Feeding behavior, natural food, and nutritional relationships of larval mosquitoes. Ann. Rev. Entomol. 37, 349–376.

Minard, G., Mavingui, P., Moro, C.V., 2013. Diversity and function of bacterial microbiota in the mosquito holobiont. Parasites Vectors 6, 146.

Minard, G., Tran, G., Dubost, A., Tran-Van, V., Mavingui, P., Moro, C.V., 2014. Pyrosequencing 16S rRNA genes of bacteria associated with wild tiger mosquito *Aedes albopictus*: a pilot study. Front. Cell. Infect. Microbiol. 4, 59.

Moll, R.M., Romoser, W.S., Modrzakowski, M.C., Moncayo, A.C., Lerdthusnee, K., 2001. Meconial peritrophic membranes and the fate of midgut bacteria during mosquito (Diptera: Culicidae) metamorphosis. J. Med. Entomol. 38, 29–32.

Muturi, E.J., Kim, C., Bara, J., Bach, E.M., Siddappaji, M.H., 2016a. *Culex pipiens* and *Culex restuans* mosquitoes harbor distinct microbiota dominated by few bacterial taxa. Parasites Vectors 9, 18.

Muturi, E.J., Bara, J.J., Rooney, A.P., Hansen, A.K., 2016b. Midgut fungal and bacterial microbiota of *Aedes triseriatus* and *Aedes japonicus* shift in response to La Crosse virus infection. Mol. Ecol 25(16), 4075–4090.

Nava, G.M., Stappenbeck, T.S., 2011. Diversity of the autochthonous colonic bacteria. Gut Microbes 2, 99–104.

Nikoh, N., Hosokawa, T., Oshima, K., Hattori, M., Fukatsu, T., 2011. Reductive evolution of bacterial genome in insect gut environment. Genome Biol. Evol. 3, 702–714.

Nicholson, J.K., Holmes, E., Kinross, J., Burcelin, R., Gibson, G., Jia, W., Pettersson, S., 2012. Host-gut microbiota metabolic interactions. Science 336, 1262–1267.

Oliveira, J.H.M., Goncalves, R.L.S., Lara, F.A., et al., 2011. Blood meal-derived heme decreases ROS levels in the midgut of *Aedes aegypti* and allows proliferation of intestinal microbiota. PLoS Pathog. 7, 1001320.

Osei-Poku, J., Mbogo, C.M., Palmer, W.J., Jiggins, F.M., 2012. Deep sequencing reveals extensive variation in the gut microbiota of wild mosquitoes from Kenya. Mol. Ecol. 21, 5138–5150.

Pan, X., Zhou, G., Wu, J., Bian, G., Lu, P., Raihel, A.S., Xi, Z., 2012. *Wolbachia* induces reactive oxygen species (ROS)-dependent activation of the Toll pathway to control dengue virus in the mosquito *Aedes aegypti*. Proc. Natl. Acad. Sci. U.S.A. 109, E23–E32.

Ponnusamy, L., Xu, N., Nojima, S., Wesson, D.M., Schal, C., Apperson, C.S., 2008. Identification of bacteria and bacteria-associated chemical cues that mediate oviposition site preferences by *Aedes aegypti*. Proc. Natl. Acad. Sci. U.S.A. 105, 9262–9267.

Ponnusamy, L., Boroczky, K., Wesson, D.M., Schal, C., Apperson, C.S., 2011. Bacteria stimulate hatching of the yellow fever mosquito eggs. PLoS One 6, e24409.

Pumpuni, C.B., Beier, M.S., Nataro, J.P., Guers, L.D., Davis, J.R., 1993. *Plasmodium falciparum*: inhibition of sporogonic development in *anopheles stephensi* by gram-negative bacteria. Exp. Parasitol. 77, 195–199.

Pumpuni, C.B., Demaio, J., Kent, M., Davis, J.R., Beier, J.C., 1996. Bacterial population dynamics in three anopheline species: the impact on *Plasmodium* sporogonic development. Am. J. Trop. Med. Hyg. 54, 214–218.

Ramirez, J.L., Short, S.M., Bahia, A.C., Saraiva, R.G., Dong, Y., Kang, S., Tripathi, A., Mlambo, G., Dimopoulos, G., 2014. Chromobacterium Csp_P reduces malaria and dengue infection in vector mosquitoes and has entomopathogenic and in vitro anti-pathogen activities. PLoS Pathog. 23, e1004398.

Rani, A., Sharma, A., Rajagopal, R., Adak, T., Bhatnagar, R.K., 2009. Bacterial diversity analysis of larvae and adult midgut microflora using culture-dependent and culture-independent methods in lab-reared and field-collected *Anopheles stephensi*-an Asian malarial vector. BMC Microbiol. 9, 96.

Reidenbach, K.R., Cook, S., Bertone, M.A., Harbach, R.E., Wiegmann, B.M., Besansky, N.J., 2009. Phylogenetic analysis and temporal diversification of mosquitoes (Diptera: Culicidae) based on nuclear genes and morphology. BMC Evol. Biol. 9, 298.

Ricci, I., Mosca, M., Valzano, M., Damiani, C., Scuppa, P., Rossi, P., Crotti, E., Cappelli, A., Ulissi, U., Capone, A., Esposito, F., Alma, A., Mandrioli, M., Sacchi, L., Bandi, C., Daffonchio, D., Favia, G., 2011. Different mosquito species host *Wicherhamomyces anomalus* (*Pichia anomala*): perspectives on vector-diseases symbiotic control. Antonie Van Leeuwehnoek 99, 43–50.

Ross, P.A., Endersby, N.M., Hoffmann, A.A., 2016. Costs of three *Wolbachia* infections on the survival of *Aedes aegypti* larvae under starvation conditions. PLoS Negl. Trop. Dis. 10, e0004320.

Rossi, P., Ricci, I., Cappelli, A., Damiani, C., Ulissi, U., Mancini, M.V., Valzano, M., Capone, A., Epis, S., Crotti, E., Chouaia, B., Scuppa, P., Joshi, D., Xi, Z., Manndrioli, M., Sacchi, L., O'Neill, S.L., Favia, G., 2015. Mutual exclusion of *Asaia* and *Wolbachia* in the reproductive organs of mosquito vectors. Parasites Vectors 8, 278.

Rozeboom, L.E., 1935. The relation of bacteria and bacterial filtrates to the development of mosquito larvae. Am. J. Hyg. 21, 167–179.

Saraiva, R.G., Kang, S., Simoes, M.L., Anglero-Rodriguez, Y.I., Dimopoulos, G., 2016. Mosquito gut antiparasitic and antiviral immunity. Dev. Comp. Immunol.64, 53–64.

Sharma, P., Sharma, S., Maurya, R.K., De, T.D., Thomas, T., Lata, S., Singh, N., Pandey, K.C., Valecha, N., Dixit, R., 2014. Salivary glands harbor more diverse microbial communities than gut in *Anopheles culicifacies*. Parasites Vectors 7, 235.

Shin, S.C., Kim, S.H., You, H., et al., 2011. *Drosophila* microbiome modulates host developmental and metabolic homeostasis via insulin signaling. Science 334, 670–674.

Terenius, O., Lindh, J.M., Eriksson-Gonzales, K., et al., 2012. Midgut bacterial dynamics in *Aedes aegypti*. FEMS Microbiol. Ecol. 80, 556–565.

Tchioffo, M.T., Boissiere, A., Abate, L., Nsango, S.E., Bayibeki, A.N., Awono-Ambene, P.H., Gimonneau, G., Morlais, I., 2016. Dynamics of bacterial community composition in the malaria mosquito's epithelia. Front. Microbiol. 5 (6), 1500.

Thibert, J., Farine, J.P., Cortot, J., Ferbeur, J.F., 2016. *Drosophila* food-associated pheromones: effect of experience, genotype and antibiotics on larval behavior. PLoS One 11, e0151451.

Volkman, A., Peters, W., 1989. Investigations on the midgut caseca of mosquito larvae-II. Functional aspects. Tissue Cell 21, 253–261.

Walker, E.D., Olds, E.J., Meritt, R.W., 1988. Gut content analysis of mosquito larvae (Diptera: Culicidae) using DAPI stain and epifluorescence microscopy. J. Med. Entomol. 25, 551–554.

Wang, Y., Gilbreath, T.M., Kukutla, P., Yan, G.Y., Xu, J.N., 2011. Dynamic gut microbiome across life history of the malaria mosquito *Anopheles gambiae* in Kenya. PLoS One 6, e24767.

Wilkerson, R.C., Linton, Y.-M., Fonseca, D.M., Schultz, T.R., Price, D.C., Strickman, D.A., 2015. Making mosquito taxonomy useful: a stable classification of tribe Aedini that balances utility with current knowledge of evolutionary relationships. PLoS One. http://dx.doi.org/10.1371/journal.pone.0133602.

Wong, A.C.N., Ng, P., Douglas, A.E., 2011. Low diversity bacterial community in the gut of the fruit fly *Drosophila melanogaster*. Environ. Microbiol. 13, 1889–1900.

Xi, Z., Ramirez, J.L., Dimopoulos, G., 2008. The *Aedes aegypti* toll pathway controls dengue virus infection. PLoS Pathog. 4, e1000098.

Yadav, K.K., Bora, A., Datta, S., Chandel, K., Gogoi, H.K., Prasad, G.B., Veer, V., 2015. Molecular characterization of midgut microbiota of *Aedes albopictus* and *Aedes aegypti* from Arunachal Pradesh, India. Parasites Vectors 18, 641.

Yun, J.-H., Roh, S.W., Whon, T.W., Jung, M.-J., Kim, M.-S., Park, D.-S., Yoon, C., Nam, Y.-D., Kim, Y.-J., Choi, J.-H., Kim, J.-Y., Shin, N.-S., Kim, S.-H., Lee, W.-J., Bae, J.-W., 2014. Insect gut bacterial diversity determined by environmental habitat, diet, developmental stage, and phylogeny of host. App. Environ. Microbiol. 80, 5254–5264.

Zouache, K., Raharimalala, F.N., Raquin, V., Tran-Van, V., Raveloson, L.H., Ravelonandro, P., Mavingui, P., 2011. Bacterial diversity of field-caught mosquitoes, *Aedes albopictus* and *Aedes aegypti*, from different geographic regions of Madagascar. FEMS Microbiol. Ecol. 75, 377–389.

Zouache, K., Michelland, R.J., Failloux, A.B., Grundmann, G.L., Mavingui, P., 2012. Chikungunya virus impacts diversity of symbiotic bacteria in mosquito vector. Mol. Ecol. 21, 2297–2309.

Further Reading

Briones, A.M., Shililu, J., Githure, J., Novak, R., Raskin, L., 2008. *Thorsellia anophelis* is the dominant bacterium in a Kenyan population of adult *Anopheles gambiae* mosquitoes. ISME J. 2, 74–82.

Cornel, A.J., Mcabee, R.D., Rasgon, J., Stanich, M.A., Scott, T.W., Coetzee, M., 2003. Differences in extent of genetic introgression between sympatric *Culex pipiens* and *Culex quinquefasciatus* (Diptera: Culicidae) in California and South Africa. J. Med. Entomol. 40, 36–51.

Favia, G., Ricci, I., Damiani, C., et al., 2007. Bacteria of the genus *Asaia* stably associate with *Anopheles stephensi*, an Asian malarial mosquito vector. Proc. Natl. Acad. Sci. U.S.A. 104, 9047–9051.

Guindon, S., Dufayard, J.F., Lefort, V., Anisimova, M., Hordijk, W., Gascuel, O., 2010. New algorithms and methods to estimate maximum-likelihood phylogenies: assessing the performance of PhyML. Syst. Biol. 59, 307–321.

Huang, S., Molaei, G., Andreadis, T.G., 2011. Reexamination of *Culex pipiens* hybridization zone in the Eastern United States by ribosomal DNA-based single nucleotide polymorphism markers. Am. J. Trop. Med. Hyg. 85 , 434–441.

Moullan, N., Mouchiroud, L., Wang, X., et al., 2015. Tetracyclines disturb mitochondrial function across eukaryotic models: a call for caution in biomedical research. Cell Rep. 10, 1681–1691.

Ramirez, J.L., Souza-Neto, J., Cosme, R.T., et al., 2012. Reciprocal tripartite interactions between the *Aedes aegypti* midgut microbiota, innate immune system and dengue virus influences vector competence. PLoS Negl. Trop. Dis. 6, e1561.

Rey, J.R., Nishimura, N., Wagner, B., Braks, M.A.H., O'Connell, S.M., Lounibos, L.P., 2006. Habitat segregation of mosquito arbovirus vectors in south Florida. J. Med. Entomol. 43, 1134–1141.

Staubach, F., Baines, J.F., Kunzel, S., Bik, E.M., Petrov, D.A., 2013. Host species and environmental effects on bacterial communities associated with *Drosophila* in the laboratory and in the natural environment. PLoS One 8, e70749.

Storelli, G., Defaye, A., Erkosar, B., Hols, P., Royet, J., Leulier, F., 2011. *Lactobacillus plantarum* promotes *Drosophila* systemic growth by modulating hormonal signals through TOR-dependent nutrient sensing. Cell Metab. 14, 403–414.

Ye, Y.X.H., Woolfit, M., Huttley, G.A., et al., 2013. Infection with a virulent strain of *Wolbachia* disrupts genome wide-patterns of cytosine methylation in the mosquito *Aedes aegypti*. PLoS One 8, e66482.

Zhang, G.M., Hussain, M., O'Neill, S.L., Asgari, S., 2013. *Wolbachia* uses a host microRNA to regulate transcripts of a methyltransferase, contributing to dengue virus inhibition in *Aedes aegypti*. Proc. Natl. Acad. Sci. U.S.A. 110, 10276–10281.

Zouache, K., Voronin, D., Tran-Van, V., Mousson, L., Failloux, A., Mavingui, P., 2009. Persistent *Wolbachia* and cultivable bacteria infection in the reproductive and somatic tissues of the mosquito vector *Aedes albopictus*. PLoS One 4, e6388.

CHAPTER

12

Targeting Dengue Virus Replication in Mosquitoes

Carol D. Blair, Ken E. Olson

Colorado State University, Fort Collins, CO, United States

INTRODUCTION: WHY TARGET DENGUE VIRUS IN MOSQUITOES?

Dengue virus (DENV) is a species that includes four serologically related but genetically distinct viruses, DENV serotype-1, DENV-2, DENV-3, and DENV-4, which are members of the family *Flaviviridae*, genus *Flavivirus*. DENVs are arthropod-borne viruses (arboviruses), indicating that they are maintained in a natural cycle involving transmission by bite of an infected arthropod vector, in this case *Aedes* spp. mosquitoes, to a vertebrate host. Unlike most other arboviruses, human infection with DENV is not considered "incidental" and does not result from "spillover" from another vertebrate reservoir; humans are the natural vertebrate host in the DENV epidemic (urban) cycle (Hanley et al., 2013). The major DENV vector, *Aedes aegypti aegypti*, prefers to feed on and live in close association with humans and frequently oviposits its desiccation-resistant eggs in small volumes of standing water in containers around human habitats, such as flower pots or discarded cans, jars, and tires. The vector takes multiple blood meals, frequently from different humans, during each gonotrophic cycle with the potential for an infected female that has survived the extrinsic incubation period to transmit virus each time she probes the skin (Gubler, 2004; Barrett and Higgs, 2007; Salazar et al., 2007). *A. aegypti* has rapidly expanded its geographic range over the last half century, and dengue is considered an emerging disease (Gubler, 1998). DENV is ubiquitous throughout the tropics, causing an estimated 390 million infections each year with 96 million of these resulting in apparent disease (Bhatt et al., 2013). Thus, DENV can be ranked the most globally important arbovirus and a rapidly increasing public health burden.

Clinical dengue disease can range from a self-limiting fever to severe, life-threatening hemorrhagic fever (Gubler, 1998). Currently, no specific therapies are available, and development of effective vaccines has been problematic for a number of reasons (Sabchareon et al., 2012; Capeding et al., 2014; Halstead, 2016). The resulting costs to human health and to global economies (Beatty et al., 2011) are major reasons to target DENV infections for reduction.

The uncomplex, efficient *Aedes*-human DENV transmission cycle presents both challenges and

Arthropod Vector: Controller of Disease Transmission, Volume 1
http://dx.doi.org/10.1016/B978-0-12-805350-8.00012-X

opportunities for dengue disease control. The fact that humans are the only vertebrate hosts known to develop the disease following DENV infection means that there is no completely accurate vertebrate animal model, which remains an impediment to development of vaccines and therapeutics. The complex immunopathology of human disease (Rothman, 2004, 2011), as well as the consideration that infection or immunization by a single DENV serotype increases the risk of antibody-dependent enhancement of disease severity during subsequent infection by a heterologous virus serotype (Halstead, 1988), has been among the additional obstacles to vaccine and immunotherapeutic development. Nevertheless, effective vaccines and specific therapeutics must ultimately be part of an integrated dengue control strategy.

The DENV transmission cycle involving only *Aedes* spp. mosquitoes and humans, with no other vertebrate reservoir, provides a rationale for targeting DENV in mosquitoes. The major DENV vector, *A. aegypti*, is an authentic and important laboratory animal model that is essential for virus maintenance and transmission. Interruption of DENV replication in and transmission by mosquitoes is at present the most accessible target for disease control. Understanding DENV–*A. aegypti* interactions is requisite for identification of DENV targets in the mosquito.

With DENV, as with most arboviruses, the outcome of infection of the natural vector is almost always a nonpathogenic viral persistence for the life of the mosquito, allowing maintenance of the cycle and ongoing transmission, in contrast to the potential for disease and eventual immune system–mediated clearance in humans. Characterizing *A. aegypti*–DENV interactions, including understanding natural mosquito defense systems and determination of why they do not result in clearance of virus but allow its persistence, could lead to identifying a target for interruption of DENV replication in its natural vector. Since *Ae. aegypti* and the secondary DENV vector *Aedes albopictus* are also responsible for transmission of a number of other human pathogens such as yellow fever, Zika, and chikungunya viruses (CHIKV), understanding vulnerable points in *A. aegypti*–DENV interactions could provide a model for addressing other mosquito-borne arbovirus diseases. Such expansive expectations, however, need to be viewed with caution; increasing evidence suggests that mosquito-by-virus interactions vary depending on the virus family, genus, and even genotype as well as the mosquito genus, species, and genotype (Armstrong and Rico-Hesse, 2001; Lambrechts et al., 2009; Bennett et al., 2002; Black et al., 2002).

MOSQUITOES NATURALLY TARGET DENGUE VIRUS REPLICATION

The ability of individual *A. aegypti* and *A. albopictus* to be infected by DENV after ingesting an infectious blood meal and to transmit virus by bite after a suitable extrinsic incubation period is defined as vector competence. Vector competence is influenced by both the external environment and intrinsic factors, including mosquito and virus genetics (Hardy et al., 1983).

Both mosquito and viral genetically determined factors are involved in determining the ability of virus introduced in a blood meal to overcome various anatomic barriers to infection and replication such as the midgut infection barrier (inability to infect and replicate in midgut epithelial cells), midgut escape barrier (inability to disseminate from midgut to infect secondary tissues), and salivary gland barriers (inability to invade and replicate in salivary glands and/or to be released into salivary ducts) (Bennett et al., 2002). In each case, genetically encoded structural components of mosquito cell plasma membranes and virus envelopes must enable interactions that allow virus attachment to a specific cell surface receptor, cellular membrane–virus envelope fusion, and nucleocapsid entry into the cytoplasm (Modis et al., 2004; van der Schaar et al., 2008; Erb et al., 2010; Butrapet et al., 2011), and intracellular components must provide the environment required for virus replication (Sessions et al., 2009; Behura et al., 2011; Fansiri et al., 2013).

Other important mosquito genetic determinants encode potential antipathogen defense mechanisms. DENVs also presumably encode genetic determinants to evade mosquito defenses and maintain their transmission cycle. Among the classical eukaryotic cell antiviral defense mechanisms displayed in mosquitoes are autophagy (Behura et al., 2014), apoptosis (Bryant et al., 2008; Clem, 2016), and canonical innate immune pathways that are also found in vertebrates such as Toll, IMD, and JAK-STAT (Waterhouse et al., 2007). Activation of some components of these classical pathways has been demonstrated during DENV infection of *A. aegypti* (Sim et al., 2012; Xi et al., 2008; Luplertlop et al., 2011), and the influence of the Toll pathway on DENV replication has been validated by silencing either positive or negative regulators (Xi et al., 2008). Involvement of Toll, IMD, and JAK-STAT signaling in antiarboviral immunity has been reviewed by Saraiva et al. (2016). For each of these classical signaling pathways, insect pattern recognition receptors (PRRs) specific for pathogen-associated molecular patterns (PAMPs) of gram-negative or gram-positive bacteria or fungi have been well characterized; however, specific mechanisms of induction of these immune responses by arboviruses have not been identified, and induced antiviral effectors potentially activated by signaling pathways are not known. The predominant, most broadly reactive and potent mosquito innate immune pathway(s) that control DENV and other arbovirus infections involve antiviral components of RNA interference (RNAi), which are unique to invertebrates (Kemp et al., 2013; Keene et al., 2004; Sánchez-Vargas et al., 2009; Blair, 2011; Blair and Olson, 2015; Olson and Blair, 2015).

RNAi comprises three major pathways used in various formats by eukaryotic cells to control gene expression. These pathways are named for the small RNA effectors that are end products of each: microRNA (miRNA), small-interfering RNA (siRNA), and Piwi-interacting RNA (piRNA). Several of the most promising strategies to interrupt DENV infection and transmission in *A. aegypti* are based on the *A. aegypti* RNAi pathways, and since we have recently reviewed mosquito RNAi (Blair and Olson, 2015; Olson and Blair, 2015), we will focus here on recent discoveries and what remains to be learned about mosquito antiviral RNAi.

The miRNA pathway. The components and mechanisms of the miRNA pathway are similar in vertebrates and invertebrates, and the constituents in *A. aegypti* have been identified by orthology to those in *Drosophila melanogaster* (Ghildiyal and Zamore, 2009). The function of miRNA is posttranscriptional regulation of gene expression by a mechanism involving imperfect hybridization between a 22–23 nt miRNA and one or more mRNAs, usually in the 3′ UTR, which results in inhibition of translation or, less frequently, mRNA degradation. Biogeneration of miRNA is initiated by nuclear transcription from the host genome of hairpin structure primary miRNAs that are processed first to 60–70 nt stem loops, then to imperfectly base-paired 22–23 nt duplex miRNAs by nuclear and cytoplasmic components, including Drosha (nucleus) and Dicer 1 (Dcr1) and R3D1 (cytoplasm) (Bartel, 2004). The "passenger strand" of each duplex is discarded and the "guide strand" is loaded into Argonaute 1 (Ago1). As part of an RNA-induced silencing complex (RISC), Ago1 directs imperfect base pairing of the guide strand to the 3′ UTR of a specific mRNA, resulting in suppression of its translation (Bartel, 2004; Czech et al., 2009). The *A. aegypti* miRNA pathway has not been shown to have a direct effect on DENV replication; however, it has been shown that changes in expression levels of a large number of miRNAs occur after DENV infection (Campbell et al., 2014). Campbell et al. (2014) performed in silico analysis to identify *A. aegypti* transcripts that contained putative 3′UTR targets of differentially expressed miRNAs. Functional groups of proteins encoded by target mRNAs involved transport, transcriptional regulation, mitochondrial function, chromatin modification, and signal transduction processes, which were also linked to previous transcriptome studies defining host factors required for DENV replication and dissemination. However, experimental validation of the roles of specific gene products in DENV infection was not performed, and mechanisms of modulation of miRNA levels were not determined.

The exo-siRNA pathway. Discovery in *Caenorhabditis elegans* and *D. melanogaster* in 1998 (Fire et al., 1998; Kennerdell and Carthew, 1998) that exposure to exogenous dsRNA led to interference with expression of cognate mRNA, and subsequent identification of required factors and mechanisms for this phenomenon, called RNAi, helped us to understand potential mechanisms underlying our earlier observations of RNA-mediated "pathogen-derived resistance" to DENV replication in *A. aegypti* and cultured mosquito cells (Gaines et al., 1996; Olson et al., 1996). Since that time, we have engaged in ongoing studies of RNAi in *Aedes* spp.

What we now know as exogenous (exo)-siRNA is triggered by introduction of "foreign" long double-stranded (ds) RNA into mosquito cells. In the case of virus infection, the dsRNA is formed as replicative intermediates during replication of positive-sense RNA viruses such as flaviviruses and alphaviruses, but could be due to extensive intrastrand secondary structures or overlapping opposite-strand transcripts of DNA viruses (Kemp et al., 2013; Bronkhorst et al., 2012; Sabin et al., 2013). The dsRNA serves as a PAMP and is bound by the PRR Dicer 2 (Dcr2), a large, multidomain RNase III-family enzyme that cleaves long dsRNA to 21-nt siRNA duplexes with perfect base pairing and 2-nt overhangs at the 3′ ends. A small dsRNA-binding protein called R2D2 complexes with Dcr2 to load an siRNA duplex onto Ago2, an endonuclease that is part of the multicomponent RISC. Ago2 cleaves and releases one strand of the siRNA duplex, retaining the second strand as a guide to hybridize to a complementary coding sequence in mRNA, which it then cleaves in the center of the complementary region (Fig. 12.1) (Campbell et al., 2008b; Galiana-Arnoux et al., 2006; Liu et al., 2003; Matranga et al., 2005;

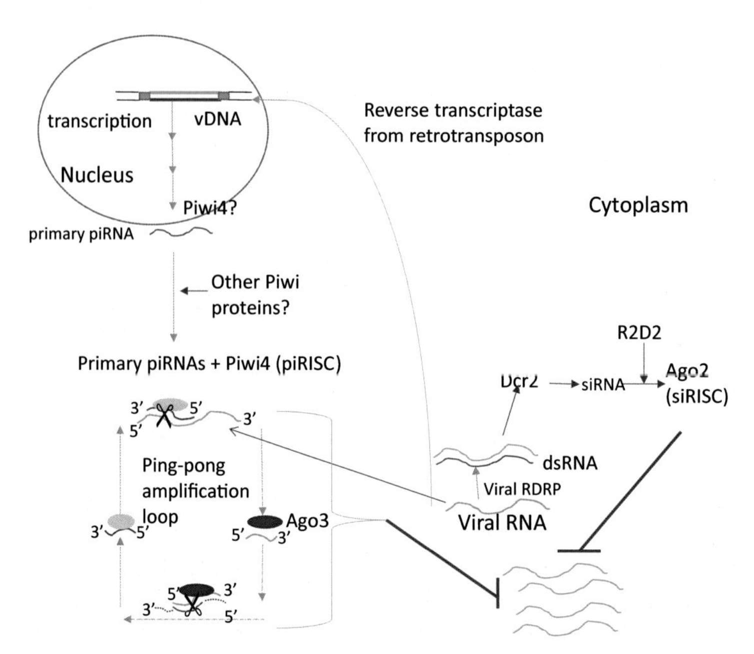

FIGURE 12.1 Antiviral small-interfering RNA (exo-siRNA) and Piwi-interacting RNA (piRNA) responses in *Aedes*–dengue virus interactions.

van Rij et al., 2006). Expression levels of these core components of the mosquito siRNA pathway do not appear to change greatly after a blood meal or arbovirus infection, suggesting that they are expressed constitutively (Bartholomay et al., 2010; Campbell et al., 2008a; Bonizzoni et al., 2011).

We first showed the replication-enhancing effect of knocking down expression of Ago2 and Dcr2 in *Anopheles gambiae* infected with o'nyong–nyong virus (ONNV, *Alphavirus*) (Keene et al., 2004). Although it is not clear that *A. gambiae* or any *Anopheles* spp. mosquito is an important vector of ONNV or any other arbovirus, the complete genome sequence of *A. gambiae* was the first full mosquito genome published (Holt et al., 2002). Our study and several subsequent studies suggested that the antiviral exo-siRNA response of *A. gambiae* is different from that of *Aedes* spp. and might be tailored for control of insect pathogens. *A. gambiae* exo-siRNA appears to be focused in tissues outside the midgut (Keene et al., 2004; Carissimo et al., 2015). Patterns of arbovirus infection and maintenance of arbovirus persistence also appear to be different from *Aedes* spp (Brault et al., 2004).

Publication of the *A. aegypti* genome sequence (Nene et al., 2007) allowed us to identify core components of the *A. aegypti* exo-siRNA pathway and show that Dcr2, Ago2, and R2D2 are required for robust antiviral defense against dengue virus type 2 (DENV-2) in its natural vector (Sánchez-Vargas et al., 2009). Importance of this innate immune pathway was shown by the observation that silencing of Dcr2 resulted in a 10-fold increase in DENV-2 titer in mosquitoes and a reduction in the extrinsic incubation period from 10 to 7 days. We also demonstrated that small RNAs derived from both the genome-sense and antisense viral RNA with a size consistent with siRNAs appeared after DENV infection (Sánchez-Vargas et al., 2009). Given our evidence that the exo-siRNA pathway was central to control of DENV in *Aedes* spp., we hypothesized that our observation that DENV replicated to 10- to 100-fold higher titers in cultured *A. albopictus* C6/36 cells than in *A. aegypti* Aag2 cells was due to a defect in the C6/36 cell exo-siRNA pathway. We characterized DENV-specific siRNAs (vsiRNAs) by next-generation sequencing and determined that DENV-2-infected Aag2 cells and *A. aegypti* mosquitoes produced a population of vsiRNAs that mapped relatively uniformly along the entire viral genome, contained approximately equal proportions of sense (genome)-strand and antisense reads, and were predominantly 21 nt in length (Scott et al., 2010), properties suggesting that they were the products of Dcr2 cleavage of replicating DENV RNA. In contrast, we showed that DENV genome-derived small RNAs in C6/36 cells were predominantly 27 nt in length, had approximately 96% positive polarity, and mapped to a few hot spots on the DENV genome, with properties suggestive of piRNAs. Importantly, we showed that C6/36 cells exhibit defective Dcr2 cleavage of DENV RNA because of a single nucleotide deletion in their *dicer2* gene, rendering it nonfunctional (Scott et al., 2010; Brackney et al., 2010). The more robust replication of DENV in exo-siRNA–deficient C6/36 mosquito cells emphasizes the importance of this pathway in *Aedes* spp. antiviral defense.

Despite the power of the exo-siRNA pathway, DENV infections are not cleared or prevented by *A. aegypti* unless the process is implemented immediately upon midgut exposure to the virus (Franz et al., 2006). Various mechanisms have been proposed for DENV evasion of RNAi. Unlike insect-pathogenic viruses, DENVs do not encode proteins with RNAi suppressor activity (VSRs). Following DENV infection of mosquito cells, rearrangement of intracellular membranes results in formation of double-membrane vesicles that enclose replication complexes, including the dsRNA-replicative intermediates (Junjhon et al., 2014). This sequestration of the PAMP that triggers the exo-siRNA response could be a strategy for counteracting mosquito defense. RNAi may drive positive selection of mutation-prone flavivirus genomes so that regions that are more highly targeted by siRNAs acquire more mutations than weakly targeted regions (Brackney et al., 2009). In addition, a highly structured noncoding subgenomic RNA (sfRNA) is generated from the 3′ UTR of DENV and other arthropod-borne flavivirus genomes after infection due to resistance to the cellular 5′-exonuclease Xrn-1 (Pijlman et al., 2008). This sfRNA could have roles in DENV RNAi evasion in *Aedes* spp. Degradation of DENV RNA after Ago2 cleavage is thought to be required for completion of exo-siRNA inhibition of DENV replication; however, if Xrn1 is retained by the nuclease-resistant sfRNA, a disruption of normal cellular mRNA stability, including virus genome degradation, ensues (Moon et al., 2012; Chapman et al., 2014). In addition, sfRNA has been implicated in directly interacting with and inhibiting the activity of Dcr2 and Ago2 (Schnettler et al., 2012; Moon et al., 2015).

The piRNA pathway. The third major RNAi pathway in mosquitoes results in biogenesis of 24–30 nt piRNAs. In *Drosophila*, production of mature piRNAs is effected by the three members of the Piwi subfamily of the Argonaute protein family, Piwi, Aubergine (Aub), and Ago3 and is Dcr independent (Brennecke et al., 2007; Gunawardane et al., 2007). These proteins are expressed almost exclusively in germ line cells, and their major function is to protect the genome by silencing transposable element (TE) activity. The substrates for production of primary piRNAs are nuclear transcripts with antisense polarity from discrete genomic loci called piRNA clusters that contain defective TE sequences. These are processed by a pathway that is not completely characterized, with the production of mature Piwi- or Aub-associated 24–27 nt primary piRNAs with 5′U (U1) and 3′O-methylation as the final step, which takes place in or near the nucleus. The antisense 24–27 nt RNA-loaded Piwi or Aub acts in a piRNA-induced silencing complex (piRISC) that hybridizes to and cleaves complementary sense-strand RNA in the cytoplasm. This "slicer" action of the piRISC can result in positive-sense, Ago3-bound piRNA to initiate a "ping-pong" amplification loop (Siomi et al., 2011).

The Piwi protein subfamily has undergone considerable expansion (to Ago3 and Piwi1-7) in *A. aegypti* (Campbell et al., 2008b), suggesting functional diversification of these gene products in mosquitoes. In addition, the piRNA pathway is active in somatic as well as germ line tissues of *A. aegypti* (Morazzani et al., 2012). Newly generated piRNAs specifically derived from the infecting virus genome occur after acute infections of *Aedes* spp. cells and mosquitoes by flaviviruses, alphaviruses, and bunyaviruses (Scott et al., 2010; Morazzani et al., 2012; Léger et al., 2013; Schnettler et al., 2013; Miesen et al., 2015, 2016). However, in contrast to siRNAs, during acute infection flavi- and alphavirus genome–derived piRNAs have almost exclusively positive (genome)-sense polarity (Scott et al., 2010; Brackney et al., 2010), and their potential role in antiviral defense is uncertain. In addition, the source of virus genome–derived piRNAs (vpiRNAs) and the mechanism(s) of their biogenesis is unknown.

We conducted recent studies based on the hypothesis that vpiRNAs, in addition to vsiRNAs, are part of the *A. aegypti* RNAi-mediated antiviral defense repertoire and that their source, like that of transposon-derived piRNAs, is nuclear transcripts (M. Kunitomi, K. Olson, C. Blair and R. Andino, unpub). We showed that transient knockdown of Piwi4 expression by intrathoracic injection of *piwi4*-cognate dsRNA into *A. aegypti* resulted in significant increases of DENV-2 genomic RNA in both the midgut and carcass at 7 and 10 days postinfection (dpi) as well as an increase of infectious virus titers in whole mosquitoes at 10 dpi, providing the first in vivo evidence of an anti-DENV function of the *A. aegypti* piRNA pathway.

The generation of antiviral vpiRNAs from cytoplasmic RNA virus genomes has been puzzling, since biogenesis of piRNA classically begins in the nucleus with the transcription of primary piRNA precursors from DNA loci defined as piRNA clusters (Iwasaki et al., 2015). A recent study by the Saleh research group (Goic et al., 2013) showed that during infection of *D. melanogaster* S2 cells, fragments of RNA virus genomes are reverse transcribed by endogenous reverse transcriptase (RT) and appear in nuclear DNA forms (vDNA). The vDNAs are templates for transcripts that are processed by RNAi machinery to generate vsiRNAs. These findings led us to examine the possibility that cytoplasmic arboviral RNA can be reverse transcribed by mosquito cell endogenous RT and deposited in the nucleus as vDNA, possibly in piRNA clusters, which could form the templates for transcription of primary vpiRNA precursors. We found that vDNA derived from both DENV and Sindbis virus (SINV) RNA genomes could be demonstrated shortly after infection of *A. aegypti* Aag2 cells by either virus and that the RT inhibitors AZT and Stavudine (d4T) blocked synthesis of infection-derived vDNA and inhibited biogenesis of vpiRNAs. In recent studies, Goic et al. (2016) have also demonstrated endogenous RT-dependent synthesis of vDNA in *A. albopictus* and *A. aegypti* after CHIKV or DENV infection and that treatment of infected cells with RT inhibitors reduced vpiRNAs. In contrast to what Goic et al. (2013) observed in *Drosophila* cells, we found that RT inhibitors did not reduce production of vsiRNAs in Aag2 cells, suggesting that biogenesis pathways and activities of antiviral small RNAs are not identical in mosquitoes and flies. Indeed, Saleh's research group has more recently shown that the piRNA pathway is not required for antiviral defense in *Drosophila* (Petit et al., 2016).

In acute DENV, as well as alphavirus, infections of *Aedes* spp. mosquitoes and mosquito cells, most vpiRNAs have positive polarity (Scott et al., 2010; Morazzani et al., 2012; Schnettler et al., 2013; Miesen et al., 2015), suggesting that they are secondary piRNAs derived from the infecting virus genome. However, in persistent infection as well as in uninfected *A. aegypti*, vpiRNAs occur at lower concentrations and have mostly antisense polarity, suggesting that they are primary piRNAs derived from transcripts of piRNA clusters (Arensburger et al., 2011). Viruslike sequences, termed endogenous viral elements (EVEs) or nonretroviral integrated RNA viruses (NIRVs), have been noted in the genomes of a number of *Aedes* spp. mosquitoes (Crochu et al., 2004; Arensburger et al., 2011; Chen et al., 2015) and have been shown to be frequently concentrated in piRNA clusters and to generate vpiRNAs (Arensburger et al., 2011; Chen et al., 2015; M. Bonizzoni, unpublished). We showed that Piwi4-dependent generation of vpiRNAs from EVEs in cultured *A. aegypti* cells suppressed the replication of the persistently infecting flavivirus cell fusing agent virus. Interestingly, a bioinformatics survey of publicly available mosquito genomes, including *A. aegypti* and 22 *Anopheles* spp., determined that EVEs occupy a higher proportion of the total genome in *A. aegypti* than in other mosquito species, especially genomes of nonarbovirus vectors (M. Kunitomi, unpub).

These studies have led us to formulate a model in which during acute infection of *A. aegypti* by DENV, vDNA is generated in somatic cells by endogenous RT, probably during a recombination event between DENV RNA and retrotransposon RNA (Olson and Blair, 2015). The vDNA is relocated to the nucleus and primary vpiRNA precursors are transcribed from this de novo synthesized vDNA, which might be episomal during acute infection. The vpiRNA precursors are processed into mature antiviral vpiRNA by components of the piRNA biogenesis pathway, including Piwi4 (Fig. 12.1). During acute infection, when DENV positive-sense genomes are abundant, these are targeted by primary, antigenome-sense piRNAs originating from vDNA transcripts to produce secondary, positive-strand vpiRNAs. As the DENV infection becomes persistent, primarily dampened by the vsiRNA response, the rate of new virus genome synthesis and the source of secondary vpiRNAs decreases, and the predominant vpiRNA population becomes negative-sense primary vpiRNAs transcribed from vDNA, which might have been integrated into piRNA clusters. Thus the vpiRNA pathway, along with the more vigorous vsiRNA pathway, is part of the *A. aegypti* antiviral defense system and might play an important role in modulating DENV replication as the infection persists. There remain a number of details of this proposed antiviral piRNA pathway, such as which additional Piwi proteins are involved in processing vpiRNAs, the mechanism and timing of conversion of vDNA to EVES, and DENV evasion strategies.

Demonstration of the antiviral role of the vpiRNA pathway, in addition to that of the vsiRNA pathway in mosquitoes, has suggested to us a resolution of an apparent contradiction in our published research on RNAi as a major antiviral defense mechanism in *Aedes* spp. mosquitoes. In earlier studies, we showed that infection of *A. aegypti* mosquitoes or *A. albopictus* C6/36 cells with a double-subgenomic SINV that expressed a 300–500 nt region of the DENV2 prM gene from its second subgenomic promoter resulted in complete inhibition of replication of subsequently infecting DENV-2 (Gaines et al., 1996; Olson et al., 1996; Adelman et al., 2001). We also showed that stable transformation of C6/36 cells with a plasmid designed to transcribe an inverted-repeat RNA (IR-RNA) derived from the genome of DENV-2, which was capable of forming dsRNA, allowed us to isolate clonal cell lines that were resistant to DENV-2 infection (Adelman et al., 2002). We attributed the DENV-2 resistance to the *A. aegypti* and C6/36 cell RNAi responses, triggered by dsRNA detection and processing in the siRNA pathway. However, our subsequent research showed that C6/36 cells have a single nucleotide deletion in their *dcr2* gene, resulting in a dysfunctional vsiRNA response to DENV-2 and other arbovirus infections (Scott et al., 2010; Brackney et al., 2010). Instead, C6/36 cells form vpiRNAs with characteristics of secondary piRNAs in response to flavivirus infection, and we hypothesized that this more robust piRNA pathway was used in antiviral defense to compensate for the ineffective vsiRNA pathway (Scott et al., 2010). We now assume that our ability to engineer DENV-resistant C6/36 cells by intracellular expression of DENV RNA–derived long dsRNA, either as part of a SINV RNA replicative intermediate or as a hairpin RNA transcript from a nuclear DNA, was due to use of the DENV RNA in RT-dependent synthesis of a vDNA template for vpiRNA biogenesis.

The reverse transcription of viral RNA genomes into vDNA upon infection also could provide mosquitoes with a mechanism of acquired adaptive immunological memory in the form of EVEs that could, if present in germ line cells, be inherited to provide antiviral immunity to mosquito progeny reminiscent of the clustered regularly interspaced short palindromic repeat (CRISPR)-Cas system in prokaryotes (Makarova et al., 2009).

STRATEGIES TO ENHANCE TARGETING OF DENGUE VIRUS REPLICATION IN MOSQUITOES

Considering the lack of effective vaccines and specific therapies, mosquito control is at present the most feasible and widely used method to limit transmission of DENV 1–4 by the mosquito vector, *A. aegypti*. Classical vector control efforts use insecticides, but these are problematic because of the possibility of adverse environmental effects and increasing insecticide resistance in mosquito populations (Saavedra-Rodriguez et al., 2015; Flores et al., 2013). Alternative technologies using strains of the bacterial endosymbiont *Wolbachia* combine a drive system (cytoplasmic incompatibility) with antiviral activity to control DENV prevalence (Hoffmann et al., 2011). This technology is promising, but it is unclear how long *Wolbachia* strains will persist in a mosquito population because of the possibility of development of resistance to the cytoplasmic incompatibility drive by target mosquitoes, or how readily targeted viruses will evolve resistance to the suppressive effects (Bull and Turelli, 2013). Combinations of these and other strategies likely will be needed to control dengue disease.

Several genetically modified mosquito (GMM) approaches have been pursued to suppress mosquito populations or control pathogen transmission. Two examples of GMMs designed to suppress vector populations are (1) release of insects with dominant lethality (RIDL) designed to kill *A. aegypti* at the pupal stage and (2) transgenic *A. gambiae* with a gene drive system based on CRISPR-associated protein 9 (Cas9) (CRISPR-Cas9) conferring a recessive phenotype that disrupts female mosquito reproduction (Hammond et al., 2016; Phuc et al., 2007). Population suppression of *A. aegypti* is an attractive control measure in the Americas because *A. aegypti* is an invasive species and has recently been involved in transmission of not only DENV and yellow fever virus, but also CHIKV and Zika viruses. However, RIDL requires repeated transient, large, inundative releases of mosquitoes (Wise de Valdez et al., 2011) and offers localized control at best. Although promising, suppression approaches using CRISPR-Cas9 have not been adapted to *A. aegypti* and are untested in the field.

Modeling of population suppression strategies with gene drive also suggests that the target mosquito population might only be reduced to an equilibrium value, determined by the balance between fitness cost of disrupting the target gene and rate of acquiring escape mutations (Hammond et al., 2016). In such cases, where suppressed target vector strains do not disappear, sufficient numbers of unmodified vectors may remain to continue disease transmission. There are also perceived general risks associated with abrupt elimination of a prevalent species from an environment whether the species is indigenous or not (Fang, 2010).

Two antiviral approaches have been successfully used to generate arbovirus-refractory transgenic *A. aegypti* in the absence of gene drive. These approaches express either dsRNAs or ribozymes that target the arbovirus genome with high specificity. The first approach expresses (from a nonautonomous transposon-based transforming plasmid) dsRNA derived from a specific virus genome sequence to direct the vector's innate exo-siRNA pathway to cleave the viral RNA at defined sequence sites. Ribozymes are RNA molecules capable of catalyzing specific biochemical reactions to cleave the target RNA virus genome. The result of these strategies is inhibition of virus replication in the vector. This is preferable to manipulating host genes controlling infection that may disrupt vector fitness. Two examples of these antiviral strategies follow.

Our research group has engineered a genetically modified mosquito line (Carb109M) designed to express an antiviral IR-RNA targeting a 568-nt region of the DENV-2 RNA genome (Franz et al., 2014). The IR-RNA was designed as a hairpin structure transcript that forms dsRNA and is processed via the mosquito exo-siRNA pathway. Expression of DENV-2-specific dsRNA in midgut epithelial cells is induced by ingestion of a blood meal, thus Carb109M mosquitoes receiving a noninfectious blood meal express the IR-RNA and generate DENV-specific siRNAs 21 nt in length that map to the target region (Franz et al., 2006, 2014). The Carb109M line has remained refractory at 98% resistance to DENV-2 for 42 generations. Additionally, the Carb109M transgene was introgressed by a series of backcrosses into the background of a genetically diverse laboratory strain (GDLS) composed of 10 *A. aegypti* populations collected in southern Mexico (Wise de Valdez et al., 2011; Franz et al., 2014; Wise de Valdez et al., 2010). This effort was significant because the transgene was active in a genetic background more representative of field populations. The GDLS population was highly competent for DENV-2 before introgression of the transgene but refractory after introgression. In cage colonies, observed allele frequencies in Carb109 M/GDLS mosquitoes after five consecutive inbred generations matched expected frequencies based on Fisher's model for natural selection (Fisher, 1950) as expected for an element without gene drive. If linked to a CRISPR-Cas9 drive, we predict that the gene frequency of the IR-RNA will increase in a non-Mendelian fashion in wild-type populations to form a DENV2 refractory population. We have clearly shown that any antiarbovirus strategy will need to be deployed in the target vector population at efficiencies approaching 100% of the wild-type mosquitoes to impact virus transmission in the field when solely relying on a population replacement strategy. Test results with Carb109 M mosquitoes, even without a gene drive, indicate that this may be feasible.

While an exo-siRNA-based antiviral strategy could potentially target and eliminate the transmission cycle for DENV2, the IR-RNAs targeting DENV2 RNA were not successful at targeting DENV-1, DENV-3, DENV-4, and other arboviruses because of their considerable sequence divergence at the target site. A second approach uses ribozymes [transsplicing Group I introns (Grp1I)], which can be targeted to destroy exposed viral RNA genomes during their replication (Franz et al., 2015; Mishra et al., 2016; Carter et al., 2015). Highly effective Grp1I catalytic RNAs are activated by a target sequence (minimum of 9 nt) shorter than siRNA strategies. Thus, a Grp1I could be more easily designed to target conserved regions represented in all four DENV serotypes. Grp1I consist of an external and an internal guide sequence, a transsplicing domain, and a 3′ exon-of-choice. The catalytic reaction of the Grp1I involves two transesterification steps, one that cleaves the target RNA and a second that ligates the 5′ region of the target sequence to the 3′ (heterologous) exon (Mishra et al., 2016). After stable transformation to introduce the Grp1I construct designed to target DENVs into cultured *A. albopictus* C6/36 cells, researchers were able to achieve complete elimination of all four DENV serotypes (Carter et al., 2010). Carter et al. (2015) also demonstrated that they could engineer a dual-acting Grp1I ribozyme that targets DENV and CHIKV RNAs in a sequence-specific manner to inhibit virus replication of both arboviruses in mosquito cells. Mishra et al. (2016) have engineered and tested a hammerhead ribozyme targeting CHIKV in transgenic *A. aegypti*, demonstrating effective suppression of arbovirus replication in transgenic mosquitoes.

CRISPR-Cas9–based gene drive is attractive for spreading antiviral genes in mosquito populations in a super-Mendelian fashion. In this concept the mosquito population would not be suppressed or eliminated but would be replaced with an arbovirus-resistant population, thus eliminating concerns about the effects of *A. aegypti* extinction on the environment. While the concern about development of escape mutations is also potentially true for approaches to genetically engineer DENV-resistant vectors, the autonomous CRISPR-Cas9 drive system could contain multiple anti-DENV gene cassettes (Gantz et al., 2015) to limit any breakdown of resistance. CRISPR-Cas9, in theory, also can be recalled in the target population should resistance appear (Esvelt et al., 2014). An advantage of a CRISPR-Cas9 gene drive is that it is not self-limiting but is expected to spread effector genes throughout the population in a relatively few generations to suppress DENV transmission without *A. aegypti* population reduction. Once the drive system is optimized, it should be straightforward to spread new effector genes into the population to target new DENV genotypes or even new species of arboviruses.

Researchers developed a very efficient CRISPR-Cas9 drive adapted from a highly effective system, mutagenic chain reaction (Gantz and Bier, 2015), to drive target-specific gene conversion at ≥99.5% efficiency in transgene heterozygotes of *Anopheles stephensi* (Gantz et al., 2015). They showed that CRISPR-Cas9, when restricted to the germ line of *A. stephensi*, copies a genetic element from one chromosome to its homolog with ≥98% efficiency while maintaining the transcriptional activity of the introgressed effector gene cassettes. Modeling of gene drive systems such as CRISPR-Cas9, which greatly exceed rates of Mendelian inheritance, shows rapid population-level transformation that could support sustainable local arboviral disease elimination at much reduced costs (Robert et al., 2014).

Gene drive systems are self-perpetuating gene editing systems that bias inheritance in populations through sexual reproduction. Endonuclease RNA–guided drives such as CRISPR-Cas9 have been described for the genetic alteration of wild-type mosquito populations (Gantz et al., 2015; Esvelt et al., 2014). Briefly, Cas9 works by cutting target sequences flanked by a protospacer-adjacent motif (PAM) at the 3′ end. Cas9 is directed to cut a specific (protospacer) sequence in the genome using only a single, chimeric guide RNA (gRNA) less than 100 base pairs in length (Jinek et al., 2012). The gRNA begins with a 17–20 nucleotide "spacer" complementary to the target gene in the genome (Fu et al., 2014). Using appropriate promoters for Cas9 expression, the drive system could be designed to work through the male germ lines of mosquitoes to enhance homology-directed repair (HDR) of the cleavage event. When a transgenic mosquito-carrying CRISPR-Cas9 drive mates with a wild-type mosquito, the Cas9 drive would be inherited by all offspring and would enable the transgene to rapidly spread through the wild-type population. The Cas9 drive is preferentially inherited because Cas9 cuts the homologous wild-type chromosome (Esvelt et al., 2014), and the break is repaired in the germ line by homologous recombination using the gene drive chromosome as a repair template, thereby copying the drive onto the wild-type chromosome. Experiments in both *Drosophila* and mosquitoes reveal that HDR greatly dominates over nonhomologous end joining (NHEJ), which impairs gene drive when repairing double-strand breaks in male germ line lineages (Gantz et al., 2015; Gantz and Bier, 2015). The system as designed could carry a relatively large gene set (~17 kb with antiviral genes) that is transcriptionally active following movement. Transgenic *A. aegypti* based on this technology could play a major role in sustaining DENV disease control.

SUMMARY AND FUTURE DIRECTIONS

DENV is the most prevalent arbovirus worldwide, estimated to cause 390 million infections annually, with about 96 million of these resulting in apparent disease. The serious burden on human health of this arboviral disease requires development of strategies for disease control/reduction. No effective vaccines or specific therapies are available, thus control of DENV transmission by mosquitoes is currently the most vulnerable target for disease reduction. *A. aegypti*, an anthropophilic mosquito, is the major vector of DENV. The epidemic DENV transmission cycle involves only *A. aegypti* and humans, with no other vertebrate reservoir.

During the last 20 years, our advances in understanding the ecology, genetics, and molecular biology of mosquito–arbovirus interactions have suggested novel strategies for targeting dengue disease in the mosquito.

A. aegypti acquires DENV in a blood meal from a viremic human, DENV infects the midgut epithelial cells, replicates, and disseminates to infect the salivary glands, and after this extrinsic incubation period, she can transmit DENV each time she takes another blood meal. Natural antiviral defense mechanisms in competent *A. aegypti* control but do not terminate DENV replication, resulting in a persistent, life-long, nonpathogenic infection. Activation of classic innate immune pathways, which occur in both mammals and insects, has some role in controlling the level of DENV replication in *A. aegypti*; however, the arthropod-specific features of the RNAi pathways are demonstrably *A. aegypti*'s most powerful antiviral defense.

The exogenous siRNA (exo-siRNA) pathway is the principal antiviral RNAi mechanism. It is activated during acute infection by intracellular recognition and processing of DENV genome-derived dsRNA, and its impairment by silencing of the key enzyme Dcr2 results in at least 10-fold higher DENV-2 titers in *A. aegypti* at 7 days after oral infection than in control mosquitoes. More recently, the piRNA pathway has also been shown to have a significant role in mosquito antiviral defense, probably in modulating persistent infection of *Aedes* spp. mosquitoes. The vDNA reverse transcribed from the RNA genome after DENV infection of *A. aegypti* acts as a nuclear template for host transcription of vpiRNA precursors and could provide heritable virus-specific immunity in germ cells. Further exploration of the *A. aegypti* antiviral piRNA system, including complete information about which members of the expanded *A. aegypti* Piwi family protein repertoire are involved and the potential role of EVEs in the *Aedes* genome in vpiRNA-mediated defense in acute as well as persistent infections remains to be carried out.

Mosquito genetic modification strategies have been devised to control pathogen transmission by either mosquito population reduction or population replacement. More persuasive arguments can be made favoring the latter strategy.

Our research group has developed genetically modified *A. aegypti* using methods that harness the mosquitoes' exo-siRNA pathway to create DENV-refractory mosquitoes. Transgenic *A. aegypti* express a DENV genome–derived long dsRNA upon ingestion of a blood meal. Processing of the transgenic dsRNA by the exo-siRNA pathway produces vsiRNAs that render the mosquitoes completely resistant to infection by DENV in the blood meal. The refractory transgene was introgressed into the background of DENV-susceptible, genetically diverse mosquitoes that are representative of field populations, and they became refractory at efficiencies predicted by natural selection models. If the transgene were linked to a gene drive such as CRISPR-Cas9, we predict introgression of the transgene into wild-type populations in a non-Mendelian fashion, at efficiencies approaching 100%, to replace the wild mosquitoes with completely DENV-2 refractory populations. However, since the exo-siRNA response is highly sequence specific, the DENV-2-resistant mosquitoes are expected to still be susceptible to infection by other DENV serotypes.

Future studies to gain a more detailed understanding of DENV–*A. aegypti* interactions, including relative importance of the piRNA pathway in maintaining the DENV transmission cycle, and DENV mechanisms for evading the antiviral RNAi response, will be needed to improve strategies for targeting DENV in the mosquito.

References

Adelman, Z.N., Blair, C.D., Carlson, J.O., Beaty, B.J., Olson, K.E., 2001. *Sindbis* virus-induced silencing of dengue viruses in mosquitoes. Insect. Mol. Biol. 10, 265–273.

Adelman, Z.N., Sanchez-Vargas, I., Travanty, E.A., Carlson, J.O., Beaty, B.J., Blair, C.D., Olson, K.E., 2002. RNA silencing of dengue virus type 2 replication in transformed C6/36 mosquito cells transcribing an inverted-repeat RNA derived from the virus genome. J. Virol. 76, 12925–12933.

Arensburger, P., Hice, R.H., Wright, J.A., Craig, N.L., Atkinson, P.W., 2011. The mosquito *Aedes aegypti* has a large genome size and high transposable element load but contains a low proportion of transposon-specific piRNAs. BMC Genomics 12, 606.

Armstrong, P.M., Rico-Hesse, R., 2001. Differential susceptibility of *Aedes aegypti* to infection by the American and Southeast Asian genotypes of dengue type 2 virus. Vector-Borne Zoonotic Dis. 1, 159–168.

Barrett, A.D.T., Higgs, S., 2007. Yellow fever: a disease that has yet to be conquered. Annu. Rev. Entomol. 52, 209–229.

Bartel, D.P., 2004. MicroRNAs: genomics, biogenesis, mechanism, and function. Cell 116, 281–297.

Bartholomay, L.C., Waterhouse, R.M., Mayhew, G.F., Campbell, C.L., Michel, K., Zou, Z., Ramirez, J.L., Das, S., Alvarez, K., Arensburger, P., Bryant, B., Chapman, S.B., Dong, Y., Erickson, S.M., Karunaratne, S.H.P.P., Kokoza, V., Kodira, C.D., Pignatelli, P., Shin, S.W., Vanlandingham, D.L., Atkinson, P.W., Birren, B., Christophides, G.K., Clem, R.J., Hemingway, J., Higgs, S., Megy, K., Ranson, H., Zdobnov, E.M., Raikhel, A.S., Christensen, B.M., Dimopoulos, G., Muskavitch, M.A.T., 2010. Pathogenomics of *Culex quinquefasciatus* and meta-analysis of infection responses to diverse pathogens. Science 330, 88–90.

Beatty, M.E., Beutels, P., Meltzer, M.I., Shepard, D.S., Hombach, J., Hutubessy, R., Dessis, D., Coudeville, L., Dervaux, B., Wichmann, O., Margolis, H.S., Kuritsky, J.N., 2011. Health economics of dengue: a systematic literature review and expert panel's assessment. Am. J. Trop. Med. Hyg. 84, 473–488.

Behura, S.K., Gomez-Machorro, C., Harker, B.W., deBruyn, B., Lovin, D.D., Hemme, R.R., Mori, A., Romero-Severson, J., Severson, D.W., 2011. Global cross-talk of genes of the mosquito *Aedes aegypti* in response to dengue virus infection. PLoS Negl. Trop. Dis. 5, e1385.

Behura, S., Gomez-Machorro, C., deBruyn, B., Lovin, D., Harker, B., Romero-Severson, J., Mori, A., Severson, D., 2014. Influence of mosquito genotype on transcriptional response to dengue virus infection. Funct. Integr. Genomics 14, 581–589.

Bennett, K., Olson, K., Muñoz, M.L., Fernandez-Salas, I., Farfan-Ale, J., Higgs, S., Black 4th, W.C., Beaty, B., 2002. Variation in vector competence for dengue 2 virus among 24 collections of *Aedes aegypti* from Mexico and the United States. Am. J. Trop. Med. Hyg. 67, 85–92.

Bhatt, S., Gething, P.W., Brady, O.J., Messina, J.P., Farlow, A.W., Moyes, C.L., Drake, J.M., Brownstein, J.S., Hoen, A.G., Sankoh, O., Myers, M.F., George, D.B., Jaenisch, T., Wint, G.R.W., Simmons, C.P., Scott, T.W., Farrar, J.J., Hay, S.I., 2013. The global distribution and burden of dengue. Nature 496, 504–507.

Black, W.C.I., Bennett, K.E., Gorrochotegui-Escalante, N., Barillas-Mury, C.V., Fernandez-Salas, I., de Lourdes Munoz, M., Farfan-Ale, J.A., Olson, K.E., Beaty, B.J., 2002. Flavivirus susceptibility in *Aedes aegypti*. Arch. Med. Res. 33, 379–388.

Blair, C.D., Olson, K.E., 2015. The role of RNA interference (RNAi) in arbovirus-vector interactions. Viruses 7, 820–843.

Blair, C.D., 2011. Mosquito RNAi is the major innate immune pathway controlling arbovirus infection and transmission. Future Microbiol. 6, 265–277.

Bonizzoni, M., Dunn, W., Campbell, C., Olson, K., Dimon, M., Marinotti, O., James, A., 2011. RNA-seq analyses of blood-induced changes in gene expression in the mosquito vector species, *Aedes aegypti*. BMC Genomics 12, 82.

Brackney, D.E., Beane, J.E., Ebel, G.D., 2009. RNAi targeting of west nile virus in mosquito midguts promotes virus diversification. PLoS Pathog. 5, e1000502.

Brackney, D.E., Scott, J.C., Sagawa, F., Woodward, J.E., Miller, N.A., Schilkey, F.D., Mudge, J., Wilusz, J., Olson, K.E., Blair, C.D., Ebel, G.D., 2010. C6/36 *Aedes albopictus* cells have a dysfunctional antiviral RNA interference response. PLoS Negl. Trop. Dis. 4, e856.

Brault, A.C., Foy, B.D., Myles, K.M., Kelly, C.L.H., Higgs, S., Weaver, S.C., Olson, K.E., Miller, B.R., Powers, A.M., 2004. Infection patterns of o'nyong nyong virus in the malaria-transmitting mosquito, *Anopheles gambiae*. Insect Mol. Biol. 13, 625–635.

Brennecke, J., Aravin, A.A., Stark, A., Dus, M., Kellis, M., Sachidanandam, R., Hannon, G.J., 2007. Discrete small RNA-generating loci as master regulators of transposon activity in *Drosophila*. Cell 128, 1089–1103.

Bronkhorst, A.W., van Cleef, K.W.R., Vodovar, N., İnce, İ.A., Blanc, H., Vlak, J.M., Saleh, M.-C., van Rij, R.P., 2012. The DNA virus invertebrate iridescent virus 6 is a target of the *Drosophila* RNAi machinery. Proc. Natl. Acad. Sci. 109, E3604–E3613.

Bryant, B., Blair, C.D., Olson, K.E., Clem, R.J., 2008. Annotation and expression profiling of apoptosis-related genes in the yellow fever mosquito, *Aedes aegypti*. Insect Biochem. Mol. Biol. 38, 331–345.

Bull, J.J., Turelli, M., 2013. *Wolbachia* versus dengue: evolutionary forecasts. Evol. Med. Public Health 2013, 197–207.

Butrapet, S., Childers, T., Moss, K.J., Erb, S.M., Luy, B.E., Calvert, A.E., Blair, C.D., Roehrig, J.T., Huang, C.Y.H., 2011. Amino acid changes within the E protein hinge region that affect dengue virus type 2 infectivity and fusion. Virology 413, 118–127.

Campbell, C.L., Keene, K.M., Brackney, D.E., Olson, K.E., Blair, C.D., Wilusz, J., Foy, B.D., 2008a. *Aedes aegypti* uses RNA interference in defense against *Sindbis* virus infection. BMC Microbiol. 8, 47.

Campbell, C.L., Black, W.C.t., Hess, A.M., Foy, B.D., 2008b. Comparative genomics of small RNA regulatory pathway components in vector mosquitoes. BMC Genomics 9, 425.

Campbell, C.L., Harrison, T., Hess, A.M., Ebel, G.D., 2014. MicroRNA levels are modulated in *Aedes aegypti* after exposure to Dengue-2. Insect Mol. Biol. 23, 132–139.

Capeding, M.R., Tran, N.H., Hadinegoro, S.R.S., Ismail, H.I.H.J.M., Chotpitayasunondh, T., Chua, M.N., Luong, C.Q., Rusmil, K., Wirawan, D.N., Nallusamy, R., Pitisuttithum, P., Thisyakorn, U., Yoon, I.-K., van der Vliet, D., Langevin, E., Laot, T., Hutagalung, Y., Frago, C., Boaz, M., Wartel, T.A., Tornieporth, N.G., Saville, M., Bouckenooghe, A., 2014. Clinical efficacy and safety of a novel tetravalent dengue vaccine in healthy children in Asia: a phase 3, randomised, observer-masked, placebo-controlled trial. Lancet 384, 1358–1365.

Carissimo, G., Pondeville, E., McFarlane, M., Dietrich, I., Mitri, C., Bischoff, E., Antoniewski, C., Bourgouin, C., Failloux, A.-B., Kohl, A., Vernick, K.D., 2015. Antiviral immunity of *Anopheles gambiae* is highly compartmentalized, with distinct roles for RNA interference and gut microbiota. Proc. Natl. Acad. Sci. 112, E176–E185.

Carter, J., Keith, J., Barde, P., Fraser, T., Fraser, M., 2010. Targeting of highly conserved dengue virus sequences with anti-dengue virus trans-splicing group I introns. BMC Mol. Biol. 11, 84.

Carter, J.R., Taylor, S., Fraser, T.S., Kucharski, C.A., Dawson, J.L., Fraser Jr., M.J., 2015. Suppression of the arboviruses dengue and chikungunya using a dual-acting Group-I intron coupled with conditional expression of the Bax C-terminal domain. PLoS One 10, e0139899.

Chapman, E.G., Costantino, D.A., Rabe, J.L., Moon, S.L., Wilusz, J., Nix, J.C., Kieft, J.S., 2014. The structural basis of pathogenic subgenomic flavivirus RNA (sfRNA) production. Science 344, 307–310.

Chen, X.-G., Jiang, X., Gu, J., Xu, M., Wu, Y., Deng, Y., Zhang, C., Bonizzoni, M., Dermauw, W., Vontas, J., Armbruster, P., Huang, X., Yang, Y., Zhang, H., He, W., Peng, H., Liu, Y., Wu, K., Chen, J., Lirakis, M., Topalis, P., Van Leeuwen, T., Hall, A.B., Jiang, X., Thorpe, C., Mueller, R.L., Sun, C., Waterhouse, R.M., Yan, G., Tu, Z.J., Fang, X., James, A.A., 2015. Genome sequence of the Asian Tiger mosquito, *Aedes albopictus*, reveals insights into its biology, genetics, and evolution. Proc. Natl. Acad. Sci. 112, E5907–E5915.

Clem, R.J., 2016. Arboviruses and apoptosis: the role of cell death in determining vector competence. J. Gen. Virol. 97, 1033–1036.

Crochu, S., Cook, S., Attoui, H., Charrel, R.N., De Chesse, R., Belhouchet, M., Lemasson, J.-J., de Micco, P., de Lamballerie, X., 2004. Sequences of flavivirus-related RNA viruses persist in DNA form integrated in the genome of *Aedes* spp. mosquitoes. J. Gen. Virol. 85, 1971–1980.

Czech, B., Zhou, R., Erlich, Y., Brennecke, J., Binari, R., Villalta, C., Gordon, A., Perrimon, N., Hannon, G.J., 2009. Hierarchical rules for argonaute loading in *Drosophila*. Mol. Cell 36, 445–456.

Erb, S.M., Butrapet, S., Moss, K.J., Luy, B.E., Childers, T., Calvert, A.E., Silengo, S.J., Roehrig, J.T., Huang, C.Y.H., Blair, C.D., 2010. Domain-III FG loop of the dengue virus type 2 envelope protein is important for infection of mammalian cells and *Aedes aegypti* mosquitoes. Virology 406, 328–335.

Esvelt, K.M., Smidler, A.L., Catteruccia, F., Church, G.M., 2014. Concerning RNA-guided gene drives for the alteration of wild populations. eLife e03401. http://dx.doi.org/10.7554/eLife.03401.

Fang, J., 2010. Ecology: a world without mosquitoes. Nature 466, 432–434.

Fansiri, T., Fontaine, A., Diancourt, L., Caro, V., Thaisomboonsuk, B., Richardson, J.H., Jarman, R.G., Ponlawat, A., Lambrechts, L., 2013. Genetic mapping of specific interactions between *Aedes aegypti* mosquitoes and dengue viruses. PLoS Genet. 9, e1003621.

Fire, A., Xu, S., Montgomery, M.K., Kostas, S.A., Driver, S.E., Mello, C.C., 1998. Potent and specific genetic interference by double-stranded RNA in *Caenorhabditis elegans*. Nature 391, 806–811.

Fisher, R.A., 1950. Gene frequencies in a cline determined by selection and diffusion. Biometrics 6, 353–361.

Flores, A.E., Ponce, G., Silva, B.G., Gutierrez, S.M., Bobadilla, C., Lopez, B., Mercado, R., Black, W.C., 2013. Wide spread cross resistance to pyrethroids in *Aedes aegypti*; (Diptera: Culicidae) from Veracruz state Mexico. J. Econ. Entomol. 106, 959–969.

Franz, A.W.E., Sanchez-Vargas, I., Adelman, Z.N., Blair, C.D., Beaty, B.J., James, A.A., Olson, K.E., 2006. Engineering RNA interference-based resistance to dengue virus type 2 in genetically modified *Aedes aegypti*. Proc. Natl. Acad. Sci. U.S.A. 103, 4198–4203.

Franz, A.W.E., Sanchez-Vargas, I., Raban, R.R., Black, W.C.I., James, A.A., Olson, K.E., 2014. Fitness impact and stability of a transgene conferring resistance to Dengue-2 virus following introgression into a genetically diverse *Aedes aegypti* strain. PLoS Negl. Trop. Dis. 8, e2833.

Franz, A.W.E., Balaraman, V., Fraser Jr., M.J., 2015. Disruption of dengue virus transmission by mosquitoes. Curr. Opin. Insect Sci. 8, 88–96.

Fu, Y., Sander, J.D., Reyon, D., Cascio, V.M., Joung, J.K., 2014. Improving CRISPR-Cas nuclease specificity using truncated guide RNAs. Nat. Biotech. 32, 279–284.

Gaines, P.J., Olson, K.E., Higgs, S., Powers, A.M., Beaty, B.J., Blair, C.D., 1996. Pathogen-derived resistance to dengue type 2 virus in mosquito cells by expression of the premembrane coding region of the viral genome. J. Virol. 70, 2132–2137.

Galiana-Arnoux, D., Dostert, C., Schneemann, A., Hoffmann, J.A., Imler, J.-L., 2006. Essential function in vivo for Dicer-2 in host defense against RNA viruses in drosophila. Nat. Immunol. 7, 590–597.

Gantz, V.M., Bier, E., 2015. The mutagenic chain reaction: a method for converting heterozygous to homozygous mutations. Science 348, 442–444.

Gantz, V.M., Jasinskiene, N., Tatarenkova, O., Fazekas, A., Macias, V.M., Bier, E., James, A.A., 2015. Highly efficient Cas9-mediated gene drive for population modification of the malaria vector mosquito *Anopheles stephensi*. Proc. Natl. Acad. Sci. 112, E6736–E6743.

Ghildiyal, M., Zamore, P.D., 2009. Small silencing RNAs: an expanding universe. Nat. Rev. Genet. 10.

Goic, B., Vodovar, N., Mondotte, J.A., Monot, C., Frangeul, L., Blanc, H., Gausson, V., Vera-Otarola, J., Cristofari, G., Saleh, M.-C., 2013. RNA-mediated interference and reverse transcription control the persistence of RNA viruses in the insect model *Drosophila*. Nat. Immunol. 14, 396–403.

Goic, B., Stapleford, K.A., Frangeul, L., Doucet, A.J., Gausson, V., Blanc, H., Schemmel-Jofre, N., Cristofari, G., Lambrechts, L., Vignuzzi, M., Saleh, M.-C., 2016. Virus-derived DNA drives mosquito vector tolerance to arboviral infection. Nat. Commun. 7, 12410.

Gubler, D.J., 1998. Dengue and dengue hemorrhagic fever. Clin. Microbiol. Rev. 11, 480–496.

Gubler, D.J., 2004. The changing epidemiology of yellow fever and dengue, 1900 to 2003: full circle? Comparative immunology. Microbiology and Infectious Dis. 27, 319–330.

Gunawardane, L.S., Saito, K., Nishida, K.M., Miyoshi, K., Kawamura, Y., Nagami, T., Siomi, H., Siomi, M.C., 2007. A slicer-mediated mechanism for repeat-associated siRNA 5′ end formation in *Drosophila*. Science 315, 1587–1590.

Halstead, S., 1988. Pathogenesis of dengue: challenges to molecular biology. Science 239, 476–481.

Halstead, S.B., 2016. Licensed dengue vaccine: public health conundrum and scientific challenge. Am. J. Trop. Med. Hyg. 95, 741–745

Hammond, A., Galizi, R., Kyrou, K., Simoni, A., Siniscalchi, C., Katsanos, D., Gribble, M., Baker, D., Marois, E., Russell, S., Burt, A., Windbichler, N., Crisanti, A., Nolan, T., 2016. A CRISPR-Cas9 gene drive system targeting female reproduction in the malaria mosquito vector *Anopheles gambiae*. Nat. Biotech. 34, 78–83.

Hanley, K.A., Monath, T.P., Weaver, S.C., Rossi, S.L., Richman, R.L., Vasilakis, N., 2013. Fever versus fever: the role of host and vector susceptibility and interspecific competition in shaping the current and future distributions of the sylvatic cycles of dengue virus and yellow fever virus. Infect. Genet. Evol. 19, 292–311.

Hardy, J.L., Houk, E.J., Kramer, L.D., Reeves, W.C., 1983. Intrinsic factors affecting vector competence of mosquitoes for arboviruses. Annu. Rev. Entomol. 28, 229–262.

Hoffmann, A.A., Montgomery, B.L., Popovici, J., Iturbe-Ormaetxe, I., Johnson, P.H., Muzzi, F., Greenfield, M., Durkan, M., Leong, Y.S., Dong, Y., Cook, H., Axford, J., Callahan, A.G., Kenny, N., Omodei, C., McGraw, E.A., Ryan, P.A., Ritchie, S.A., Turelli, M., O'Neill, S.L., 2011. Successful establishment of *Wolbachia* in *Aedes* populations to suppress dengue transmission. Nature 476, 454–457.

Holt, R.A., Subramanian, G.M., Halpern, A., Sutton, G.G., Charlab, R., Nusskern, D.R., Wincker, P., Clark, A.G., Ribeiro, J.M.C., Wides, R., Salzberg, S.L., Loftus, B., Yandell, M., Majoros, W.H., Rusch, D.B., Lai, Z., Kraft, C.L., Abril, J.F., Anthouard, V., Arensburger, P., Atkinson, P.W., Baden, H., de Berardinis, V., Baldwin, D., Benes, V., Biedler, J., Blass, C., Bolanos, R., Boscus, D., Barnstead, M., Cai, S., Center, A., Chatuverdi, K., Christophides, G.K., Chrystal, M.A., Clamp, M., Cravchik, A., Curwen, V., Dana, A., Delcher, A., Dew, I., Evans, C.A., Flanigan, M., Grundschober-Freimoser, A., Friedli, L., Gu, Z., Guan, P., Guigo, R., Hillenmeyer, M.E., Hladun, S.L., Hogan, J.R., Hong, Y.S., Hoover, J., Jaillon, O., Ke, Z., Kodira, C., Kokoza, E., Koutsos, A., Letunic, I., Levitsky, A., Liang, Y., Lin, J.-J., Lobo, N.F., Lopez, J.R., Malek, J.A., McIntosh, T.C., Meister, S., Miller, J., Mobarry, C., Mongin, E., Murphy, S.D., O'Brochta, D.A., Pfannkoch, C., Qi, R., Regier, M.A., Remington, K., Shao, H., Sharakhova, M.V., Sitter, C.D., Shetty, J., Smith, T.J., Strong, R., Sun, J., Thomasova, D., Ton, L.Q., Topalis, P., Tu, Z., Unger, M.F., Walenz, B., Wang, A., Wang, J., Wang, M., Wang, X., Woodford, K.J., Wortman, J.R., Wu, M., Yao, A., Zdobnov, E.M., Zhang, H., Zhao, Q., Zhao, S., Zhu, S.C., Zhimulev, I., Coluzzi, M., della Torre, A., Roth, C.W., Louis, C., Kalush, F., Mural, R.J., Myers, E.W., Adams, M.D., Smith, H.O., Broder, S., Gardner, M.J., Fraser, C.M., Birney, E., Bork, P., Brey, P.T., Venter, J.C., Weissenbach, J., Kafatos, F.C., Collins, F.H., Hoffman, S.L., 2002. The genome sequence of the malaria mosquito *Anopheles gambiae*. Science 298, 129–149.

Iwasaki, Y.W., Siomi, M.C., Siomi, H., 2015. Piwi-interacting RNA: its biogenesis and functions. Annu. Rev. Biochem. 84, 405–433.

Jinek, M., Chylinski, K., Fonfara, I., Hauer, M., Doudna, J.A., Charpentier, E., 2012. A programmable dual-RNA–guided DNA endonuclease in adaptive bacterial immunity. Science 337, 816–821.

Junjhon, J., Pennington, J.G., Edwards, T.J., Perera, R., Lanman, J., Kuhn, R.J., 2014. Ultrastructural characterization and three-dimensional architecture of replication sites in dengue virus-infected mosquito cells. J. Virol. 88, 4687–4697.

Keene, K.M., Foy, B.D., Sanchez-Vargas, I., Beaty, B.J., Blair, C.D., Olson, K.E., 2004. RNA interference acts as a natural antiviral response to O'nyong-nyong virus (*Alphavirus; Togaviridae*) infection of *Anopheles gambiae*. Proc. Natl. Acad. Sci. U.S.A. 101, 17240–17245.

Kemp, C., Mueller, S., Goto, A., Barbier, V., Paro, S., Bonnay, F., Dostert, C., Troxler, L., Hetru, C., Meignin, C., Pfeffer, S., Hoffmann, J.A., Imler, J.-L., 2013. Broad RNA interference–mediated antiviral immunity and virus-specific inducible responses in *Drosophila*. J. Immunol. 190, 650–658.

Kennerdell, J.R., Carthew, R.W., 1998. Use of dsRNA-mediated genetic interference to demonstrate that frizzled and frizzled 2 act in the Wingless pathway. Cell 95, 1017–1026.

Lambrechts, L., Chevillon, C., Albright, R., Thaisomboonsuk, B., Richardson, J., Jarman, R., Scott, T., 2009. Genetic specificity and potential for local adaptation between dengue viruses and mosquito vectors. BMC Evol. Biol. 9, 160.

Léger, P., Lara, E., Jagla, B., Sismeiro, O., Mansuroglu, Z., Coppée, J.Y., Bonnefoy, E., Bouloy, M., 2013. Dicer-2- and piwi-mediated RNA interference in Rift valley fever virus-infected mosquito cells. J. Virol. 87, 1631–1648.

Liu, Q., Rand, T.A., Kalidas, S., Du, F., Kim, H.-E., Smith, D.P., Wang, X., 2003. R2D2, a bridge between the initiation and effector steps of the *Drosophila* RNAi pathway. Science 301, 1921–1925.

Luplertlop, N., Surasombatpattana, P., Patramool, S., Dumas, E., Wasinpiyamongkol, L., Saune, L., Hamel, R., Bernard, E., Sereno, D., Thomas, F., Piquemal, D., Yssel, H., Briant, L., Missé, D., 2011. Induction of a peptide with activity against a broad spectrum of pathogens in the *Aedes aegypti* salivary gland, following infection with dengue virus. PLoS Pathog. 7, e1001252.

Makarova, K., Wolf, Y., van der Oost, J., Koonin, E., 2009. Prokaryotic homologs of Argonaute proteins are predicted to function as key components of a novel system of defense against mobile genetic elements. Biol. Direct. 4, 29.

Matranga, C., Tomari, Y., Shin, C., Bartel, D.P., Zamore, P.D., 2005. Passenger-strand cleavage facilitates assembly of siRNA into Ago2-containing RNAi enzyme complexes. Cell 123, 607–620.

Miesen, P., Girardi, E., van Rij, R.P., 2015. Distinct sets of PIWI proteins produce arbovirus and transposon-derived piRNAs in *Aedes aegypti* mosquito cells. Nucleic Acids Res. 43, 6545–6556.

Miesen, P., Ivens, A., Buck, A.H., van Rij, R.P., 2016. Small RNA profiling in dengue virus 2-infected *Aedes* mosquito cells reveals viral piRNAs and novel host miRNAs. PLoS Negl. Trop. Dis. 10, e0004452.

Mishra, P., Furey, C., Balaraman, V., Fraser, M., 2016. Antiviral hammerhead ribozymes are effective for developing transgenic suppression of chikungunya virus in *Aedes aegypti* mosquitoes. Viruses 8, 163.

Modis, Y., Ogata, S., Clements, D., Harrison, S.C., 2004. Structure of the dengue virus envelope protein after membrane fusion. Nature 427, 313–319.

Moon, S.L., Anderson, J.R., Kumagai, Y., Wilusz, C.J., Akira, S., Khromykh, A.A., Wilusz, J., 2012. A noncoding RNA produced by arthropod-borne flaviviruses inhibits the cellular exoribonuclease XRN1 and alters host mRNA stability. RNA 18, 2029–2040.

Moon, S.L., Dodd, B.J.T., Brackney, D.E., Wilusz, C.J., Ebel, G.D., Wilusz, J., 2015. Flavivirus sfRNA suppresses antiviral RNA interference in cultured cells and mosquitoes and directly interacts with the RNAi machinery. Virology 485, 322–329.

Morazzani, E.M., Wiley, M.R., Murreddu, M.G., Adelman, Z.N., Myles, K.M., 2012. Production of virus-derived ping-pong-dependent piRNA-like small RNAs in the mosquito soma. PLoS Pathog. 8, e1002470.

Nene, V., Wortman, J.R., Lawson, D., Haas, B., Kodira, C., Tu, Z., Loftus, B., Xi, Z., Megy, K., Grabherr, M., Ren, Q., Zdobnov, E.M., Lobo, N.F., Campbell, K.S., Brown, S.E., Bonaldo, M.F., Zhu, J., Sinkins, S.P., Hogenkamp, D.G., Amedeo, P., Arensburger, P., Atkinson, P.W., Bidwell, S., Biedler, J., Birney, E., Bruggner, R.V., Costas, J., Coy, M.R., Crabtree, J., Crawford, M., deBruyn, B., DeCaprio, D., Eiglmeier, K., Eisenstadt, E., El-Dorry, H., Gelbart, W.M., Gomes, S.L., Hammond, M., Hannick, L.I., Hogan, J.R., Holmes, M.H., Jaffe, D., Johnston, J.S., Kennedy, R.C., Koo, H., Kravitz, S., Kriventseva, E.V., Kulp, D., LaButti, K., Lee, E., Li, S., Lovin, D.D., Mao, C., Mauceli, E., Menck, C.F.M., Miller, J.R., Montgomery, P., Mori, A., Nascimento, A.L., Naveira, H.F., Nusbaum, C., O'Leary, S., Orvis, J., Pertea, M., Quesneville, H., Reidenbach, K.R., Rogers, Y.-H., Roth, C.W., Schneider, J.R., Schatz, M., Shumway, M., Stanke, M., Stinson, E.O., Tubio, J.M.C., VanZee, J.P., Verjovski-Almeida, S., Werner, D., White, O., Wyder, S., Zeng, Q., Zhao, Q., Zhao, Y., Hill, C.A., Raikhel, A.S., Soares, M.B., Knudson, D.L., Lee, N.H., Galagan, J., Salzberg, S.L., Paulsen, I.T., Dimopoulos, G., Collins, F.H., Birren, B., Fraser-Liggett, C.M., Severson, D.W., 2007. Genome sequence of *Aedes aegypti*, a major arbovirus vector. Science 316, 1718–1723.

Olson, K.E., Blair, C.D., 2015. Arbovirus–mosquito interactions: RNAi pathway. Curr. Opin. Virol. 15, 119–126.

Olson, K.E., Higgs, S., Gaines, P.J., Powers, A.M., Davis, B.S., Kamrud, K.I., Carlson, J.O., Blair, C.D., Beaty, B.J., 1996. Genetically engineered resistance to Dengue-2 virus transmission in mosquitoes. Science 272, 884–886.

Petit, M., Mongelli, V., Frangeul, L., Blanc, H., Jiggins, F., Saleh, M.-C., 2016. piRNA pathway is not required for antiviral defense in *Drosophila melanogaster*. Proc. Natl. Acad. Sci. 113, E4218–E4227.

Phuc, H.K., Andreasen, M.H., Burton, R.S., Vass, C., Epton, M.J., Pape, G., Fu, G., Condon, K.C., Scaife, S., Donnelly, C.A., Coleman, P.G., White-Cooper, H., Alphey, L., 2007. Late-acting dominant lethal genetic systems and mosquito control. BMC Biol. 5, 1–11.

Pijlman, G.P., Funk, A., Kondratieva, N., Leung, J., Torres, S., van der Aa, L., Liu, W.J., Palmenberg, A.C., Shi, P.-Y., Hall, R.A., Khromykh, A.A., 2008. A highly structured, nuclease-resistant, noncoding RNA produced by flaviviruses is required for pathogenicity. Cell Host Microbe 4, 579–591.

Robert, M.A., Okamoto, K.W., Gould, F., Lloyd, A.L., 2014. Antipathogen genes and the replacement of disease-vectoring mosquito populations: a model-based evaluation. Evol. Appl. 7, 1238–1251.

Rothman, A.L., 2004. Dengue: defining protective versus pathologic immunity. J. Clin. Invest. 113, 946–951.

Rothman, A.L., 2011. Immunity to dengue virus: a tale of original antigenic sin and tropical cytokine storms. Nat. Rev. Immunol. 11, 532–543.

Saavedra-Rodriguez, K., Beaty, M., Lozano-Fuentes, S., Denham, S., Garcia-Rejon, J., Reyes-Solis, G., Machain-Williams, C., Loroño-Pino, M.A., Flores-Suarez, A., Ponce-Garcia, G., Beaty, B., Eisen, L., Black, W.C., 2015. Local evolution of pyrethroid resistance offsets gene flow among *Aedes aegypti* collections in Yucatan state, Mexico. Am. J. Trop. Med. Hyg. 92, 201–209.

Sabchareon, A., Wallace, D., Sirivichayakul, C., Limkittikul, K., Chanthavanich, P., Suvannadabba, S., Jiwariyavej, V., Dulyachai, W., Pengsaa, K., Wartel, T.A., Moureau, A., Saville, M., Bouckenooghe, A., Viviani, S., Tornieporth, N.G., Lang, J., 2012. Protective efficacy of the recombinant, live-attenuated, CYD tetravalent dengue vaccine in Thai schoolchildren: a randomised, controlled phase 2b trial. Lancet 380, 1559–1567.

Sabin, L.R., Zheng, Q., Thekkat, P., Yang, J., Hannon, G.J., Gregory, B.D., Tudor, M., Cherry, S., 2013. Dicer-2 processes diverse viral RNA species. PLoS One 8, e55458.

Salazar, M.I., Richardson, J.H., Sanchez-Vargas, I., Olson, K.E., Beaty, B.J., 2007. Dengue virus type 2: replication and tropisms in orally infected *Aedes aegypti* mosquitoes. BMC Microbiol. 7, 9.

Sánchez-Vargas, I., Scott, J.C., Poole-Smith, B.K., Franz, A.W., Barbosa-Solomieu, V., Wilusz, J., Olson, K.E., Blair, C.D., 2009. Dengue virus type 2 infections of *Aedes aegypti* are modulated by the mosquito's RNA interference pathway. PLoS Pathog. 5, e1000299.

Saraiva, R.G., Kang, S., Simões, M.L., Angleró-Rodríguez, Y.I., Dimopoulos, G., 2016. Mosquito gut antiparasitic and antiviral immunity. Dev. Comp. Immunol. 64, 53–64.

Schnettler, E., Sterken, M.G., Leung, J.Y., Metz, S.W., Geertsema, C., Goldbach, R.W., Vlak, J.M., Kohl, A., Khromykh, A.A., Pijlman, G.P., 2012. Noncoding flavivirus RNA displays RNA interference suppressor activity in insect and mammalian cells. J. Virol. 86, 13486–13500.

Schnettler, E., Donald, C.L., Human, S., Watson, M., Siu, R.W.C., McFarlane, M., Fazakerley, J.K., Kohl, A., Fragkoudis, R., 2013. Knockdown of piRNA pathway proteins results in enhanced Semliki forest virus production in mosquito cells. J. Gen. Virol. 94, 1680–1689.

Scott, J.C., Brackney, D.E., Campbell, C.L., Bondu-Hawkins, V., Hjelle, B., Ebel, G.D., Olson, K.E., Blair, C.D., 2010. Comparison of dengue virus type 2-specific small RNAs from RNA interference-competent and -incompetent mosquito cells. PLoS Negl. Trop. Dis. 4, e848.

Sessions, O.M., Barrows, N.J., Souza-Neto, J.A., Robinson, T.J., Hershey, C.L., Rodgers, M.A., Ramirez, J.L., Dimopoulos, G., Yang, P.L., Pearson, J.L., Garcia-Blanco, M.A., 2009. Discovery of insect and human dengue virus host factors. Nature 458, 1047–1050.

Sim, S., Ramirez, J.L., Dimopoulos, G., 2012. Dengue virus infection of the *Aedes aegypti* salivary gland and chemosensory apparatus induces genes that modulate infection and blood-feeding behavior. PLoS Pathog. 8, e1002631.

Siomi, M.C., Sato, K., Pezic, D., Aravin, A.A., 2011. PIWI-interacting small RNAs: the vanguard of genome defence. Nat. Rev. Mol. Cell Biol. 12, 246–258.

van der Schaar, H.M., Rust, M.J., Chen, C., van der Ende-Metselaar, H., Wilschut, J., Zhuang, X., Smit, J.M., 2008. Dissecting the cell entry pathway of dengue virus by single-particle tracking in living cells. PLoS Pathog. 4, e1000244.

van Rij, R.P., Saleh, M.-C., Berry, B., Foo, C., Houk, A., Antoniewski, C., Andino, R., 2006. The RNA silencing endonuclease Argonaute 2 mediates specific antiviral immunity in *Drosophila melanogaster*. Genes Dev. 20, 2985–2995.

Waterhouse, R.M., Kriventseva, E.V., Meister, S., Xi, Z., Alvarez, K.S., Bartholomay, L.C., Barillas-Mury, C., Bian, G., Blandin, S., Christensen, B.M., Dong, Y., Jiang, H., Kanost, M.R., Koutsos, A.C., Levashina, E.A., Li, J., Ligoxygakis, P., MacCallum, R.M., Mayhew, G.F., Mendes, A., Michel, K., Osta, M.A., Paskewitz, S., Shin, S.W., Vlachou, D., Wang, L., Wei, W., Zheng, L., Zou, Z., Severson, D.W., Raikhel, A.S., Kafatos, F.C., Dimopoulos, G., Zdobnov, E.M., Christophides, G.K., 2007. Evolutionary dynamics of immune-related genes and pathways in disease-vector mosquitoes. Science 316, 1738–1743.

Wise de Valdez, M.R., Suchman, E.L., Carlson, J.O., Black, W.C., 2010. A large scale laboratory cage trial of *Aedes densonucleosis* virus (AeDNV). J. Med. Entomol. 47, 392.

Wise de Valdez, M.R., Nimmo, D., Betz, J., Gong, H.-F., James, A.A., Alphey, L., Black, W.C., 2011. Genetic elimination of dengue vector mosquitoes. Proc. Natl. Acad. Sci. 108, 4772–4775.

Xi, Z., Ramirez, J., Dimopoulos, G., 2008. The *Aedes aegypti* Toll pathway controls dengue virus infection. PLoS Pathog. 4, e1000098.

CHAPTER

13

Paratransgenesis Applications: Fighting Malaria With Engineered Mosquito Symbiotic Bacteria

Sibao Wang[1], *Marcelo Jacobs-Lorena*[2]

[1]Chinese Academy of Sciences, Shanghai, China; [2]Johns Hopkins Bloomberg School of Public Health, Baltimore, MD, United States

INTRODUCTION

Mosquitoes transmit many infectious diseases, including malaria, lymphatic filariasis, yellow fever, dengue fever, chikungunya, and Zika. Malaria is one of the most devastating infectious diseases worldwide. Malaria occurs mostly in poor, tropical, and subtropical countries. It is endemic in over 106 countries accounting for nearly 200 million malaria cases and around 627,000 deaths annually, especially young children in sub-Saharan Africa (WHO, 2014). Around 3.2 billion people (half of the world's population) live in areas at risk of malaria infection. Clearly, the current tools are not sufficient for malaria control.

Malaria is caused by *Plasmodium* parasites and is transmitted to people through the bite of infected female anopheline mosquitoes. Therefore, eliminating the mosquito or interfering with its ability to support the parasite development will block malaria transmission. Vector control via use of insecticide-treated bed nets and indoor residual spraying represent the front-line tools to combat transmission and have helped alleviate the malaria burden in many endemic areas (Greenwood et al., 2008). However, the emergence and rapid spreading of insecticide-resistant mosquitoes and drug-resistant parasites, combined with the lack of an effective malaria vaccine severely undermine current control efforts to fight the disease (Enayati and Hemingway, 2010; Trape et al., 2011). Another rarely considered but equally important limitation is that insecticides leave intact the biological niche where mosquitoes reproduce, meaning that mosquito populations quickly revert to original density as soon as insecticide treatment stops or when mosquitoes become resistant to the insecticide. Recently, the malERA consultative group stressed that malaria eradication cannot be achieved without introduction of novel control tools (Baum et al.,

Arthropod Vector: Controller of Disease Transmission, Volume 1
http://dx.doi.org/10.1016/B978-0-12-805350-8.00013-1

2011). Therefore, new approaches for disease control are urgently needed.

Paratransgenesis is a "population replacement" strategy for interfering with pathogen development via genetic modification of symbiotic microbes to produce antipathogen effector molecules in the host. In this chapter, we provide an overview on (1) means for interfering with malaria parasite development in the mosquito gut; (2) the *Anopheles* mosquito microbiome and its impact on mosquito physiology and pathogen transmission; (3) recent advances related to the paratransgenesis approach; and (4) the challenges for translation of laboratory findings to field applications aiming at reducing mosquito vectorial competence.

GENETIC MANIPULATION OF MOSQUITO VECTORIAL COMPETENCE

Transmission of the malaria parasite requires the completion of a complex developmental cycle in vector mosquitoes (Ghosh et al., 2000). Based on this premise, a promising approach being considered is not to kill the mosquito, but instead to convert it into an ineffective malaria vector. In other words, fill the biological niche with mosquitoes that cannot transmit the parasite.

The most vulnerable stages of *Plasmodium* development occur in the lumen of the mosquito midgut, making the mosquito midgut a prime target for parasite intervention (Abraham and Jacobs-Lorena, 2004; Drexler et al., 2008). One option for interfering with parasite transmission is to genetically modify the mosquito to promote midgut expression of "effector genes" whose products inhibit parasite development (Wang and Jacobs-Lorena, 2013; Cui et al., 2015). This proof of concept was tested for the first time by genetically engineering *Anopheles stephensi* for midgut expression of the SM1 peptide that binds to a putative ookinete receptor on the luminal surface of the midgut epithelium and strongly inhibits ookinete midgut invasion (Ghosh et al., 2001; Ito et al., 2002). Subsequent reports from different laboratories making use of a variety of effector molecules reached similar conclusions (Moreira et al., 2002; Yoshida et al., 2007; Dong et al., 2011; Corby-Harris et al., 2010; Isaacs et al., 2011; Abraham et al., 2005), suggesting that this strategy works successfully in the laboratory (Corby-Harris et al., 2010; Isaacs et al., 2011; Marrelli et al., 2007). However, one unresolved challenge is how to spread transgenes into wild mosquito populations. Progress has been made in the development of a genetic drive, but additional issues remain to be addressed (Gantz et al., 2015; McLean and Jacobs-Lorena, 2016; Hammond et al., 2016). An additional consideration for any genetic drive approach is that there are about 430 anopheline species, about 30–40 of which are natural vectors for human malaria (Arino et al., 2007) and very few of them have been shown to be amenable to genetic manipulation (Coutinho-Abreu et al., 2010). Moreover, an important additional challenge faced by genetic drive approaches in general is that anopheline vectors frequently exist as reproductively isolated populations (cryptic species) (Powell et al., 1999), thus preventing gene flow from one population to another. In addition, fitness load imposed by refractory gene(s), insertional mutagenesis, and positional effects need to be considered (Corby-Harris et al., 2010; Catteruccia et al., 2003; Li et al., 2008). Once these issues are resolved, transgenesis could provide a powerful tool to combat malaria.

An alternative strategy for delivering anti-*Plasmodium* effector molecules is paratransgenesis, which consists of introducing into vector mosquitoes bacteria genetically engineered to secrete antipathogen compounds (Hurwitz et al., 2011). A key factor relating to this strategy is that the mosquito microbiome resides in the same compartment where the most vulnerable stages of malaria parasite development occur (Drexler et al., 2008). Furthermore, bacteria

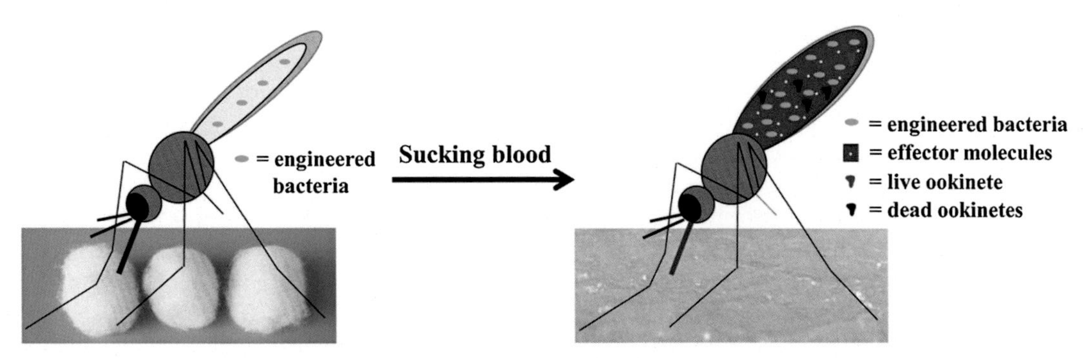

FIGURE 13.1 **Paratransgenesis to block malaria transmission in the mosquito.** In this strategy, a symbiotic bacterium native to the mosquito is genetically engineered in vitro to export anti-*Plasmodium* effector molecules that interfere with parasite transmission. Genetically modified bacteria are then reintroduced into mosquitoes, e.g., by providing mosquitoes with cotton balls soaked in a mixture of bacteria and sugar solution (left panel). After a blood meal, the engineered bacteria proliferate by hundred-fold thus dramatically increasing the output of effector molecules. The effector molecules kill *Plasmodium* parasites or influence the mosquito's ability to transmit malaria, thwarting vector competence.

numbers increase dramatically (hundreds of fold) after ingestion of a blood meal (Whitten et al., 2006) and output of anti-*Plasmodium* effector molecules produced by recombinant bacteria can be expected to increase proportionally, hindering the mosquito's ability to support parasite development (Fig. 13.1).

ANOPHELES GUT MICROBIOTA

Insects are the most diverse and abundant group of organisms dominating terrestrial habitats, in terms of number of species (Basset et al., 2012). The evolutionary success of insects and their diversification into a wide range of ecological niches is dependent in part on the beneficial members of their associated microbiome (Engel and Moran, 2013). The gut of insects is colonized by diverse communities of symbiotic microorganisms (the microbiota), which vary from bacteria to viruses, yeasts, and protists (Engel and Moran, 2013; Minard et al., 2013). These communities are thought to be beneficial to the insect hosts in many ways, including provision of nutritional supplements, enhancement of digestive mechanism, tolerance of environmental perturbations, and manipulation of host immune homeostasis (Engel and Moran, 2013; Weiss and Aksoy, 2011; Coon et al., 2014). Conversely, the insect host can affect the microbial community.

The association between gut symbionts and mosquitoes has attracted considerable attention for its use in development of the novel transmission blocking tools against vectorborne diseases (Durvasula et al., 1997; Chang et al., 2003; Favia et al., 2007; Rao et al., 2005; Riehle et al., 2007; Wang et al., 2012). The gut microbiota composition and structure have been investigated in several anophelines using culture-dependent and culture-independent approaches (Rani et al., 2009). Both laboratory-reared and field-caught *Anopheles* mosquitoes carry a broad diversity of bacteria, mostly Gram-negative proteobacteria and enterobacteria (Straif et al., 1998; Wang et al., 2011). Common gut symbiotic bacteria include members of the genera *Serratia*, *Pseudomonas*, *Enterobacter*, and *Pantoea*. Despite substantial differences in the microbiota composition between laboratory-reared and field-collected mosquitoes, the most frequent genera present in the laboratory-reared adult mosquitoes are similar to those of field-collected mosquitoes, suggesting that anopheline mosquitoes retain their microbiome in a selective way (Wang et al., 2011). Many studies have revealed that the mosquito gut microbiota is highly dynamic, varies greatly among individuals and species, and even fluctuates within the same individual on account of factors such as age, diet, physiology, and geographic location (Osei-Poku et al., 2012). During pupation, the gut contents (including the microbiota) are wrapped by the peritrophic matrix, to form the meconium. When the adult mosquito emerges from the pupa, the meconium is discarded within 24–48 h after emergence, resulting in the loss of the larval gut bacteria (Romoser et al., 2000; Moll et al., 2001). A laboratory study found that mosquitoes can take up bacteria from the water they emerged from and transfer the same bacteria to the water they laid eggs in (Lindh et al., 2008). However, it is not yet clear how wild, adult mosquitoes acquire their microbiota. *Pantoea agglomerans* is a dominant gut bacterium that colonizes the gut of different wild mosquito species in Kenya and Mali (Straif et al., 1998) and laboratory-reared *A. stephensi, Anopheles gambiae,* and *Anopheles albimanus* (Riehle et al., 2007; Pumpuni et al., 1996). *P. agglomerans* is also commonly identified on plant surfaces, fruit, and blossoms (Andrews and Harris, 2000; Pusey, 1997, 2002), suggesting that flower nectar is a possible source of the mosquito microbiota in the field. A study indicated that sugar feeding is a practical way to reintroduce the engineered symbiotic bacteria into wild mosquitoes (Lindh et al., 2006). Cotton balls soaked with sugar solution and bacteria were placed in clay jar refuges and the jars are placed at different locations around houses. A second option is to introduce bacteria into the larval breeding sites. The larvae are expected to ingest the bacteria and the emerged adult might also take up the bacteria from the water. A third option is to spray a mixture of bacteria and sugar meals on fresh grasses around mosquito larval sites. By spraying an attractive toxic sugar bait (ATSB) solution on tender grasses around larval sites, Beier et al. (2012) could locally decimate (more than 95% of the population) mosquito populations in both sugar-rich and sugar-poor environments.

Stable colonization of the mosquito gut and efficient transmission to the next generation are crucial properties required for the successful introduction of recombinant bacteria into field mosquito populations. Bacteria from the genus *Asaia* were identified as a stable symbiont in laboratory *A. stephensi* colonies and in wild *A. gambiae* populations. *Asaia* sp. was also observed in several mosquito organs, including salivary glands and ovaries (Favia et al., 2007). Importantly, *Asaia* sp. is able to be vertically transmitted from female to larval progeny, venereally transmitted from male to female during mating, and transstadially transmitted from larva to adult (Favia et al., 2007; Damiani et al., 2008; Favia et al., 2008). These features favor dissemination and should be helpful when considering potential introduction of genetically modified bacteria into mosquito populations in the field. However, bacteria from the genus *Asaia* appear to be absent in field-collected or laboratory-reared Asian malaria mosquitoes *A. stephensi* and *Anopheles maculipennis* in Iran and India (Djadid et al., 2011; Rani et al., 2009). There are also no records of *Asaia* sp. in American anopheline mosquitoes.

The intracellular bacterium *Wolbachia* infects 40%–69% of arthropod species (Zug and Hammerstein, 2012; Hilgenboecker et al., 2008), including many mosquito species (Moreira et al., 2009; Kambris et al., 2009; Bian et al., 2010). These bacteria infect the insect germ cells, thus assuring their transmission from one generation to the next. Importantly, many *Wolbachia* can spread through insect populations by inducing cytoplasmic incompatibility, whereby the progeny of certain crosses (*Wolbachia*-positive males X *Wolbachia*-negative females) do not survive (Werren et al., 2008). A crucial finding was that when *Wolbachia* was introduced into *Aedes aegypti*, the ability of this mosquito to vector the dengue, chikungunya, and Zika viruses was dramatically reduced (Moreira et al., 2009; Dutra et al., 2016). These two properties—ability to spread through mosquito populations

(Hoffmann et al., 2011) and ability to thwart virus replication—are being explored as a population replacement strategy for the fight against arboviral diseases. There is evidence that *Wolbachia* may also interfere with *Plasmodium* development in the mosquito (Hughes et al., 2011; Bian et al., 2013). Previous efforts to identify *Wolbachia* in over 30 species of *Anopheles* from four continents were unsuccessful (Walker and Moreira, 2011), until a novel *Wolbachia* strain—*w*Anga—was identified in natural populations of two *Anopheles* species in Burkina Faso, West Africa (Baldini et al., 2014). *w*Anga infections significantly accelerate egg laying without inducing CI or sex ratio distortion. A negative correlation between the presence of *Plasmodium* parasites and *Wolbachia* infection was observed in natural *Anopheles coluzzii* populations, suggesting that infection with these endosymbionts may reduce the vectorial competence of these mosquitoes (Shaw et al., 2016).

Fungi are underappreciated components of *Anopheles* microbiota compared to the bacterial kingdom. Fungi are of special interest, as fungal microorganisms may affect the mosquito physiology, and directly or indirectly interact with gut bacterial microbiota and pathogens. Several yeasts such as *Candida* sp., *Pichia* sp., and *Wickerhamomyces anomalus* were identified in the gut of laboratory-reared *Anopheles* mosquitoes (Ignatova et al., 1996; Ricci et al., 2011a,b). *W. anomalus* was also found in wild mosquitoes, having a wide tissue distribution in adult, including the female midgut and the male gonoduct (Ricci et al., 2011a,b), suggesting the possible use of symbiotic yeasts for the control of vectorborne diseases.

Linear single-stranded DNA densoviruses thought to be widely distributed among arthropods have been found to infect several important vector mosquitoes [*A. aegypti* (Ward et al., 2001), *A. gambiae* (Ren et al., 2008), and *Culex pipiens* (Ma et al., 2011)], and to be able to be vertically transmitted (Ren et al., 2008; Carlson et al., 2006). The densovirus AgDNV was found to efficiently infect *A. gambiae* larvae and to spread to the adult midgut, fat body, and ovaries (Ren et al., 2008). This virus was also found to be vertically transmitted to subsequent mosquito generations (Ren et al., 2008). These properties suggest that densoviruses could be used to produce effector molecules in host mosquitoes (Carlson et al., 2006). However, the limited length of foreign DNA that these viruses can carry may be a limiting factor.

IMPACT OF MICROBIOTA ON *ANOPHELES* PHYSIOLOGY AND PATHOGEN TRANSMISSION

Insects coexist with a myriad of commensal microorganisms in a symbiotic relationship that play important roles in host immunity, metabolic homeostasis, and development. The insect gut and its microbiota is akin to an internal organ. The insect host provides a nutrient-rich environment for the commensal microbiota and in turn, commensal microbes profoundly affect multiple aspects of insect biology, including secretion of metabolites, prevention of infections by pathogenic microbes, and signaling of immune responses. Thus, the gut microbiota has the potential to alter mosquito vector competence. The complex interactions of the commensal microbiota with the immune system is a topic of substantial interest, since antimicrobial defenses can cross-react and strongly affect the survival of pathogens ingested by the mosquito (Dong et al., 2009). Understanding the interactions and mechanisms through which commensal microbes actively shape host immunity may yield new insights into the interactions between insect hosts and pathogens, and could help in the development of new insect-based disease interventions.

Only a small proportion of parasites ingested with the infectious blood meal succeed in invading the mosquito midgut epithelium for further propagation (the midgut bottleneck). This is due to many factors, including formation of the peritrophic matrix, secretion of gut enzymes, and vector immunity. The gut microbiota also plays an important role as determinants of parasite survival and development and therefore contributes to the modulation of vector competence. Vectorial competence is modulated either through direct microbiota–*Plasmodium* interactions (Cirimotich et al., 2011), or indirectly by modulating vector immune defenses (Dong et al., 2009; Meister et al., 2009; Meister et al., 2005). The bacterial population in the mosquito gut increases by hundreds of times within 24h after a blood meal (Wang et al., 2012; Pumpuni et al., 1996) and this rapid proliferation stimulates mosquito immune responses that limit malaria parasites infection (Dong et al., 2009; Meister et al., 2005). Reduction of the gut microbiota with antibiotics renders the mosquito more susceptible to *Plasmodium* infection. Conversely, coinfection of bacteria with *Plasmodium* gametocytes reduces the oocyst load (Dong et al., 2009). Co-feeding infectious blood and an *Enterobacter* bacterium strain Esp_Z, isolated from wild *Anopheles arabiensis* mosquitoes in Zambia, was shown to arrest *Plasmodium* development via synthesis of reactive oxygen species (Cirimotich et al., 2011). Another study indicated that the dominant commensal *Enterobacteriaceae* favors mosquito *Plasmodium* infection, suggesting that *Enterobacteriaceae* might also play a positive role in the natural infection of *Plasmodium falciparum* (Boissiere et al., 2012).

FIGHTING MALARIA TRANSMISSION WITH PARATRANSGENESIS

Paratransgenesis is a "Trojan Horse" approach to interfere with pathogen development and differs from the genetic modification of the insect itself (Wang and Jacobs-Lorena, 2013; Wang et al., 2012; Yoshida et al., 2001). The principle of the approach is to engineer an insect symbiont to secrete antipathogen molecules and introduce the engineered symbiont into the vector (Fig. 13.1). This approach seeks to block pathogen transmission and, importantly, avoid any adverse effects to the insect vector. At the heart of the paratransgenesis strategy for malaria control is the fact that the mosquito microbiota and *Plasmodium* share the same compartment—the midgut—where the most vulnerable stage of the parasite development occurs (Abraham and Jacobs-Lorena, 2004; Drexler et al., 2008). In other words, in this compartment bacteria and pathogens reside next to each other. The first step is to identify effector proteins that interfere with pathogen transmission. The genes coding for these proteins are then genetically inserted into a symbiont that is native to disease-transmitting vectors. The final step in the strategy is to reintroduce these genetically altered symbionts into vector populations in the wild.

Paratransgenesis has already been proposed as a means to control other insect-borne diseases. The parasitic protozoan *Trypanosoma cruzi*, the causative agent of Chagas disease, is transmitted by the triatomid bug *Rhodnius prolixus* (Durvasula et al., 1997). The *Rhodnius* obligate Gram-positive bacterium *Rhodococcus rhodnii* was genetically engineered to produce the antimicrobial peptide cecropin A, and fed to naïve *R. prolixus* nymphs. Durvasula et al. (1997) showed that expression of the anti-parasite peptide by the genetically modified symbionts significantly reduces *T. cruzi's* ability to survive in the bug.

The paratransgenesis approach has a number of attractive features and could help overcome some of the limitations of transgenesis (Wang and Jacobs-Lorena, 2013; Riehle and Jacobs-Lorena, 2005). (1) Mosquitoes carry a diverse microbiota in their midgut, the site where a severe bottleneck of *Plasmodium* development occurs, making this compartment a prime target for intervention (Abraham and Jacobs-Lorena, 2004; Drexler et al., 2008).

(2) The gut bacterial number increases dramatically, by hundreds of fold, after ingestion of a blood meal (Whitten et al., 2006), correspondingly increasing the output of effector molecules produced by recombinant bacteria (Wang et al., 2012). (3) Genetic manipulation of bacteria is much simpler and faster than genetic manipulation of mosquitoes. (4) Given that the use of multiple effector proteins is essential to maximize blocking and avoid parasite resistance, it is straightforward to construct strains of bacteria that express multiple effector molecules or simply introduce into mosquitoes a mixture of GM bacteria, each expressing a different effector gene. (5) Transgenes are much easier to spread into mosquito populations via engineered bacteria than via transgenic mosquitoes. Importantly, this approach bypasses genetic barriers of reproductively isolated mosquito populations that commonly occur in areas of high malaria transmission and that hinder transgene introgression. Paratransgenesis may well be "universal," effective with multiple mosquito and parasite species. (6) Bacteria can be mass-produced easily and cheaply in large scale, also in malaria-affected countries. (7) The paratransgenesis approach is compatible with current mosquito control strategies and integrated pest management programs, including those that use insecticides. It is also compatible with mosquito transgenesis.

Basic Requirements for Paratransgenesis

There are several basic requirements for the implementation of the paratransgenesis strategy. (1) The symbiotic bacterium should preferably originate from the disease-transmitting vector and have a stable symbiotic relationship with the vector; (2) The symbiotic bacterium can be cultured; (3) The symbiotic bacterium can be genetically manipulated; (4) The effector gene product should not impair symbiont and vector fitness; (5) The effector gene product should be secreted to assure interaction with the target pathogen; (6) An efficient means of introducing the engineered symbiont into field vector populations must be devised.

Effector Molecules

Identification of potent effector genes that confer resistance to *Plasmodium* is the first requirement for the generation of a refractory mosquito via transgenesis or paratransgenesis. The effector gene product should not impair symbiont and vector fitness and should not be toxic to other organisms such as mammals. Combination of multiple anti-*Plasmodium* effector proteins with different modes of action will be necessary for a successful intervention both to maximize the effectiveness of interference with parasite development (additive or synergistic effects) and to reduce the probability that resistance to effector proteins emerges in the *Plasmodium* population.

Anti-*Plasmodium* effector molecules can be grouped into four classes based on their modes of action against *Plasmodium* parasites (Table 13.1). (1) **Parasite killing**. This class includes host antibacterial peptides such as defensins (Kokoza et al., 2010), gambicin (Vizioli et al., 2001) and cecropins (Kim et al., 2004), and peptides from other sources that lyse parasites, such as scorpine (Wang et al., 2012; Conde et al., 2000), Hemolytic C-Type Lectin CEL-III (Moreira et al., 2002), angiotensin II (Maciel et al., 2008), magainins (Gwadz et al., 1989), synthetic anti-parasitic lytic peptides Shiva 1 and Shiva 3 (Possani et al., 1998), and gomesin (Moreira et al., 2007). (2) **Interaction with parasites**. EPIP, a *Plasmodium* Enolase–Plasminogen Interaction Peptide, is a peptide that inhibits mosquito midgut invasion by preventing plasminogen binding to the ookinete surface (Ghosh et al., 2011). Other effector molecules for generating transgenic symbionts that are refractory to *Plasmodium* are single-chain monoclonal antibodies (scFvs) that bind to ookinete or sporozoite surface or secreted proteins. For

TABLE 13.1 Anti-*Plasmodium* Effector Molecules

Effector	Properties	Target Parasite	Parasite Stage(s)	Function or Mechanism	References
PARASITE KILLING					
Scorpine	Scorpion *Pandinus imperator* venom peptide	*Plasmodium falciparum* *Plasmodium berghei*	Gametocyte to ookinete	Cecropin and defensin-like lytic peptide	Conde et al. (2000)
Shiva1	Cecropin-like synthetic peptide	*P. berghei* *P. falciparum*	Gametocyte to ookinete	Lyses the parasite	Jaynes et al. (1988)
Shiva 3	Cecropin-like synthetic peptide	*P. falciparum* *P. berghei*	Gametocyte to ookinete	Lyses the parasite	Possani et al. (1998)
Cec A	*Anopheles gambiae* cecropin A	*P. berghei*	Ookinete	Lyses the parasite	Kim et al. (2004)
Cec B	*A. gambiae* cecropin B	*P. falciparum*	Oocyst	Lyses the parasite	Gwadz et al. (1989)
DEF1 A	*A. gambiae* defensin A	*P. falciparum* *P. berghei*	Ookinete	Lyses the parasite	Kokoza et al. (2010)
Gambicin	*A. gambiae* antimicrobial peptide	*P. falciparum* *P. berghei*	Ookinete	Lyses the parasite	Vizioli et al. (2001)
Angiotensin II		*P. gallinaceum*	Sporozoite	Lyses the parasite	Maciel et al. (2008)
Magainins	Peptides from the African clawed frog *Xenopus laevis* skin	*P. falciparum* *P. knowlesi* *P. cynomolgi*	All mosquito stages	Lyses the parasite	Gwadz et al. (1989)
Gomesin	A antimicrobial peptide from spider	*P. falciparum* *P. berghei*	All mosquito stages	Lyses the parasite	Moreira et al. (2007)
CEL-III	Hemolytic C-Type lectin	*P. falciparum* *P. berghei*	Ookinete, oocyst	Lyses the parasite	Moreira et al. (2002)
TP10	Wasp venom peptide	*P. falciparum*	Gametocyte to ookinete	Lyses the parasite	Arrighi (2008)
AdDLP	*Anaeromyxobacter dehalogenans* defensin-like peptide	*P. berghei*	Ookinete	Lyses the parasite	Gao et al. (2009)
Meucin-25	Scorpion *Mesobuthus eupeus* venom gland	*P. berghei* *P. falciparum*	Gametocyte	Antimicrobial linear cationic peptide	Gao et al. (2010)
Drosomycin	An inducible antifungal peptide initially isolated from the *Drosophila melanogaster* hemolymph	*P. berghei*	Ookinete	Lyses the parasite	Tian et al. (2008)

INTERACTION WITH PARASITES					
EPIP	Enolase–plasminogen interaction peptide	*P. falciparum* *P. berghei*	Ookinete	Inhibits mosquito midgut invasion by preventing plasminogen binding to the ookinete surface	Ghosh et al. (2011)
Pro:EPIP	A fusion peptide composed of a chitinase propeptide (pro) and EPIP	*P. falciparum* *P. berghei*	Ookinete	Blocks ookinete traversal of the mosquito peritrophic matrix and prevents plasminogen binding to the ookinete surface	Wang et al. (2012)
Pbs21scFv-Shiva1	Single-chain immunotoxin	*P. falciparum* *P. berghei*	Gametocyte to oocyst	A single-chain monoclonal antibody (scFv) targeting the major ookinete surface protein Pbs21 and linked to the lytic peptide Shiva1	Yoshida et al. (2001)
scFv 4B7	A single-chain antibody	*P. falciparum*	Ookinete	Binds to *P. falciparum* ookinete surface protein Pfs25	Corby-Harris et al. (2010) and de Lara Capurro et al. (2000)
scFv 2A10	Single-chain antibody	*P. falciparum*	Sporozoite	Targets the *P. falciparum* circumsporozoite protein (CSP)	Corby-Harris et al. (2010) and de Lara Capurro et al. (2000)
PfNPNA-1	Single-chain antibody	*P. falciparum*	Sporozoite	Recognizes the repeat region (Asn-Pro-Asn-Ala) of The *P. falciparum* surface circumsporozoite protein	Fang et al. (2011)
scFv 1C3	Single-chain antibody	*P. falciparum*	Ookinete	Binds a *P. falciparum* secreted enzyme chitinase 1	Corby-Harris et al. (2010)
INTERACTION WITH MOSQUITO MIDGUT OR SALIVARY GLAND EPITHELIA					
SM1	Salivary gland and midgut peptide 1	*P. berghei* *P. falciparum*	Ookinete, sporozoite	Blocks ookinete invasion of the midgut epithelium and sporozoite invasion of the salivary gland epithelium	Ghosh et al. (2001)
MP2	Midgut peptide 2	*P. berghei* *P. falciparum*	Ookinete	Blocks ookinete invasion of the midgut epithelium and invasion of the salivary gland epithelium	Vega-Rodriguez et al. (2014)

Continued

TABLE 13.1 Anti-*Plasmodium* Effector Molecules—cont'd

Effector	Properties	Target Parasite	Parasite Stage(s)	Function or Mechanism	References
mPLA2	Bee venom phospholipase	*P. falciparum* *P. berghei*	Ookinete	Inhibits ookinete midgut invasion, probably by modifying the properties of the midgut epithelial membrane	Ito et al. (2002)
Pro	A chitinase propeptide	*P. falciparum* *P. berghei*	Ookinete	Inhibits the enzyme chitinase and blocks ookinete traversal of the mosquito peritrophic matrix	Bhatnagar et al. (2003)
Pchtscfv	Single-chain antibody	*P. falciparum*	Ookinete	Inhibits the *P. falciparum* chitinase and blocks ookinete traversal of the mosquito peritrophic matrix	Li et al. (2005)
MANIPULATION OF MOSQUITO IMMUNE SYSTEM					
Akt	A protein kinase	*P. falciparum* *P. berghei*	Ookinete	Akt boosts mosquito innate immunity via insulin signaling	Dong et al. (2011)
Rel2	*Anopheles* Imd pathway transcription factor	*P. falciparum* *P. berghei*	Ookinete, sporozoite	Rel2 overexpression enhances the mosquito Imd pathway	Yoshida et al. (2007)

instance, scFv 4B7 binds to *P. falciparum* ookinete surface protein Pfs25, 2A10 targets the *P. falciparum* circumsporozoite protein (CSP) (Isaacs et al., 2011; de Lara Capurro et al., 2000), anti-Pbs21 single-chain antibody targets the *Plasmodium berghei* major ookinete surface protein Pbs21 (Yoshida et al., 2001), and scFv 1C3 binds a *P. falciparum* secreted enzyme chitinase 1 (Isaacs et al., 2011). Monoclonal antibodies have the disadvantage that a parasite mutation leading to a single amino acid change in the target epitope may reduce its effectiveness of interfering with parasite development. (3) **Interaction with mosquito midgut or salivary gland epithelia**. To complete its life cycle in the mosquito, *Plasmodium* has to traverse the midgut and salivary gland epithelia and this traversal is dependent on parasite interaction with surface receptors (Ghosh et al., 2009; Vega-Rodriguez et al., 2014). Therefore, a protein or peptide that binds specifically to the surface receptor will inhibit *Plasmodium* invasion of the corresponding organs. Examples of this class are SM1—a 12-amino acid Salivary gland and Midgut Peptide 1—that binds to a putative receptor on the luminal surface of the mosquito midgut and basal surface of the salivary gland epithelia, blocking ookinete and sporozoite invasion (Ghosh et al., 2001; Fang et al., 2011) and MP2—a 12-amino acid Midgut Peptide 2—that binds to a putative receptor on the surface of the mosquito midgut, blocking ookinete invasion (Vega-Rodriguez et al., 2014); mPLA2 is a mutant phospholipase A2 that inhibits ookinete invasion, possibly by modifying the properties of the midgut epithelial membrane (Moreira et al., 2002; Wang et al., 2012; Zieler et al., 2001); and a chitinase propeptide that inhibits this enzyme and in this way hinders ookinete traversal of the mosquito peritrophic matrix (PM) (Bhatnagar et al., 2003). The PM is a chitin-based extracellular structure that surrounds the entire blood meal and must be crossed by the ookinete to reach the mosquito midgut (Shao et al., 2001). (4) **Manipulation of mosquito immune system**. The mosquito's innate immune system plays an important role in inhibiting the *Plasmodium* parasite development in the mosquito. Thus, boosting mosquito immune-related genes can lead to reduced mosquito vectorial competence. Blood meal–induced expression of Akt, a key signaling component in the insulin signaling pathway renders the mosquito refractory to *Plasmodium* infection (Dong et al., 2011). Overexpression of Imd pathway–mediated transcription factor Rel2 renders the mosquito resistant to *Plasmodium* infection (Dong et al., 2011). Manipulation of mosquito immune pathway using RNA interference or "smart sprays" enhances mosquito antimicrobial response (Christophides et al., 2004; Brown et al., 2003).

Fighting Malaria With Engineered Symbionts

Early reports on the use of paratransgenesis to interfere with *Plasmodium* transmission were based on a recombinant laboratory bacterium *Escherichia coli* expressing a single-chain immunotoxin (Yoshida et al., 1999), a dimer of the SM1 peptide or a modified phospholipase A2 (Riehle et al., 2007). Although these recombinant bacteria cause a significant decrease of *P. berghei* oocyst numbers in *A. stephensi*, inhibition of parasite development was modest. One limitation of these earlier studies was that *E. coli*, an attenuated laboratory bacterium, survives poorly in the mosquito gut (Riehle et al., 2007). An additional limitation was that the recombinant effector molecules either remained attached to the bacterial surface (Riehle et al., 2007) or formed an insoluble inclusion body within the bacterial cells (Yoshida et al., 2001). In either case, the effector molecules could not diffuse to their intended parasite or mosquito midgut targets.

Recently, a substantially improved strategy to deliver effector molecules was developed, which consists of engineering a natural symbiotic bacterium *P. agglomerans* to produce and secrete antimalaria proteins in the mosquito midgut (Wang et al., 2012). The *E. coli* hemolysin A secretion system was used to promote the secretion of a variety of anti-*Plasmodium* effector proteins that either inhibit midgut invasion by *Plasmodium*, such as the *Plasmodium* Enolase–Plasminogen Interaction Peptide $[EPIP]_4$ that prevents plasminogen binding to the ookinete surface (Ghosh et al., 2011), or by directly targeting the parasite, such as the scorpion-derived antiplasmodial scorpine. These engineered *P. agglomerans* strains inhibited development of both, the human malaria parasite *P. falciparum* and the rodent malaria parasite *P. berghei*, up to 98%. The engineered bacteria were as effective when carried by an African mosquito vector (*A. gambiae*) as by an Asian vector (*A. stephensi*). Significantly, the proportion of mosquitoes carrying parasites (prevalence) decreased down to 84% (Wang et al., 2012). This strong reduction in the proportion of infected mosquitoes should translate into important reduction of transmission in the field. Moreover, the use of multiple effector molecules, each acting by a different mechanism, should greatly reduce the probability of selecting resistant parasites. Therefore, the paratransgenesis strategy may well turn out to be "universal," being effective with multiple mosquito species and parasite species. These promising laboratory results will next need to be translated to field applications. A major outstanding issue is how to efficiently introduce the engineered bacteria into wild mosquito populations.

CONCLUSION AND REMARKS

Despite vast efforts and expenditures in the past few decades, mosquito-transmitted infectious diseases remain a leading cause of morbidity and mortality. Current insecticide-based vector control strategies have the disadvantage that they create an "empty ecological niche" that is readily repopulated as soon as intervention stops. Thus, any population suppression strategy needs to be implemented forever. Genetic modification of the mosquito or its symbiotic microbes to render the vector incapable to transmit the malaria parasite is a promising approach that needs to be explored. Unlike population suppression strategies, transgenesis and paratransgenesis are "population replacement" strategies and should require much less follow-up effort.

Paratransgenesis is low-tech and independent of mosquito or parasite species. However, the genetic modification strategies discussed in this chapter even if ultimately successful, cannot by themselves provide a final solution in the fight against malaria. Rather, transgenesis and/or paratransgenesis should be considered as a complement to existing and future control measures. Vectorborne diseases can only be conquered by the coordinated deployment of as many weapons as possible. The remaining roadblocks need to be resolved before paratransgenesis can be implemented to block malaria transmission in the field. These include the development of means to introduce the engineered bacteria into field mosquito populations, study of the bacteria dynamics in natural mosquito populations to ensure effective propagation of the engineered bacteria, and testing of effectors against multiple parasite species/isolates and in multiple vector species. Another major roadblock refers to solving regulatory and ethical issues related to the release of genetically modified organisms in nature.

Acknowledgments

Work was supported by a grant from the Strategic Priority Research Program of Chinese Academy of Sciences (grant no. XDB11010500), the National Nature Science Foundation of China (grant no. 31472044), Pujiang talent project of Shanghai (grant no. 14PJ1410200), and National Institute of Allergy and Infectious Diseases grant AI 031478.

References

Abraham, E.G., et al., 2005. Driving midgut-specific expression and secretion of a foreign protein in transgenic mosquitoes with AgAper1 regulatory elements. Insect Mol. Biol. 14 (3), 271–279.

Abraham, E.G., Jacobs-Lorena, M., 2004. Mosquito midgut barriers to malaria parasite development. Insect Biochem. Mol. 34 (7), 667–671.

Andrews, J.H., Harris, R.F., 2000. The ecology and biogeography of microorganisms on plant surfaces. Annu. Rev. Phytopathol. 38, 145–180.

Arino, J., Bowman, C., Gumel, A., Portet, S., 2007. Effect of pathogen-resistant vectors on the transmission dynamics of a vector-borne disease. J. Biol. Dyn. 1 (4), 320–346.

Arrighi, R.B., 2008. Cell-penetrating peptide TP10 shows broad-spectrum activity against both *Plasmodium falciparum* and *Trypanosoma brucei brucei*. Antimicrob. Agents Chemother. 52, 3414–3417.

Baldini, F., et al., 2014. Evidence of natural *Wolbachia* infections in field populations of *Anopheles gambiae*. Nat. Commun. 5, 3985.

Basset, Y., et al., 2012. Arthropod diversity in a tropical forest. Science 338 (6113), 1481–1484.

Baum, J., et al., 2011. A research agenda for malaria eradication: basic science and enabling technologies. PLoS Med. 8 (1).

Beier, J.C., Muller, G.C., Gu, W.D., Arheart, K.L., Schlein, Y., 2012. Attractive toxic sugar bait (ATSB) methods decimate populations of *Anopheles* malaria vectors in arid environments regardless of the local availability of favoured sugar-source blossoms. Malar. J. 11, 31.

Bhatnagar, R.K., et al., 2003. Synthetic propeptide inhibits mosquito midgut chitinase and blocks sporogonic development of malaria parasite. Biochem. Biophys. Res. Commun. 304 (4), 783–787.

Bian, G., et al., 2013. *Wolbachia* invades *Anopheles stephensi* populations and induces refractoriness to *Plasmodium* infection. Science 340 (6133), 748–751.

Bian, G., Xu, Y., Lu, P., Xie, Y., Xi, Z., 2010. The endosymbiotic bacterium *Wolbachia* induces resistance to dengue virus in *Aedes aegypti*. PLoS Pathog. 6 (4), e1000833.

Boissiere, A., et al., 2012. Midgut microbiota of the malaria mosquito vector *Anopheles gambiae* and interactions with *Plasmodium falciparum* infection. PLoS Pathog. 8 (5), e1002742.

Brown, A.E., Bugeon, L., Crisanti, A., Catteruccia, F., 2003. Stable and heritable gene silencing in the malaria vector *Anopheles stephensi*. Nucleic Acids Res. 31 (15), e85.

Carlson, J., Suchman, E., Buchatsky, L., 2006. Densoviruses for control and genetic manipulation of mosquitoes. Adv. Virus Res. 68, 361–392.

Catteruccia, F., Godfray, H.C., Crisanti, A., 2003. Impact of genetic manipulation on the fitness of *Anopheles stephensi* mosquitoes. Science 299 (5610), 1225–1227.

Chang, T.L., et al., 2003. Inhibition of HIV infectivity by a natural human isolate of *Lactobacillus jensenii* engineered to express functional two-domain CD4. Proc. Natl. Acad. Sci. U.S.A. 100 (20), 11672–11677.

Christophides, G.K., Vlachou, D., Kafatos, F.C., 2004. Comparative and functional genomics of the innate immune system in the malaria vector *Anopheles gambiae*. Immunol. Rev. 198, 127–148.

Cirimotich, C.M., et al., 2011. Natural microbe-mediated refractoriness to *Plasmodium* infection in *Anopheles gambiae*. Science 332 (6031), 855–858.

Conde, R., Zamudio, F.Z., Rodriguez, M.H., Possani, L.D., 2000. Scorpine, an anti-malaria and anti-bacterial agent purified from scorpion venom. FEBS Lett. 471 (2–3), 165–168.

Coon, K.L., Vogel, K.J., Brown, M.R., Strand, M.R., 2014. Mosquitoes rely on their gut microbiota for development. Mol. Ecol. 23 (11), 2727–2739.

Corby-Harris, V., et al., 2010. Activation of Akt signaling reduces the prevalence and intensity of malaria parasite infection and lifespan in *Anopheles stephensi* mosquitoes. PLoS Pathog. 6 (7), e1001003.

Coutinho-Abreu, I.V., Zhu, K.Y., Ramalho-Ortigao, M., 2010. Transgenesis and paratransgenesis to control insect-borne diseases: current status and future challenges. Parasitol. Int. 59 (1), 1–8.

Cui, C., Chen, J., Wang, S., 2015. Genetic control and paratransgenesis of mosquito-borne diseases. Chin. J. Appl. Entomol. 52, 1061–1071.

Damiani, C., et al., 2008. Paternal transmission of symbiotic bacteria in malaria vectors. Curr. Biol. 18 (23), R1087–R1088.

de Lara Capurro, M., et al., 2000. Virus-expressed, recombinant single-chain antibody blocks sporozoite infection of salivary glands in *Plasmodium gallinaceum*-infected *Aedes aegypti*. Am. J. Trop. Med. Hyg. 62 (4), 427–433.

Djadid, N.D., et al., 2011. Identification of the midgut microbiota of *An. stephensi* and *An. maculipennis* for their application as a paratransgenic tool against malaria. PLoS One 6 (12).

Dong, Y., et al., 2011. Engineered anopheles immunity to *Plasmodium* infection. PLoS Pathog. 7 (12), e1002458.

Dong, Y., Manfredini, F., Dimopoulos, G., 2009. Implication of the mosquito midgut microbiota in the defense against malaria parasites. PLoS Pathog. 5 (5), e1000423.

Drexler, A.L., Vodovotz, Y., Luckhart, S., 2008. *Plasmodium* development in the mosquito: biology bottlenecks and opportunities for mathematical modeling. Trends Parasitol. 24 (8), 333–336.

Durvasula, R.V., et al., 1997. Prevention of insect-borne disease: an approach using transgenic symbiotic bacteria. Proc. Natl. Acad. Sci. U.S.A. 94 (7), 3274–3278.

Dutra, H.L., et al., 2016. *Wolbachia* blocks currently circulating zika virus isolates in Brazilian *Aedes aegypti* mosquitoes. Cell Host Microbe 19 (6), 771–774.

Enayati, A., Hemingway, J., 2010. Malaria management: past, present, and future. Annu. Rev. Entomol. 55, 569–591.

Engel, P., Moran, N.A., 2013. The gut microbiota of insects - diversity in structure and function. FEMS Microbiol. Rev. 37 (5), 699–735.

Fang, W.G., et al., 2011. Development of transgenic fungi that kill human malaria parasites in mosquitoes. Science 331 (6020), 1074–1077.

Favia, G., et al., 2007. Bacteria of the genus *Asaia* stably associate with *Anopheles stephensi*, an Asian malarial mosquito vector. Proc. Natl. Acad. Sci. U.S.A. 104 (21), 9047–9051.

Favia, G., et al., 2008. Bacteria of the genus *Asaia*: a potential paratransgenic weapon against malaria. Adv. Exp. Med. Biol. 627, 49–59.

Gantz, V.M., et al., 2015. Highly efficient Cas9-mediated gene drive for population modification of the malaria vector mosquito *Anopheles stephensi*. Proc. Natl. Acad. Sci. U.S.A. 112 (49), E6736–E6743.

Gao, B., et al., 2010. Characterization of two linear cationic antimalarial peptides in the scorpion *Mesobuthus eupeus*. Biochimie 92 (4), 350–359.

Gao, B., Rodriguez Mdel, C., Lanz-Mendoza, H., Zhu, S., 2009. AdDLP, a bacterial defensin-like peptide, exhibits anti-Plasmodium activity. Biochem. Biophys. Res. Commun. 387 (2), 393–398.

Ghosh, A.K., et al., 2009. Malaria parasite invasion of the mosquito salivary gland requires interaction between the *Plasmodium* TRAP and the *Anopheles* saglin proteins. PLoS Pathog. 5 (1).

Ghosh, A., Edwards, M.J., Jacobs-Lorena, M., 2000. The journey of the malaria parasite in the mosquito: hopes for the new century. Parasitol. Today 16 (5), 196–201.

Ghosh, A.K., Ribolla, P.E.M., Jacobs-Lorena, M., 2001. Targeting *Plasmodium* ligands on mosquito salivary glands and midgut with a phage display peptide library. Proc. Natl. Acad. Sci. U.S.A. 98 (23), 13278–13281.

Ghosh, A.K., Coppens, I., Gardsvoll, H., Ploug, M., Jacobs-Lorena, M., 2011. *Plasmodium* ookinetes coopt mammalian plasminogen to invade the mosquito midgut. Proc. Natl. Acad. Sci. U.S.A. 108 (41), 17153–17158.

Greenwood, B.M., et al., 2008. Malaria: progress, perils, and prospects for eradication. J. Clin. Invest. 118 (4), 1266–1276.

Gwadz, R.W., et al., 1989. Effects of magainins and cecropins on the sporogonic development of malaria parasites in mosquitoes. Infect. Immun. 57 (9), 2628–2633.

Hammond, A., et al., 2016. A CRISPR-Cas9 gene drive system targeting female reproduction in the malaria mosquito vector *Anopheles gambiae*. Nat. Biotechnol. 34 (1), 78–83.

Hilgenboecker, K., Hammerstein, P., Schlattmann, P., Telschow, A., Werren, J.H., 2008. How many species are infected with *Wolbachia*?–A statistical analysis of current data. FEMS Microbiol. Lett. 281 (2), 215–220.

Hoffmann, A.A., et al., 2011. Successful establishment of *Wolbachia* in *Aedes* populations to suppress dengue transmission. Nature 476 (7361), 454–457.

Hughes, G.L., Koga, R., Xue, P., Fukatsu, T., Rasgon, J.L., 2011. *Wolbachia* infections are virulent and inhibit the human malaria parasite *Plasmodium falciparum* in *Anopheles gambiae*. PLoS Pathog. 7 (5), e1002043.

Hurwitz, I., et al., 2011. Paratransgenic control of vector borne diseases. Int. J. Biol. Sci. 7 (9), 1334–1344.

Ignatova, E.A., Nagornaia, S.S., Povazhnaia, T.N., Ianishevskaia, G.S., 1996. The yeast flora of blood-sucking mosquitoes. Mikrobiol Z. 58 (2), 12–15.

Isaacs, A.T., et al., 2011. Engineered resistance to *Plasmodium falciparum* development in transgenic *Anopheles stephensi*. PLoS Pathog. 7 (4), e1002017.

Ito, J., Ghosh, A., Moreira, L.A., Wimmer, E.A., Jacobs-Lorena, M., 2002. Transgenic anopheline mosquitoes impaired in transmission of a malaria parasite. Nature 417 (6887), 452–455.

Jaynes, J.M., et al., 1988. In vitro cytocidal effect of novel lytic peptides on *Plasmodium falciparum* and *Trypanosoma cruzi*. FASEB J. 2 (13), 2878–2883.

Kambris, Z., Cook, P.E., Phuc, H.K., Sinkins, S.P., 2009. Immune activation by life-shortening *Wolbachia* and reduced filarial competence in mosquitoes. Science 326 (5949), 134–136.

Kim, W., et al., 2004. Ectopic expression of a cecropin transgene in the human malaria vector mosquito *Anopheles gambiae* (Diptera: Culicidae): effects on susceptibility to *Plasmodium*. J. Med. Entomol. 41 (3), 447–455.

Kokoza, V., et al., 2010. Blocking of *Plasmodium* transmission by cooperative action of Cecropin A and Defensin A in transgenic *Aedes aegypti* mosquitoes. Proc. Natl. Acad. Sci. U.S.A. 107 (18), 8111–8116.

Li, F., Patra, K.P., Vinetz, J.M., 2005. An anti-Chitinase malaria transmission-blocking single-chain antibody as an effector molecule for creating a *Plasmodium falciparum*-refractory mosquito. J. Infect Dis. 192 (5), 878–887.

Li, C., Marrelli, M.T., Yan, G., Jacobs-Lorena, M., 2008. Fitness of transgenic *Anopheles stephensi* mosquitoes expressing the SM1 peptide under the control of a vitellogenin promoter. J. Hered. 99 (3), 275–282.

Lindh, J.M., Terenius, O., Eriksson-Gonzales, K., Knols, B.G.J., Faye, I., 2006. Re-introducing bacteria in mosquitoes - a method for determination of mosquito feeding preferences based on coloured sugar solutions. Acta Tropica 99 (2–3), 173–183.

Lindh, J.M., Borg-Karlson, A.K., Faye, I., 2008. Transstadial and horizontal transfer of bacteria within a colony of *Anopheles gambiae* (Diptera: Culicidae) and oviposition response to bacteria-containing water. Acta Tropica 107 (3), 242–250.

Ma, M., et al., 2011. Discovery of DNA viruses in wild-caught mosquitoes using small RNA high throughput sequencing. PLoS One 6 (9), e24758.

Maciel, C., et al., 2008. Anti-plasmodium activity of angiotensin II and related synthetic peptides. PLoS One 3 (9), e3296.

Marrelli, M.T., Li, C.Y., Rasgon, J.L., Jacobs-Lorena, M., 2007. Transgenic malaria-resistant mosquitoes have a fitness advantage when feeding on *Plasmodium*-infected blood. Proc. Natl. Acad. Sci. U.S.A. 104 (13), 5580–5583.

McLean, K.J., Jacobs-Lorena, M., 2016. Genetic control of malaria mosquitoes. Trends Parasitol. 32 (3), 174–176.

Meister, S., et al., 2005. Immune signaling pathways regulating bacterial and malaria parasite infection of the mosquito *Anopheles gambiae*. Proc. Natl. Acad. Sci. U.S.A. 102 (32), 11420–11425.

Meister, S., et al., 2009. *Anopheles gambiae* PGRPLC-mediated defense against bacteria modulates infections with malaria parasites. PLoS Pathog. 5 (8), e1000542.

Minard, G., Mavingui, P., Moro, C.V., 2013. Diversity and function of bacterial microbiota in the mosquito holobiont. Parasites Vectors 6, 146.

Moll, R.M., Romoser, W.S., Modrzakowski, M.C., Moncayo, A.C., Lerdthusnee, K., 2001. Meconial peritrophic membranes and the fate of midgut bacteria during mosquito (Diptera: Culicidae) metamorphosis. J. Med. Entomol. 38 (1), 29–32.

Moreira, L.A., et al., 2002. Bee venom phospholipase inhibits malaria parasite development in transgenic mosquitoes. J. Biol. Chem. 277 (43), 40839–40843.

Moreira, C.K., et al., 2007. Effect of the antimicrobial peptide gomesin against different life stages of *Plasmodium* spp. Exp. Parasitol. 116 (4), 346–353.

Moreira, L.A., et al., 2009. A *Wolbachia* symbiont in *Aedes aegypti* limits infection with *dengue*, *Chikungunya*, and *Plasmodium*. Cell 139 (7), 1268–1278.

Osei-Poku, J., Mbogo, C.M., Palmer, W.J., Jiggins, F.M., 2012. Deep sequencing reveals extensive variation in the gut microbiota of wild mosquitoes from Kenya. Mol. Ecol. 21 (20), 5138–5150.

Possani, L.D., Zurita, M., Delepierre, M., Hernandez, F.H., Rodriguez, M.H., 1998. From noxiustoxin to Shiva-3, a peptide toxic to the sporogonic development of *Plasmodium berghei*. Toxicon 36 (11), 1683–1692.

Powell, J.R., Petrarca, V., della Torre, A., Caccone, A., Coluzzi, M., 1999. Population structure, speciation, and introgression in the *Anopheles gambiae* complex. Parasitology 41 (1–3), 101–113.

Pumpuni, C.B., Demaio, J., Kent, M., Davis, J.R., Beier, J.C., 1996. Bacterial population dynamics in three anopheline species: the impact on *Plasmodium* sporogonic development. Am. J. Trop. Med. Hyg. 54 (2), 214–218.

Pusey, P.L., 1997. Crab apple blossoms as a model for research on biological control of fire blight. Phytopathology 87 (11), 1096–1102.

Pusey, P.L., 2002. Biological control agents for fire blight of apple compared under conditions limiting natural dispersal. Plant Dis. 86 (6), 639–644.

Rani, A., Sharma, A., Rajagopal, R., Adak, T., Bhatnagar, R.K., 2009. Bacterial diversity analysis of larvae and adult midgut microflora using culture-dependent and culture-independent methods in lab-reared and field-collected *Anopheles stephensi*-an Asian malarial vector. BMC Microbiol. 9, 96.

Rao, S., et al., 2005. Toward a live microbial microbicide for HIV: commensal bacteria secreting an HIV fusion inhibitor peptide. Proc. Natl. Acad. Sci. U.S.A. 102 (34), 11993–11998.

Ren, X., Hoiczyk, E., Rasgon, J.L., 2008. Viral paratransgenesis in the malaria vector *Anopheles gambiae*. PLoS Pathog. 4 (8), e1000135.

Ricci, I., et al., 2011a. The yeast *Wickerhamomyces anomalus* (*Pichia anomala*) inhabits the midgut and reproductive system of the Asian malaria vector *Anopheles stephensi*. Environ. Microbiol. 13 (4), 911–921.

Ricci, I., et al., 2011b. Different mosquito species host *Wickerhamomyces anomalus* (*Pichia anomala*): perspectives on vector-borne diseases symbiotic control. Antonie Van Leeuwenhoek 99 (1), 43–50.

Riehle, M.A., Jacobs-Lorena, M., 2005. Using bacteria to express and display anti-parasite molecules in mosquitoes: current and future strategies. Insect Biochem. Mol. 35 (7), 699–707.

Riehle, M.A., Moreira, C.K., Lampe, D., Lauzon, C., Jacobs-Lorena, M., 2007. Using bacteria to express and display anti-*Plasmodium* molecules in the mosquito midgut. Int. J. Parasitol. 37 (6), 595–603.

Romoser, W.S., Moll, R.M., Moncayo, A.C., Lerdthusnee, K., 2000. The occurrence and fate of the meconium and meconial peritrophic membranes in pupal and adult mosquitoes (Diptera: Culicidae). J. Med. Entomol. 37 (6), 893–896.

Shao, L., Devenport, M., Jacobs-Lorena, M., 2001. The peritrophic matrix of hematophagous insects. Arch. Insect Biochem. Physiol. 47 (2), 119–125.

Shaw, W.R., et al., 2016. *Wolbachia* infections in natural *Anopheles* populations affect egg laying and negatively correlate with *Plasmodium* development. Nat. Commun. 7, 11772.

Straif, S.C., et al., 1998. Midgut bacteria in *Anopheles gambiae* and *An. funestus* (Diptera: Culicidae) from Kenya and Mali. J. Med. Entomol. 35 (3), 222–226.

Tian, C., et al., 2008. Gene expression, antiparasitic activity, and functional evolution of the drosomycin family. Mol. Immunol. 45 (15), 3909–3916.

Trape, J.F., et al., 2011. Malaria morbidity and pyrethroid resistance after the introduction of insecticide-treated bednets and artemisinin-based combination therapies: a longitudinal study. Lancet Infect Dis. 11 (12), 925–932.

Vega-Rodriguez, J., et al., 2014. Multiple pathways for *Plasmodium* ookinete invasion of the mosquito midgut. Proc. Natl. Acad. Sci. U.S.A. 111 (4), E492–E500.

Vizioli, J., et al., 2001. Gambicin: a novel immune responsive antimicrobial peptide from the malaria vector *Anopheles gambiae*. Proc. Natl. Acad. Sci. U.S.A. 98 (22), 12630–12635.

Walker, T., Moreira, L.A., 2011. Can *Wolbachia* be used to control malaria? Mem. Oswaldo Cruz 106, 212–217.

Wang, S., et al., 2012. Fighting malaria with engineered symbiotic bacteria from vector mosquitoes. Proc. Natl. Acad. Sci. U.S.A. 109 (31), 12734–12739.

Wang, S., Jacobs-Lorena, M., 2013. Genetic approaches to interfere with malaria transmission by vector mosquitoes. Trends Biotechnol. 31 (3), 185–193.

Wang, Y., Gilbreath 3rd, T.M., Kukutla, P., Yan, G., Xu, J., 2011. Dynamic gut microbiome across life history of the malaria mosquito *Anopheles gambiae* in Kenya. PLoS One 6 (9), e24767.

Ward, T.W., et al., 2001. *Aedes aegypti* transducing densovirus pathogenesis and expression in *Aedes aegypti* and *Anopheles gambiae* larvae. Insect Mol. Biol. 10 (5), 397–405.

Weiss, B., Aksoy, S., 2011. Microbiome influences on insect host vector competence. Trends Parasitol. 27 (11), 514–522.

Werren, J.H., Baldo, L., Clark, M.E., 2008. *Wolbachia*: master manipulators of invertebrate biology. Nat. Rev. Microbiol. 6 (10), 741–751.

Whitten, M.M.A., Shiao, S.H., Levashina, E.A., 2006. Mosquito midguts and malaria: cell biology, compartmentalization and immunology. Parasite Immunol. 28 (4), 121–130.

WHO, 2014. World Malaria Report (World Health Organization).

Yoshida, S., et al., 1999. A single-chain antibody fragment specific for the *Plasmodium berghei* ookinete protein Pbs21 confers transmission blockade in the mosquito midgut. Mol. Biochem. Parasitol. 104 (2), 195–204.

Yoshida, S., et al., 2007. Hemolytic C-type lectin CEL-III from sea cucumber expressed in transgenic mosquitoes impairs malaria parasite development. PLoS Pathog. 3 (12), e192.

Yoshida, S., Ioka, D., Matsuoka, H., Endo, H., Ishii, A., 2001. Bacteria expressing single-chain immunotoxin inhibit malaria parasite development in mosquitoes. Mol. Biochem. Parasitol. 113 (1), 89–96.

Zieler, H., Keister, D.B., Dvorak, J.A., Ribeiro, J.M., 2001. A snake venom phospholipase A(2) blocks malaria parasite development in the mosquito midgut by inhibiting ookinete association with the midgut surface. J. Exp. Biol. 204 (Pt 23), 4157–4167.

Zug, R., Hammerstein, P., 2012. Still a host of hosts for *Wolbachia*: analysis of recent data suggests that 40% of terrestrial arthropod species are infected. PLoS One 7 (6), e38544.

CHAPTER

14

Insulin-Like Peptides Regulate *Plasmodium falciparum* Infection in *Anopheles stephensi*

Jose E. Pietri[1], *Shirley Luckhart*[2]

[1]University of California, Santa Cruz, CA, United States; [2]University of California, Davis, CA, United States

INTRODUCTION

Insulin-like peptides (ILPs) have long been studied in vertebrate organisms due to their roles in human health. Most notably, human insulin has been at the forefront of biomedical research for decades and understanding the mechanisms that regulate its function has led to significant advances in the treatment of metabolic diseases such as diabetes (Le Roith and Zick, 2001). Other human ILPs are also of interest in biomedicine. For example, insulin-like growth factor 1 (IGF-1) has been implicated in various types of cancers and cardiovascular disease (Laughlin et al., 2004; Renehan et al., 2004), while relaxin has been targeted to treat fibrosis and regulate fertility (Van Der Westhuizen et al., 2007).

The first invertebrate ILP was discovered in the silkmoth *Bombyx mori* (Nagasawa et al., 1984). Since then, countless ILPs have been identified in a variety of mollusk and insect species (Wu and Brown, 2006). In these organisms, ILPs act through highly conserved signaling and regulatory pathways to mediate diverse physiological functions (Antonova et al., 2012; Antonova-Koch et al., 2013; Nassel, 2012; Nassel et al., 2013, 2015; Nassel and Vanden Broeck, 2016). Consequently, there exists a large literature on the roles of ILPs in the regulation of invertebrate behavior, metabolism, and reproduction. Now, invertebrate ILPs have also emerged as key mediators of infection and immunity (Luckhart and Riehle, 2007). Much of this work comes from invertebrate models, where studies of ILP biology have focused on the link between insulin signaling and microbial pathogenesis to inform analogous processes in humans. Though these studies have provided some mechanistic information, no relationship between invertebrate ILPs and infection has been more extensively characterized than in the mosquito *Anopheles stephensi*. In *A. stephensi*, multiple ILPs are upregulated during infection with the malaria parasite *Plasmodium falciparum* via parasite manipulation of specific host cell signaling

Arthropod Vector: Controller of Disease Transmission, Volume 1
http://dx.doi.org/10.1016/B978-0-12-805350-8.00014-3

pathways (Marquez et al., 2011; Pietri et al., 2015). These ILPs then activate signal transduction and gene expression to produce distinct, networked effects on mosquito physiology that ultimately alter the course of parasite development and transmission, making them of particular importance to global health (Cator et al., 2015; Pietri et al., 2016a).

In the first half of this chapter, the basic biology of ILP structure, function, and synthesis is discussed. In the second half of the chapter, we detail current understanding of the interplay between *A. stephensi* ILP signaling and *P. falciparum* infection at a mechanistic level.

THE BIOLOGY OF THE INSULIN-LIKE PEPTIDES

Structure and Function of Insulin-Like Peptides

The structure and function of ILPs is highly conserved across the organisms that produce them (Antonova et al., 2012). However, the molecular mechanisms of ILP synthesis and processing are not fully understood in insects. In vertebrates, mature ILPs are comprised of two small polypeptide chains, the B chain and the A chain, which are held together by intramolecular and intermolecular disulfide bonds that are essential for bioactivity. The synthesis of ILPs begins in the cytoplasm where prepro-ILP is synthesized as a single polypeptide with three distinct domains, the B, C, and A chains (Fig. 14.1). This single chain of amino acids contains an N-terminal signal peptide that targets the peptide for translocation into the endoplasmic reticulum. Here, the peptide is folded and three specific disulfide bonds are formed. Two of these bonds are intermolecular, formed between highly conserved cysteine residues of the A and B chains of the peptide, while the third is an intramolecular bond formed between neighboring cysteine residues on the A chain. Following the formation of these disulfide bonds, the signal peptide is cleaved and pro-ILP is exported to the Golgi apparatus for packaging into secretory vesicles. Once in these secretory vesicles, the C chain of the peptide undergoes proteolytic cleavage, yielding mature and biologically active ILP for secretion from the cell. This synthesis and the biological effects of the ILPs are mediated by tightly controlled cell signaling pathways (Fig. 14.2).

Once secreted from the cell, insect ILPs act primarily through the insulin receptor (IR) (Antonova et al., 2012; Antonova-Koch et al., 2013). The IRs are receptor tyrosine kinases (RTKs) that form homodimers and undergo autophosphorylation upon ligand binding. Autophosphorylation of RTKs results in the activation of multiple downstream signaling pathways (Fig. 14.2). However, IR affinities can vary and secondary receptors may support ILP signaling as G-protein coupled receptors (GPCRs) do in vertebrates (Bathgate et al., 2013; Wen et al., 2010). In particular, several orphan GPCRs that are possible orthologs of vertebrate relaxin receptors have been identified in mosquitoes (Vogel et al., 2013). Ultimately, activation of IGF-1 signaling (IIS) results in the regulation of transcriptional responses that modulate an array of physiological effects (Antonova et al., 2012; Antonova-Koch et al., 2013; Nassel, 2012). For instance, in invertebrates IIS regulates growth and development, macronutrient provisioning and metabolism, feeding, sleep, reproduction, and egg laying (Antonova et al., 2012; Antonova-Koch et al., 2013; Brown et al., 2008; Cong et al., 2015; Cator et al., 2015; Defferrari et al., 2016; Liu et al., 2015; Luo et al., 2014; Pietri et al., 2016a; Post and Tatar, 2016; Wigby et al., 2011). Notably, the pathways activated by ILPs also regulate *ILP* transcription (Fig. 14.2) indicating that the synthesis of ILPs is controlled by a feed-forward loop in which the production and secretion of ILPs activates cell signaling that increases ILP production (Bouche et al., 2010; Khoo et al., 2003; Leibiger et al., 1998). ILP synthesis and secretion is not only regulated by nutritional signals such as glucose, but also by a variety of

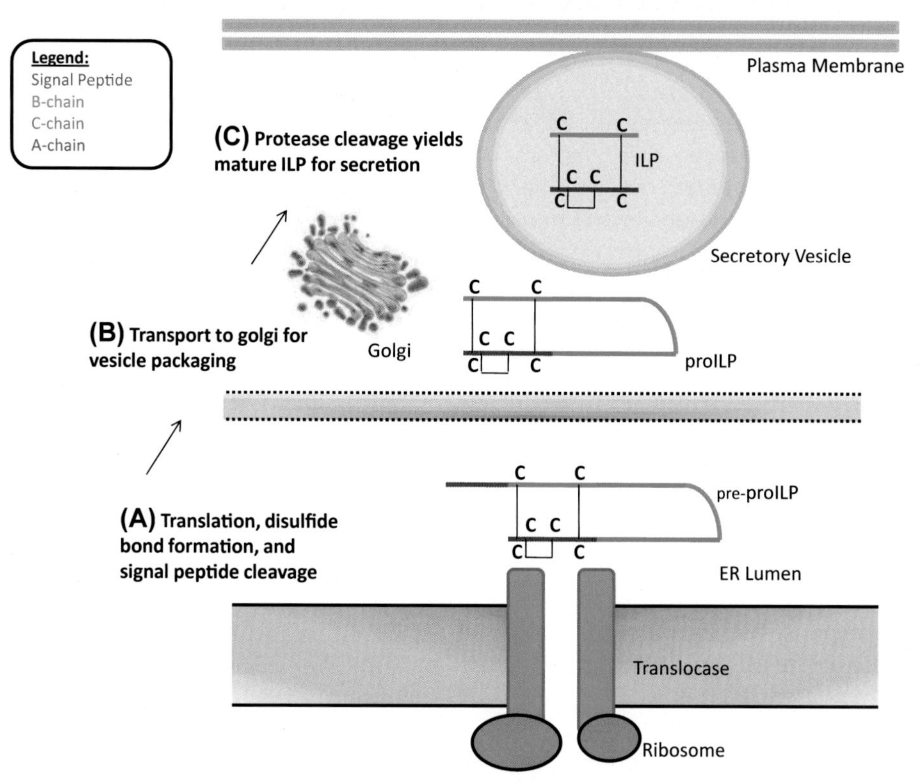

FIGURE 14.1 **Synthesis and structure of insulin-like peptides (ILPs).** ILPs are characterized by a highly conserved structure and mode of synthesis. In the endoplasmic reticulum of the cell, prepro-ILPs are translated as a single chain of amino acids consisting of a signal peptide and three domains (A–C). Following folding, three specific disulfide bonds are formed and the signal peptide is cleaved. In the Golgi, the pro-ILP is packaged into a secretory vesicle where it undergoes further proteolytic processing to remove the C-chain, yielding mature ILP that can then be secreted into the extracellular environment.

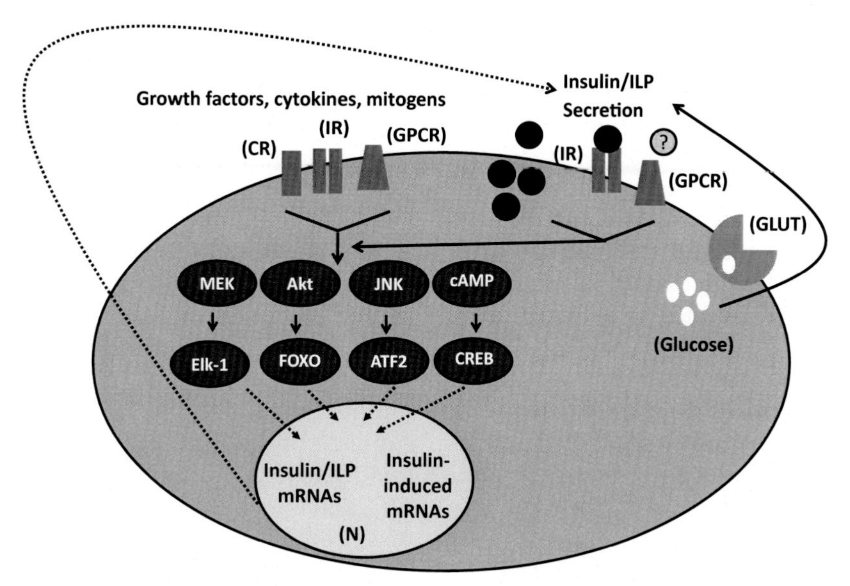

FIGURE 14.2 **Pathways that regulate insulin synthesis and response to insulin stimulation.** Pathways that regulate insulin transcription and secretion and responses to insulin stimulation are widely conserved across vertebrates and invertebrates. *CR*, cytokine receptor; *GLUT*, glucose transporter; *GPCR*, G-protein coupled receptor; (?), insulin-like ligands; *IR*, insulin receptor; *JNK*, c-Jun N-terminal kinase; *N*, nucleus; *RTK*, receptor tyrosine kinase.

other important stimuli such as growth factors, mitogens, and cytokines (Leibiger et al., 2002; Melloul et al., 2002). In insects, these factors include: multiple ILPs, serotonin, octopamine, GABA, neuropeptide F, corazonin, tachykinin, the glucose-sensitive hormone adiponectin, and juvenile hormone (Nassel et al., 2013; Nassel and Vanden Broeck, 2016).

Insulin-Like Peptides, Infection, and Immunity

To date, most research on invertebrate ILPs has focused on their neuroendocrine functions. However, more recently ILPs have emerged as regulators of invertebrate immunity and host responses to infection, balancing metabolism and the immune response for host defense. In the fruit fly *Drosophila melanogaster*, IIS and immunity are inversely related. That is, activation of the pathogen sensing Toll signaling pathway by an immune challenge reduces endogenous insulin signaling, likely as a means to divert energy resources to the mounting of an efficient immune response (DiAngelo et al., 2009). Further, activation of IIS during *Mycobacterium* infection can contribute to pathological wasting in *D. melanogaster*, suggesting that metabolic provisioning by IIS is critical for the host response to infection (Dionne et al., 2006). In the nematode *Caenorhabditis elegans*, virulence factor-mediated ILP synthesis reduces the host immune response, resulting in increased bacterial growth (Evans et al., 2008a). The role of IIS in immunosuppression in the nematode is further supported by the fact that DAF-2 (IR) and DAF-16 (downstream forkhead transcription factor-FOXO) mutants are resistant to bacterial infection (Evans et al., 2008b; Garsin et al., 2003). Bacterial infection in the razor clam and endoparasitism of the cabbage moth by a Hymenopteran wasp also alter host ILP expression, suggesting that the relationship between insulin signaling and immunity is conserved across invertebrate phyla (Kumar et al., 2016; Niu et al., 2016).

In mosquitoes, IIS regulates responses to infection in a manner similar to that observed in other invertebrates, with the important distinction that blood-feeding creates a direct interface between ingested host insulin and IGF-1 with mosquito midgut IIS. In this context, ingested human insulin (Pakpour et al., 2012) and human IGF-1 (Drexler et al., 2013, 2014) control resistance to *P. falciparum* in *A. stephensi*. Specifically, insulin decreases parasite resistance by dampening nuclear factor kappa-light-chain-enhancer (NF-kB) mediated immune responses in the midgut (Pakpour et al., 2012), while IGF-1 appears to increase resistance by enhancing the production of reactive oxygen species and altering mitochondrial integrity and midgut homeostasis (Drexler et al., 2014). In addition to these observations, human insulin drives a feed-forward loop of ILP synthesis in *A. stephensi* that is dependent on mosquito IIS (Marquez et al., 2011; Pietri et al., 2015), suggesting that the host blood-feeding interface provides important context for understanding the regulation and function of mosquito ILPs. This context and biology are a focus of the remainder of this chapter.

REGULATION OF INSULIN-LIKE PEPTIDE SYNTHESIS DURING *PLASMODIUM* INFECTION

Tissue-Specific Production of Insulin-Like Peptides

The *A. stephensi* genome encodes five ILPs (Marquez et al., 2011) that are expressed in multiple tissues (Fig. 14.3). In particular, mRNAs for *ILP1-5* have been detected to varying degrees in the head, thorax, abdomen, and midgut (Marquez et al., 2011). *ILP4* and *ILP5* are expressed at the highest levels (up to 0.4–0.5-fold relative to *ribosomal protein s7* RNA), followed by *ILP2* and *ILP3* which are expressed to an intermediate degree (up to ~0.1-fold relative to *ribosomal*

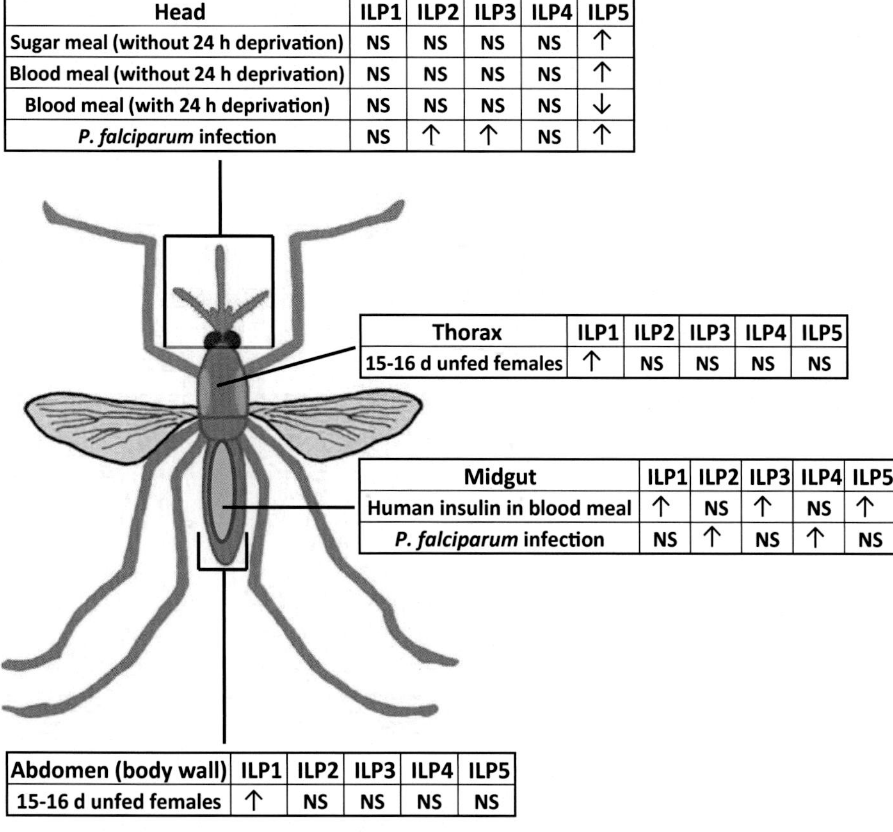

FIGURE 14.3 Expression of insulin-like peptide (ILP) genes in *Anopheles stephensi* is dynamic and tightly controlled. In *A. stephensi*, tissue-specific ILP expression is minimally responsive to dietary changes or aging, but responds robustly to activation of insulin signaling and infection with *Plasmodium falciparum*. Experimental manipulations that produced significant increases in the expression of a particular *ILP* in a tissue are denoted by *upward facing arrows*, while significant decreases are denoted as *downward facing arrows*. *NS*, no significant change. *Re-published with permission from Marquez, A.G., Pietri, J.E., Smithers, H.M., Nuss, A., Antonova, Y., Drexler, A.L., Riehle, M.A., Brown, M.R., Luckhart, S., 2011. Insulin-like peptides in the mosquito* Anopheles stephensi: *identification and expression in response to diet and infection with* Plasmodium falciparum. *Gen. Comp. Endocrinol.* http://dx.doi.org/10.1016/j.ygcen.2011.06.005.

protein s7 RNA). *ILP1* is expressed at much lower levels (0.0–0.001-fold relative to *ribosomal protein s7* RNA), and in many cases is difficult to detect by quantitative real-time polymerase chain reaction (qRT-PCR). Generally, expression of all five *ILPs* is higher in the head and midgut than in the thorax and abdomen. As determined by immunostaining, neurosecretory cells (NSCs) of the *A. stephensi* brain are the primary cellular sources of ILPs in the head and peptides produced here can be transferred to the thorax and abdomen through the axons of NSCs (Marquez et al., 2011). However, ILPs are also found in ganglia and muscle bands in the abdomen, as well as along the midgut wall (Marquez et al., 2011). Though no immunoreactivity has been observed in the midgut endocrine cells of *A. stephensi*, anti-ILP antibodies stain these cells in the yellow fever mosquito, *Aedes aegypti* (Riehle et al., 2006), and in *D. melanogaster*, gut endocrine cells generate

ILPs (Nassel, 2012), suggesting that midgut endocrine cells may be undetected sources of ILPs in *A. stephensi*. Given this tissue distribution in *A. stephensi*, ILPs are localized to impact the development and transmission of *P. falciparum*.

The Effects of Diet and Age on Insulin-Like Peptide Synthesis

Insulin signaling is an important regulator of both growth and metabolism. However, expression of ILPs in adult female *A. stephensi* is only modestly sensitive to age and dietary changes (Fig. 14.3). In particular, examination of young (3–4 days postemergence) and old (15–16 days postemergence) female mosquitoes by qRT-PCR revealed that expression of only *ILP1* changed with age and this was reflected as increased expression in the thorax and abdomen relative to younger females (Marquez et al., 2011). Though the significance of this phenomenon is unknown, ILP1 is likely involved in regulating homeostatic mechanisms important to aging and development, as is the case in *Drosophila* where *ILP1* in regulated by juvenile hormone (Liu et al., 2016). As with aging, only minor changes in *ILP* expression were observed in response to dietary variation in female *A. stephensi* (Marquez et al., 2011). Expression of *ILP1–4* was not responsive to starvation, sugar-feeding, or blood-feeding, while *ILP5* expression increased in response to either food source. Notably, this response occurred only in the head and was blunted by a 24-h period of starvation before feeding, suggesting that *ILP5* may control neuronal responses to food, such as satiety and the sensing of olfactory cues, in a manner similar to that observed in other insects and for other ILPs (Carlsson et al., 2010; Cator et al., 2015; Liu et al., 2012; Root et al., 2011). Although a lack of marked *ILP* transcriptional responses to diet and female age was somewhat unexpected, this does not preclude ILP control of nutrition, growth, and metabolic homeostasis as these processes may be dependent on post-transcriptional regulation of *ILP*s as is the case in *Drosophila* (Nassel et al., 2015). Collectively, however, these observations suggested an important possibility that *ILP* expression is fine-tuned to alternative stimuli that regulate mosquito physiology beyond central metabolism, such as the host response to infection.

Fine-Tuning of Insulin-Like Peptide Expression to Parasite Infection Through the Insulin/Insulin-Like Growth Factor Signaling Pathway

Infection of *A. stephensi* via ingestion of a blood meal with *P. falciparum*-infected red blood cells significantly and notably induced the expression of *ILP2* and *ILP4* in the midgut, while expression of *ILP2*, *3*, and *5* was induced in the head (Marquez et al., 2011). Interestingly, induction of *ILP* expression by parasites was more rapid in the midgut [6 h postinfection (PI)] than in the head (24–48 h PI), suggesting that the former tissue is a key insulin signaling hub that mediates ILP-dependent responses throughout the body. The response to live parasites has been recapitulated in both the midgut and in immortalized *A. stephensi* cells in vitro by stimulation with purified *P. falciparum* products provided in a blood meal or added directly to cell culture medium, respectively (Pietri et al., 2015). In these experiments, the response occurred much more rapidly (1–3 h post stimulation) suggesting that parasites must be digested or given time to secrete intracellular factors in order to induce *ILP* expression during a natural infection. Further, *ILP* expression following treatment with purified *P. falciparum* products suggested that host immune challenge might act as a general stimulus for the production of ILPs. However, *ILP* expression was not induced by fungal zymosan or bacterial lipopolysaccharide (Pietri et al., 2015). Further studies revealed that *P. falciparum* activates mosquito *ILP* expression by hijacking the host IIS. Specifically, soluble products from *P. falciparum* likely act as insulin mimics to induce phosphorylation of extracellular signal-regulated kinase (ERK), and this activity is required to achieve induction of *ILP* expression (Pietri et al., 2015).

Though the parasite molecules responsible for IIS activation have not been determined, glycosylphosphatidylinositols (GPIs) can mimic the action of insulin in the midgut of *A. stephensi* by inducing ERK activation (Lim et al., 2005). In addition to ERK signaling, basal activity of phosphoinositide-3 kinase (PI3K), which signals an alternate pathway of IIS, is required for ILP synthesis (Pietri et al., 2015). As such, ILP synthesis in the midgut during *P. falciparum* infection is regulated by a combination of both parasite-induced and host-driven factors.

INSULIN-LIKE PEPTIDE REGULATION OF *ANOPHELES STEPHENSI* PHYSIOLOGY DURING *PLASMODIUM* INFECTION

Insulin-Like Peptide Regulation of Cell Signaling

The induction of ILP synthesis in the midgut during *P. falciparum* infection can initiate a myriad of physiological responses that are mediated by changes in host cell signaling. While the signaling pathways activated by ILPs are widely conserved across phyla (Antonova et al., 2012; Antonova-Koch et al., 2013), there nonetheless exist significant species, stimulus, and ILP-specific variation. The ILPs of *A. stephensi* are no exception. In vertebrates, insulin binding to the IR results in activation of both ERK and PI3K-Akt signaling, as well as activation of downstream factors such as p70S6K and FOXO (Fig. 14.2). However, ILPs in the relaxin family can signal through alternate RTKs and GPCRs to activate other kinases, such as c-Jun N-terminal kinase (JNK) and p38 mitogen-activated protein kinase (MAPK) (Siddle, 2011). Mosquitoes synthesize multiple ILPs, but possess only a single, genome-encoded mosquito IR homolog (MIR) (Wu and Brown, 2006). In *A. aegypti*, endogenous ILPs have been shown to bind two receptors (Wen et al., 2010), one of which is the MIR and the other an uncharacterized transmembrane protein. In *A. stephensi*, both ingested human insulin and ingested IGF-1 can activate MIR and IIS (Drexler et al., 2013, 2014; Pakpour et al., 2012). Unlike human insulin, however, *A. stephensi* ILP3 and ILP4 do not activate IIS-associated Akt and FOXO and do not activate p70S6K signaling in the mosquito midgut (Pietri et al., 2016a). Given that p70S6K can be activated by amino acid sensing through the target of rapamycin pathway, the lack of p70S6K activation may be due to a saturation of signaling from amino acids from ingested blood (Hansen et al., 2004). Notably, ILP3 alone induced phosphorylation of JNK in the *A. stephensi* midgut, a difference that is likely involved in mediating the differential effects of ILP3 and ILP4 on the midgut physiology discussed below.

Insulin-Like Peptide Effects on Nuclear Factor Kappa-Light-Chain-Enhancer-Mediated Immunity

Previous studies of invertebrate ILPs have connected IIS with host immunity, as discussed earlier in this chapter. In *A. stephensi*, we have shown that ingested human insulin dampens NF-kB-dependent immune responses in the midgut (Pakpour et al., 2012), suggesting consistency of an immunosuppressive role for IIS. Similarly, *A. stephensi* ILP3 and ILP4 are negative modulators of NF-kB-regulated immune effectors that respond to *P. falciparum* (Pietri et al., 2015, 2016a). In particular, knockdown of either peptide increased expression of the complement-like factors *APL1*, *LRIM1*, and *TEP1*, in the midgut during *P. falciparum* infection (Pietri et al., 2015). These factors have been shown to function against *P. falciparum* and are controlled by the Imd pathway, a major regulator of immunity in the mosquito (Cirimotich et al., 2009; Mitri et al., 2009). Expression of *nitric oxide synthase*, another potent antiparasite factor, was also increased by knockdown of ILP3 or ILP4 in the midgut (Pietri et al., 2015). Conversely, provision of ILP3 in the context of infection suppressed expression of these genes as did provision of ILP4, which additionally reduced expression of the antimicrobial peptide *defensin* (Pietri et al., 2016a). Interestingly, ILP3 and ILP4 appeared to modulate these responses through kinetically distinct mechanisms. That is, ILP4 regulates host immunity within 1–6h PI, primarily influencing survival of *P. falciparum* ookinetes prior to invasion of the midgut epithelium, while ILP3 controls later responses (24h PI) and impacts parasites during and after invasion (e.g., oocysts). Although human insulin can suppress mosquito NF-kB-dependent immune responses through PI3K-Akt signaling, the pathway(s) that transduce the effects of ILP3 and ILP4 on *A. stephensi* immunity have not yet been identified. Here, conservation of signaling suggests that observations in *D. melanogaster* may be informative. In the fly, the Toll and immune deficiency (Imd) pathways that converge on NF-kB transcription, as well as the immune signaling JAK-STAT pathway (Boutros et al., 2002; De Gregorio et al., 2002) are networked with PI3K-Akt and FOXO of IIS and with signaling by the MAPKs, JNK, and ERK (Becker et al., 2010; Delaney et al., 2006; Kallio et al., 2005; Nemoto et al., 1998; Park et al., 2004; Zhao and Lee, 1999). Accordingly, the differential regulation of JNK activation by ILP3 and ILP4 in the context of signaling through canonical IIS may underlie temporal differences in the regulation of *P. falciparum* development in the midgut of *A. stephensi* by ILP3 and ILP4.

Insulin-Like Peptide Effects on Metabolic Homeostasis in the Midgut

The metabolic effects of various ILPs have been extensively studied in several invertebrates (Defferrari et al., 2016; Nassel, 2012; Pietri et al., 2016a; Post and Tatar, 2016; Teleman, 2010). In these organisms, ILPs appear to play an important role in the regulation of carbohydrate, protein, and lipid metabolism. While expression of *A. stephensi ILP3* and *ILP4* did not change significantly in response to diet in adult female mosquitoes (Marquez et al., 2011), we have recently demonstrated that these peptides play significant roles in regulating carbohydrate, protein, and fatty acid metabolism in the midgut, thereby altering tissue homeostasis. In particular, ILP3 and ILP4 induced distinct metabolic shifts in *A. stephensi* (Pietri et al., 2016a). Provision of ILP3 induced a high energy metabolic state in the midgut with an increased capacity to synthesize biomolecules and increased energy production from mitochondria. Provision of ILP4, on the other hand induced an energy-deficient state in which energy demand could not be matched by metabolism. Specifically, ILP4 appeared to reduce energy production from oxidative phosphorylation, resulting in increased glycolysis and lipolysis and decreased synthesis of proteins and nucleic acids. Conversely, ILP3 increased glycolysis and lipolysis and reduced levels of free precursors for the synthesis of biomolecules (e.g., amino acids), consistent with the high energy state induced by this ILP. Importantly, analogous inducible metabolic shifts have been implicated in the regulation of immunity, inflammation, and resistance to infection in other organisms (Chambers et al., 2012; Cheng et al., 2014; DiAngelo et al., 2009; Dionne et al., 2006; Liu et al., 2012), suggesting that ILP3 and ILP4 control parasite development, in part, through changes to midgut metabolic homeostasis.

Insulin-Like Peptide Effects on *Plasmodium falciparum* Development

Knockdown and provision of ILP3 and ILP4 resulted in distinct effects on *P. falciparum* development in the midgut (Table 14.1). Specifically, knockdown of either peptide reduced the prevalence and intensity of oocyst development (Pietri et al., 2015). Notably, provision

TABLE 14.1 The Effects of Insulin-Like Peptides (ILPs) on *Plasmodium falciparum* Development in the Midgut of *Anopheles stephensi*

	Midgut Knockdown		Peptide Feeding	
	Pf Prevalence	**Pf Intensity**	**Pf Prevalence**	**Pf Intensity**
ILP3	↓	↓	↓	—
ILP4	↓	↓	↑	↑

of ILP4 increased the prevalence and intensity of oocyst infection as expected, but feeding of ILP3 decreased oocyst prevalence and had no effect on the intensity of infection (Pietri et al., 2016a). In essence, these observations suggested that ILP4 is both necessary for parasite development and sufficient to increase parasite growth, but while ILP3 is necessary for parasite development, it is not sufficient to increase parasite growth. Both peptides are negative modulators of NF-kB-dependent gene expression in *A. stephensi* (Pietri et al., 2015, 2016a), indicating that ILP-dependent effects on host immunity alone cannot explain the differential effects of ILP3 and ILP4 on infection.

The shifts in midgut metabolism, however, induced by ILP3 and ILP4—to high energy and to energy-deficient states, respectively—suggested a direct, alternative explanation for effects on infection that could be tested directly through manipulation of key enzymatic targets responsible for these metabolic changes. That is, treatment to reverse the metabolic shifts induced by ILP3 or ILP4 could reverse their respective effects on the prevalence and intensity of *P. falciparum* infection. Notably, treatment with heptelidic acid (HA), a small molecule inhibitor of glyceraldehyde 3-phosphate dehydrogenase (GAPDH), induced energy deficiency in ILP3-provisioned mosquitoes and increased the prevalence of infection to control levels (Pietri et al., 2016a). Similarly, provision of 4-hydroxy-L-phenylglycine (4H-PG), an inhibitor of carnitine palmitoyltransferase-1 (CPT1), blocked ILP4-dependent fatty acid oxidation, increased energy availability, and decreased infection levels to control levels (Pietri et al., 2016a). Neither HA nor 4H-PG altered immune gene expression, confirming independent effects of these ILP-induced metabolic shifts on *P. falciparum* development in *A. stephensi* (Pietri et al., 2016a). As such, ILP regulation of infection appears to be primarily driven by the availability of metabolic fuel for the response to infection. Proximally, the effects of JNK signaling and metabolic shifts could also alter the physical barrier of the midgut to infection as midgut dysplasia mediates the effects of IGF-1 on *P. falciparum* infection in *A. stephensi* (Drexler et al., 2014).

INSULIN-LIKE PEPTIDE REGULATION OF *ANOPHELES STEPHENSI* BEHAVIOR AND *PLASMODIUM FALCIPARUM* TRANSMISSION

Invertebrate ILPs have long held the interest of insect endocrinologists and neurobiologists for their ability to modulate behavioral responses. In particular, ILPs control sleep/wake cycles and general activity, as well as feeding (Carlsson et al., 2010; Cong et al., 2015; Liu et al., 2015; Pietri et al., 2016a; Root et al., 2011). Accordingly, it was not surprising that ILP3 and ILP4 could impact neurotransmitter levels and alter behavior in *A. stephensi*. Specifically, changes in host-seeking by *P. falciparum*-infected *A. stephensi* were correlated with low and high levels of *ILP3* and *ILP4* expression in the midgut, suggesting that ILPs are involved in the natural stimulation of blood-feeding during infection

(Cator et al., 2015). Here, during *P. falciparum* oocyst development, when mosquitoes are not infectious to humans, ILP levels, host-seeking behavior, and the ability to respond to olfactory cues were reduced relative to uninfected mosquitoes (Cator et al., 2013). However, when salivary gland sporozoites were present and mosquitoes were infectious, ILP levels, host-seeking, and olfactory responses to host odors were greatly increased relative to uninfected mosquitoes (Cator et al., 2013). Manipulation of ILP levels, however, indicated further complexity in this biology. Specifically, knockdown of ILP3 and ILP4 reduced successful engorgement of a first blood meal in *A. stephensi* (Cator et al., 2015), while provisioning of ILP3 or ILP4 increased the propensity of mosquitoes to take a second blood meal 3 days later (Pietri et al., 2016a). Further, metabolic changes suggested that ILP3 stimulated an orexigenic response to increasing hunger signals (e.g., neurotransmitters), while ILP4-induced feeding was likely driven by an energy-deficient state and unmet energy demands (Pietri et al., 2016a). Provisioning of ILP4, but not ILP3, also reduced female *A. stephensi* flight activity 24 h after a blood meal, an effect that was also likely related to the energy-deficient state induced by ILP4. Hence, the effects of ILPs on *A. stephensi* behavior are likely mediated by a combination of their impacts on metabolism, as discussed above, but also by changes in midgut neurotransmitter levels (Pietri et al., 2016a). Feeding of ILP3 triggered rapid, short-lived increases in Tyr (a precursor for dopamine), Gly, and 5-methoxytryptamine (5MT; a derivative closely associated with serotonin and melatonin). Meanwhile, feeding of ILP4 triggered rapid increases in taurine levels, but also delayed increases in Gly, Tyr, Asp, and gamma-hydroxybutyrate (GHB, a derivative of GABA). Thus, up- and downregulation of inhibitory and excitatory neurotransmitters in the midgut preceded and likely contributed to behavioral alterations associated with ILP3 and ILP4.

CONCLUSIONS AND FUTURE DIRECTIONS

As discussed in this chapter, ILP3 and ILP4 are key mediators of an array of physiological processes in *A. stephensi* that influence *P. falciparum* development and transmission (Table 14.2). As such, it may be possible to leverage ILP biology in several ways to create transgenic mosquitoes for the control of malaria parasite transmission. It has been established that knockdown of ILP transcripts in the midgut can increase resistance to *P. falciparum*, while also reducing blood-feeding propensity (Cator et al., 2015; Pietri et al., 2015). ILPs have also been shown to regulate egg production in *A. aegypti* (Brown et al., 2008; Gulia-Nuss et al., 2011). However, this physiology has not been examined in *A. stephensi* or in the context of genetic manipulation of ILPs. Similarly, ingested human insulin and IGF-1 can modulate *A. stephensi* life span (Drexler et al., 2013), as can ILP6 in *Drosophila* (Bai et al., 2012) suggesting that the endogenous mosquito ILPs may play a

TABLE 14.2 ILP-Mediated Effects on the Midgut Physiology and Behavior of *Anopheles stephensi*

	ILP3 Feeding	ILP4 Feeding
ERK activation	Yes	Yes
JNK activation	Yes	No
NK-kB immunity	Decrease	Decrease
Glycolysis	High	Low
Fat deposits	Decrease	Decrease
Bloodmeal-seeking	Increase	Increase
Nutrient intake	Increased protein	No change
Flight	No change	Decrease
Neuro transmitters	↑Excitatory at 1 h	↑Excitatory at 24 h

ERK, extracellular signal-regulated kinase; *ILP*, insulin-like peptides; *JNK*, c-Jun N-terminal kinase.

similar role in the regulation of longevity, though this area also remains unexplored. Therefore, inquiries into these aspects of ILP biology in *A. stephensi* should be a future priority as such studies may reveal ways to precisely manipulate ILPs to generate modified mosquitoes that exhibit not only enhanced resistance and reduced host-seeking, but also enhanced life span and reproductive fitness for sustainable malaria transmission control. This could be potentially achieved through transgenic approaches, by directly altering *ILP* gene expression in the mosquito, or via paratransgenesis, by using symbiotic microbes to express ILPs (Wilke and Marrelli, 2015). One promising candidate for a paratransgenic approach is the bacterium *Wolbachia* which, in addition to the germline, can localize to somatic tissues such as the midgut and may be amenable to modification that allows it to secrete antiparasite factors (Pietri et al., 2016b). While our knowledge of the molecular mechanisms that govern the relationships among ILPs, mosquito physiology, and *P. falciparum* infection continues to grow, the kinetics of ILP production and signaling are not well understood. To date, only ILP3 and ILP4 have been studied extensively in *A. stephensi*, and little is known about the remaining three ILPs, which are also produced during parasite infection. Further, ILP3 and ILP4 have only been studied in isolation and the spatial and temporal interactions between these and other ILPs remain almost completely unknown. Given that both synergy and redundancy between ILPs is commonplace in *Drosophila* (Gronke et al., 2010), these interactions are likely to be equally complex in the mosquito. Studies of *A. aegypti* ILPs have begun to elucidate such interactions by testing receptor competition between multiple ILPs and correlating these properties with differences in the physiological effects of ILPs administered separately or together (Wen et al., 2010). Experiments of a similar nature with *A. stephensi* that also incorporate the remaining ILPs would be beneficial. Similarly, a more detailed understanding of the timing of ILP production during infection in multiple tissues would provide a biologically relevant map of ILP interactions. These experiments pose unique technical challenges, but overcoming them will enhance our current understanding and likely pave the way toward the successful integration of ILP biology and transgenic/paratransgenic mosquito technology for the control of malaria transmission.

References

Antonova, Y., Arik, A.J., Brusca, W.M., Riehle, M.A., Brown, M.R., 2012. Insulin-like peptides: structure, synthesis & function. Insect Endocrinol. Elsevier. http://dx.doi.org/10.1016/B978-0-12-384749-2.10002-0.

Antonova-Koch, Y., Riehle, M.A., Arik, A.J., Brown, M.R., 2013. Insulin-like peptides. Handbook of Biologically Active Peptides. Elsevier. http://dx.doi.org/10.1016/B978-0-12-385095-9.00038-5.

Bai, H., Kang, P., Tatar, M., 2012. *Drosophila* insulin-like peptide-6 (dilp6) expression from fat body extends lifespan and represses secretion of *Drosophila* insulin-like peptide-2 from the brain. Aging Cell 11, 978–985.

Bathgate, R.A., Halls, M.L., Vander Westhuizen, E.T., Callander, G.E., Kocan, M., Summers, R.J., 2013. Relaxin family peptides and their receptors. Physiol. Rev. 93, 405–480.

Becker, T., Loch, G., Beyer, M., Zinke, I., Aschenbrenner, A.C., Carrera, P., Inhester, T., Schultze, J.L., Hoch, M., 2010. FOXO-dependent regulation of innate immune homeostasis. Nature 463, 369–373.

Bouche, C., Lopez, X., Fleischman, A., Cypess, A.M., O'Shea, S., Stefanovski, D., Bergman, R.N., Rogatsky, E., Stein, D.T., Kahn, C.R., Kulkarni, R.N., Goldfine, A.B., 2010. Insulin enhances glucose-stimulated insulin secretion in healthy humans. Proc. Natl. Acad. Sci. U.S.A. 107, 4770–4775.

Boutros, M., Agaisse, H., Perrimon, N., 2002. Sequential activation of signaling pathways during innate immune responses in *Drosophila*. Dev. Cell 3, 711–722.

Brown, M.R., Clark, K.D., Gulia, M., Zhao, Z., Garczynski, S.F., Crim, J.W., Suderman, R.J., Strand, M.R., 2008. An insulin-like peptide regulates egg maturation and metabolism in the mosquito *Aedes aegypti*. Proc. Natl. Acad. Sci. 105, 5716–5721.

Carlsson, M.A., Diesner, M., Schasctner, J., Nassel, D.R., 2010. Multiple neuropeptides in the *Drosophila* antennal lobe suggest complex modulatory circuits. J. Comp. Neurol. 518, 3359–3380.

Cator, L.J., George, J., Blanford, S., Murdock, C.C., Baker, T.C., Read, A.F., Thomas, M.B., 2013. Manipulation without the parasite: altered feeding behavior of mosquitoes is not dependent on infection with malaria parasites. Proc. Biol. Sci. 1763, 20130711.

Cator, L.J., Pietri, J.E., Murdock, C.C., Ohm, J.R., Lewis, E.E., Read, A.F., Luckhart, S., Thomas, M.B., 2015. Immune response and insulin signalling alter mosquito feeding behaviour to enhance malaria transmission potential. Sci. Rep. 5, 11947. http://dx.doi.org/10.1038/srep11947.

Chambers, M.C., Song, K.H., Schneider, D., 2012. *Listeria monocytogenes* infection causes metabolic shifts in *Drosophila melanogaster*. PLoS One 7 (12), e50679. http://dx.doi.org/10.1371/journal.pone.0050679.

Cheng, S., Joosten, L.A., Netea, M.G., 2014. The interplay between central metabolism and innate immune responses. Cytokine Growth Factor Rev. 6, 707–713.

Cirimotich, C.M., Dong, Y., Garver, L.S., Sim, S., Dimopoulos, G., 2009. Mosquito immune defenses against *Plasmodium* infection. Dev. Comp. Immunol. 34, 387–395.

Cong, X., Wang, H., Liu, Z., He, C., An, C., Zhao, Z., 2015. Regulation of sleep by insulin-like peptide system in *Drosophila melanogaster*. Sleep 38, 1075–1083.

De Gregorio, E., Spellman, P.T., Tzou, P., Rubin, G.M., Lemaitre, B., 2002. The Toll and Imd pathways are the major regulators of the immune response in *Drosophila*. EMBO J. 21, 2568–2579.

Defferrari, M.S., Orchard, I., Lange, A.B., 2016. Identification of the first insulin-like peptide in the disease vector *Rhodnius prolixus*: involvement in metabolic homeostasis of lipids and carbohydrates. Insect Biochem. Mol. Biol. 70, 148–159.

Delaney, J.R., Stoven, S., Uvell, H., Anderson, K.V., Engstrom, Y., Mlodzik, M., 2006. Cooperative control of *Drosophila* immune responses by the JNK and NF-kappaB signaling pathways. EMBO J. 25, 3068–3077.

DiAngelo, J.R., Bland, M.L., Bambina, S., Cherry, S., Birnbaum, M.J., 2009. The immune response attenuates growth and nutrient storage in *Drosophila* by reducing insulin signaling. Proc. Natl. Acad. Sci. U.S.A. 106, 20853–20858.

Dionne, M.S., Pham, L.N., Shirasu-Hiza, M., Schneider, D.S., 2006. Akt and FOXO dysregulation contribute to infection-induced wasting in *Drosophila*. Curr. Biol. 16, 1977–1985.

Drexler, A., Nuss, A., Hauck, E., Glennon, E., Cheung, K., Brown, M., Luckhart, S., 2013. Human IGF1 extends lifespan and enhances resistance to *Plasmodium falciparum* infection in the malaria vector *Anopheles stephensi*. J. Exp. Biol. 216, 208–217.

Drexler, A.L., Pietri, J.E., Pakpour, N., Hauck, E., Wang, B., Glennon, E.K., Georgis, M., Riehle, M.A., Luckhart, S., 2014. Human IGF1 regulates midgut oxidative stress and epithelial homeostasis to balance lifespan and *Plasmodium falciparum* resistance in *Anopheles stephensi*. PLoS Pathog. 10, e1004231. http://dx.doi.org/10.1371/journal.ppat.1004231.

Evans, E.A., Chen, W.C., Tan, M.W., 2008a. The DAF-2 insulin-like signaling pathway independently regulates aging and immunity in *C. elegans*. Aging Cell 7, 879–893.

Evans, E.A., Kawli, T., Tan, M.W., 2008b. *Pseudomonas aeruginosa* suppresses host immunity by activating the DAF-2 insulin-like signaling pathway in *Caenorhabditis elegans*. PLoS Pathog. 4, e1000175.

Garsin, D.A., Villanueva, J.M., Begun, J., Kim, D.H., Sifri, C.D., Calderwood, S.B., Ruvkun, G., Ausubel, F.M., 2003. Long-lived *C. elegans* daf-2 mutants are resistant to bacterial pathogens. Science. 300, (5627), 1921. http://dx.doi.org/10.1126/science.1080147.

Gronke, S., Clarke, D.F., Broughton, S., Andrews, T.D., Partridge, L., 2010. Molecular evolution and functional characterization of *Drosophila* insulin-like peptides. PLoS Genet. 6, e1000857.

Gulia-Nuss, M., Robertson, A.E., Brown, M.R., Strand, M.R., 2011. Insulin-like peptides and the target of rapamycin pathway coordinately regulate blood digestion and egg maturation in the mosquito *Aedes aegypti*. PLoS One 6, e20401.

Hansen, I.A., Attardo, G.M., Park, J.H., Peng, Q., Raikhel, A.S., 2004. Target of rapamycin-mediated amino acid signaling in mosquito anautogeny. Proc. Natl. Acad. Sci. U.S.A. 101, 10626–10631.

Kallio, J., Leinonen, A., Ulvila, J., Valanne, S., Ezekowitz, R.A., Ramet, M., 2005. Functional analysis of immune response genes in *Drosophila* identifies JNK pathway as a regulator of antimicrobial peptide gene expression in S2 cells. Microbes Infect. 7, 811–819.

Khoo, S., Griffen, S.C., Xia, Y., Baer, R.J., German, M.S., Cobb, M.H., 2003. Regulation of insulin gene transcription by ERK1 and ERK2 in pancreatic beta cells. J. Biol. Chem. 278, 32969–32977.

Kumar, S., Gu, X., Kim, Y., 2016. A viral histone H4 suppresses insect insulin signal and delays host development. Dev. Comp. Immunol. 63, 66–77.

Laughlin, G.A., Barrett-Connor, E., Criqui, M.H., Kritz-Silverstein, D., 2004. The prospective association of serum insulin-like growth factor I (IGF-I) and IGF-binding protein-1 levels with all cause and cardiovascular disease mortality in older adults: the Rancho Bernardo Study. J. Clin. Endocrinol. Metab. 89, 114–120.

Le Roith, D., Zick, Y., 2001. Recent advances in our understanding of insulin action and insulin resistance. Diabetes Care 24, 588–597.

Leibiger, I.B., Leibiger, B., Moede, T., Berggren, P.O., 1998. Exocytosis of insulin promotes insulin gene transcription via the insulin receptor/PI-3 kinase/p70 s6 kinase and CaM kinase pathways. Mol. Cell 1, 933–938.

Leibiger, B., Moede, T., Uhles, S., Berggren, P.O., Leibiger, I.B., 2002. Short-term regulation of insulin gene transcription. Biochem. Soc. Trans. 30, 312–317.

Lim, J., Gowda, D.C., Krishnegowda, G., Luckhart, S., 2005. Induction of nitric oxide synthase in *Anopheles stephensi* by *Plasmodium falciparum*: mechanism of signaling and the role of parasite glycosylphosphatidylinositols. Infect Immun. 73, 2778–2789.

Liu, T.F., Brown, C.M., El Gazzar, M., McPhail, L., Millet, P., Rao, A., Vachharajani, V.T., Yoza, B.K., McCall, C.E., 2012. Fueling the flame: bioenergy couples metabolism and inflammation. J. Leukoc. Biol. 3, 499–507.

Liu, Y., Luo, J., Carlsson, M.A., Nassel, D.R., 2015. Serotonin and insulin-like peptides modulate leucokinin -producing neurons that affect feeding and water homeostasis in *Drosophila*. J. Comp. Neurol. 523, 1840–1863.

Liu, Y., Liao, S., Veenstra, J.A., Nassel, D.R., 2016. *Drosophila* insulin-like peptide 1 (DILP1) is transiently expressed during non-feeding stages and reproductive dormancy. Sci. Rep. 6, 26620.

Luckhart, S., Riehle, M.A., 2007. The insulin signaling cascade from nematodes to mammals: insights into innate immunity of *Anopheles* mosquitoes to malaria parasite infection. Dev. Comp. Immunol. 31, 647–656.

Luo, J., Lushchak, O.V., Goergen, P., Williams, M.J., Nassel, D.R., 2014. *Drosophila* insulin-producing cells are differentially modulated by serotonin and octopamine receptors and affect social behavior. PLoS One 9, e99732.

Marquez, A.G., Pietri, J.E., Smithers, H.M., Nuss, A., Antonova, Y., Drexler, A.L., Riehle, M.A., Brown, M.R., Luckhart, S., 2011. Insulin-like peptides in the mosquito *Anopheles stephensi*: identification and expression in response to diet and infection with *Plasmodium falciparum*. Gen. Comp. Endocrinol. 173 (2), 303–312. http://dx.doi.org/10.1016/j.ygcen.2011.06.005.

Melloul, D., Marshak, S., Cerasi, E., 2002. Regulation of insulin gene transcription. Diabetologia 45, 309–326.

Mitri, C., Jacques, J., Thiery, I., Riehle, M.M., Xu, J., 2009. Fine pathogen discrimination within the APL1 gene family protects *Anopheles gambiae* against human and rodent malaria species. PLoS Pathog. 5, e1000576.

Nagasawa, H., Kataoka, H., Isogai, A., Tamura, S., Suzuki, A., Ishizaki, H., Mizoguchi, A., Fujiwara, Y., 1984. Amino-terminal amino acid sequence of the silkworm prothoracicotropic hormone: homology with insulin. Science 226, 1344–1345.

Nassel, D.R., Vanden Broeck, J., 2016. Insulin/IGF signaling in *Drosophila* and other insects: factors that regulate production, release, and post-release action of the insulin-like peptides. Cell Mol. Life Sci. 73, 271–290.

Nassel, D.R., Kubrak, O., Liu, Y., Luo, J., Lushchak, O.V., 2013. Factors that regulate insulin producing cells and their output in *Drosophila*. Front Physiol. 4, 252.

Nassel, D.R., Liu, Y., Luo, J., 2015. Insulin/IGF signaling and its regulation in *Drosophila*. Gen. Comp. Endocrinol. 221, 255–266. http://dx.doi.org/10.1016/j.ygcen.2014.11.021.

Nassel, D.R., 2012. Insulin-producing cells and their regulation in physiology and behavior of *Drosophila*. Can. J. Zool. 90, 476–488.

Nemoto, S., DiDonato, J.A., Lin, A., 1998. Coordinate regulation of IkappaB kinases by mitogen-activated protein kinase kinase kinase 1 and NF-kappaB-inducing kinase. Mol. Cell Biol. 18, 7336–7343.

Niu, D., Wang, F., Zhao, H., Wang, Z., Li, J., 2016. Identification, expression, and innate immune responses of two insulin-like peptide genes in the razor clam *Sinovacula constricta*. Fish Shellfish Immunol. 41, 401–404.

Pakpour, N., Corby-Harris, V., Green, G.P., Smithers, H.M., Cheung, K.W., Riehle, M.A., Luckhart, S., 2012. Ingested human insulin inhibits the mosquito NF-kappaB-dependent immune response to *Plasmodium falciparum*. Infect Immun. 80, 2141–2149.

Park, J.M., Brady, H., Ruocco, M.G., Sun, H., Williams, D., Lee, S.J., Kato Jr., T., Richards, N., Chan, K., Mercurio, F., Karin, M., Wasserman, S.A., 2004. Targeting of TAK1 by the NF-kappa B protein Relish regulates the JNK-mediated immune response in *Drosophila*. Genes Dev. 18, 584–594.

Pietri, J.E., Pietri, E.J., Potts, R., Riehle, M.A., Luckhart, S., 2015. *Plasmodium falciparum* suppresses the host immune response by inducing the synthesis of insulin-like peptides (ILPs) in the mosquito *Anopheles stephensi*. Dev. Comp. Immunol. 53, 134–144.

Pietri, J.E., Pakpour, N., Napoli, E., Song, G., Pietri, E.J., Potts, R., Cheung, K.W., Walker, G.T., Riehle, M.A., Starcevich, H., Giulivi, C., Lewis, E.E., Luckhart, S., 2016a. Two insulin-like peptides regulate malaria parasite infection in the mosquito through effects on intermediary metabolism. Biochem. J. 473 (20), 3487–3503.

Pietri, J.E., Debruhl, H., Sullivan, W., 2016b. The rich somatic life of *Wolbachia*. MicrobiologyOpen. 5 (6), 923–936. http://dx.doi.org/10.1002/mbo3.390.

Post, S., Tatar, M., 2016. Nutritional geometric profiles of Insulin/IGF expression in *Drosophila melanogaster*. PLoS One 11, e0155628.

Renehan, A.G., Zwahlen, M., Minder, C., O'Dwyer, S.T., Shalet, S.M., Egger, M., 2004. Insulin-like growth factor (IGF)-I, IGF binding protein-3, and cancer risk: systematic review and meta-regression analysis. Lancet (London, England) 363, 1346–1353.

Riehle, M.A., Fan, Y., Cao, C., Brown, M.R., 2006. Molecular characterization of insulin-like peptides in the yellow fever mosquito, *Aedes aegypti*: expression, cellular localization, and phylogeny. Peptides 27, 2547–2560.

Root, C.M., Ko, K., Jafari, A., Wang, J.W., 2011. Presynaptic facilitation by neuropeptide signaling mediates odor driven food search. Cell 141, 133–144.

Siddle, K., 2011. Signalling by insulin and IGF receptors: supporting acts and new players. J. Mol. Endocrinol. 47, R1–R10.

Teleman, A., 2010. Molecular mechanisms of metabolic regulation by insulin in *Drosophila*. Biochem. J. 425, 13–26.

Van Der Westhuizen, E.T., Summers, R.J., Halls, M.L., Bathgate, R.A.D., Sexton, P.M., 2007. Relaxin receptors–new drug targets for multiple disease states. Curr. Drug Targets 8, 91–104.

Vogel, K.J., Brown, M.R., Strand, M.R., 2013. Phylogenetic investigation of peptide hormone and growth factor receptors in five dipteran genomes. Front Microbiol. 4, 193.

Wen, Z., Gulia, M., Clark, K.D., Dhara, A., Crim, J.W., Strand, M.R., Brown, M.R., 2010. Two insulin-like peptide family members from the mosquito *Aedes aegypti* exhibit differential biological and receptor binding activities. Mol. Cell Endocrinol. 328, 47–55.

Wigby, S., Slack, C., Gronke, S., Martinez, P., Calboli, F.C., Chapman, T., Partridge, L., 2011. Insulin signaling regulates remating in female *Drosophila*. Proc. Biol. Sci. 278, 424–431.

Wilke, A.B.B., Marrelli, M.T., 2015. Paratransgenesis: a promising new strategy for mosquito vector control. Parasites Vectors 8, 342.

Wu, Q., Brown, M.R., 2006. Signaling and function of insulin-like peptides in insects. Annu. Rev. Entomol. 51, 1–24.

Zhao, Q., Lee, F.S., 1999. Mitogen-activated protein kinase/ERK kinase kinases 2 and 3 activate nuclear factor-kappaB through IkappaB kinase-alpha and IkappaB kinase-beta. J. Biol. Chem. 274, 8355–8358.

Index

'*Note*: Page numbers followed by "f" indicate figures, "t" indicate tables, and "b" indicate boxes.'

Printed in the United States
By Bookmasters